Advances in Navigability and Mooring

Advances in Navigability and Mooring

Guest Editor

Marko Perkovic

MDPI

Basel • Beijing • Wuhan • Barcelona • Belgrade • Novi Sad • Cluj • Manchester

Guest Editor
Marko Perkovic
University of Ljubljana
Portorož
Slovenia

Editorial Office
MDPI AG
Grosspeteranlage 5
4052 Basel, Switzerland

This is a reprint of the Special Issue, published open access by the journal *Journal of Marine Science and Engineering* (ISSN 2077-1312), freely accessible at: https://www.mdpi.com/journal/jmse/special_issues/zara_navigability_mooring.

For citation purposes, cite each article independently as indicated on the article page online and as indicated below:

Lastname, A.A.; Lastname, B.B. Article Title. *Journal Name* **Year**, *Volume Number*, Page Range.

ISBN 978-3-7258-2685-8 (Hbk)
ISBN 978-3-7258-2686-5 (PDF)
https://doi.org/10.3390/books978-3-7258-2686-5

Cover image courtesy of Marko Perkovic

Contents

Preface

This collection of research articles covers innovative methods in and solutions to current challenges aimed at improving the efficiency and safety of maritime operations. Motivated by the urgent need for improved navigational safety, efficient port logistics, and robust cybersecurity measures, this collection promotes a deeper understanding of these critical issues among researchers, practitioners, and policymakers; yet, at the same time, it is available to a broad audience, including marine engineers, naval architects, and maritime security professionals, providing valuable insights that inform both academic discourse and practical applications in the current environment. I want to thank the authors for their expertise and dedication and the reviewers for their critical insights, which have enriched the quality of the articles. I would also like to thank the editorial team for their support throughout the publication. I invite readers to engage with the research findings presented here and hope they will inspire further

Marko Perkovic
Guest Editor

Journal of
Marine Science and Engineering

MDPI

Editorial

Advances in Navigability and Mooring

Marko Perkovič

Faculty of Maritime Studies and Transport, University of Ljubljana, 6320 Portorož, Slovenia;
marko.perkovic@fpp.uni-lj.si

Abstract: Considerable technological progress has been made in ship handling and mooring in recent years, especially progress generated by the needs imposed by the introduction of ever larger ships. These advancements exploit the economic scale and environmental efficiency of larger vessels, but also present unique challenges, particularly in narrow waterways and harbour approaches. Precise navigation in these environments requires highly accurate hydrographic measurements, high-quality electronic charts, and advanced navigation systems, such as modern electronic chart display and information systems (ECDIS). Safe and efficient port operations also depend on the optimised allocation of port resources and comprehensive queuing strategies. Modern ships are increasingly susceptible to interference with Global Navigation Satellite Systems (GNSS) and Automatic Identification Systems (AIS), necessitating the development of resilient technologies and procedures to ensure navigational safety. In addition, climate change is exacerbating the challenges of ship handling in ports, as larger vessels are particularly vulnerable to sudden gusts of wind and have difficulty maintaining their position in the quay in strong crosswinds. Training and simulation are crucial to overcoming these challenges. Ship-handling simulators are invaluable for training purposes, but development is still needed to accurately simulate tilt and lean effects, especially when ships are sailing in narrow channels with following currents and changing winds. Improving the accuracy of these simulators will improve the preparation of seafarers for real-life conditions and ultimately contribute to safer and more efficient ship operations.

Citation: Perkovič, M. Advances in Navigability and Mooring. *J. Mar. Sci. Eng.* **2024**, *12*, 1601. https://doi.org/10.3390/jmse12091601

Received: 29 July 2024
Accepted: 4 September 2024
Published: 10 September 2024

1. Introduction

The increasing size of vessels has significantly heightened the complexity of maritime navigation, particularly in narrow canals and port approaches [1]. These challenges include the manoeuvrability of large ships, the limited space available in constrained waterways, and the potential for significant economic and environmental repercussions in the event of an accident [2]. Large ships require more space to turn and to slow down. The interaction with the canal or harbour walls can create additional hydrodynamic effects, such as suction and squatting, which complicate navigation [3]. Furthermore, berthing large vessels is a critical challenge that requires precise coordination and control. The size of these vessels means they have a larger windage area, making them more susceptible to wind forces [4]. Additionally, tidal currents and other hydrodynamic effects can complicate berthing operations. The limited space in ports designed for smaller vessels exacerbates these issues, leading to increased risks of collision and grounding. Recent high-profile incidents highlight the difficulties in handling large container vessels. In March 2021, the Ever Given, a 400 m long container ship, was grounded in the Suez Canal, blocking one of the world's most vital maritime routes for six days [5]. This incident resulted in significant economic losses and disruptions to global supply chains. Another notable incident occurred in January 2022 when the container vessel Milano Bridge collided with a crane at the Port of Busan, causing extensive damage to the container terminal and the vessel [6]. These incidents underscore the need for enhanced navigational technologies, improved training for ship handlers, and better infrastructure to accommodate the growing size of vessels.

To mitigate these challenges, several measures can be recommended; these include implementing more sophisticated navigational aids, such as real-time monitoring systems and automated docking technologies, upgrading port facilities to handle larger vessels, utilising more robust berthing systems, investing in advanced simulation technology for training pilots and skippers to improve their ability to navigate large vessels in constrained environments.

2. Contemporary Challenges in Navigability and Mooring

The safe and efficient handling of large vessels is a major challenge, especially when navigating in narrow channels (without tug assistance), in shallow waters, when approaching ports, when positioning along quays, and remaining safely at berth in unfavourable weather conditions. Ports often struggle to provide safe berths for large vessels exposed to strong winds, surges and passing vessels and are often equipped with undersized bollards and fenders. Precision positioning systems have evolved significantly, especially since the Panama Canal mandated precision navigation sensors for all large vessels. Pilots around the world now use high-precision equipment that is often superior to on-board systems. Onboard equipment can be less reliable due to questionable gyrocompass-derived courses and GNSS positions that are inaccurately aligned with the ship's Consistent Common Reference Point (CCRP). Given the availability of precision equipment on large vessels, minimal investment is required to use it outside Panama—primarily a portable display and licensing of electronic charts in critical areas are all that is required, which are relatively low-cost solutions compared to the risks. However, even modern portable pilot units (PPU) operating in real-time kinematic (RTK) mode are susceptible to technical errors such as multipath interference and intentional jamming. Multipath interference occurs when GNSS signals reflect off surfaces such as buildings, cranes, or other structures before reaching the receiver, resulting in inaccurate position calculations [7]. To overcome the limitations of GNSS-based systems and mitigate the multipath effects, researchers emphasise the need for LiDAR measurements as a complementary technology (*Contributions 1 and 14*). LiDAR scanners provide precise distance measurements between the vessel and the berth and offer a reliable alternative in areas where GNSS-based systems are affected by multipath effects or intentional interference. The researchers in [8] have developed an interface in which numerical data measured by two distance-measuring devices mounted on a large ship are visualised by a 3D model of Rviz to facilitate the docking of the ship for all associated structures and actors, such as the quay, the pilot, and tug operators. This integration of technologies can significantly improve the accuracy and safety of vessel positioning during the berthing process.

The maritime sector is increasingly vulnerable to cyber threats that can severely compromise security, disrupt operations and compromise data security [9–12]. This research looks at the pervasive cyber threats facing the industry. Specifically, it highlights vulnerabilities in satellite and GNSS systems, which are essential for navigation, and radio-based AIS systems, which are used to identify and track vessels. These vulnerabilities can be exploited to manipulate ship positions or transmit false information, leading to navigational hazards and security breaches. To counter these threats, this study highlights the urgent need for robust cybersecurity strategies and technologies, including preventative measures, advanced detection and response systems, and resilient recovery plans to ensure the safety and integrity of maritime operations.

In relation to navigability, maritime autonomous surface ships (MASS) represent a significant advancement in marine technology, offering improved safety, efficiency, and sustainability in maritime operations. The development of MASS involves sophisticated navigation systems, AI integration, and rigorous testing to ensure reliability and safety. The gradual market adoption of MASS is driven by regulatory support, technological advancements, and the growing demand for innovative maritime solutions, paving the way for a transformative impact on the global shipping industry. Addressing these challenges through innovative methodologies, interdisciplinary collaborations, and continuous research efforts is essential for the enhancement of the safety and security of MASS operations and the effective mitigation of collision risks. Collision risk analysis faces key challenges identified in the existing literature [13–16], including the scarcity of historical data on accidents, human factors such as coding errors and inadequate interfaces, cybersecurity risks such as hacking and communication failures, complex system interactions, uncertainty in autonomous operations and environmental conditions, and evolving regulatory frameworks [17]. These challenges underscore the need for advanced methodologies to address human errors, cybersecurity threats, system complexities, uncertain data, and regulatory compliance in collision risk assessments for MASS, emphasising the importance of interdisciplinary approaches and continuous research efforts to enhance safety and security in autonomous maritime operations. State-of-the-art methods include the expert scoring approach based on fuzzy values to gather expert opinions, the noisy-OR Technique for calculating conditional probabilities in Bayesian networks, the innovative FTA-FBN model combining fault tree analysis and Bayesian networks for quantitative risk assessment, the utilisation of machine learning algorithms, and the development of new quantitative risk analysis models to address factors like equipment failure, cyber threats, adverse environmental conditions, and even human factors.

The need for evolution of vessel traffic service (VTS) based on the maritime traffic management (STM) vision is critical as maritime systems become increasingly complex and the demand for real-time vessel traffic management increases exponentially. Future VTS systems will require sophisticated data-driven solutions to process large amounts of data from onboard sensors, AIS, radar and weather systems. Key functions include improving operational efficiency and safety, predictive risk management, route optimisation, abnormal behaviour detection, and compliance monitoring [18]. Integrating advanced technologies such as AI, machine learning, and high-speed computing will enable next-generation VTS to support fully automated ports and provide real-time data analytics and secure communication over maritime 5G. With the increasing proliferation of autonomous surface vessels, developing advanced VTMS will be essential for safe navigation and efficient traffic management [19].

3. Published Articles

This collection of 15 scientific articles comprehensively explores various critical aspects of the maritime industry. Among these papers, two specifically tackle the intricate challenges of docking increasingly larger ships in confined harbour basins. These studies delve into innovative technical solutions, such as integrating AIS data with lidar measurements and implementing reference points to enhance the efficiency and safety of docking operations. Furthermore, a detailed analysis of the cognitive load experienced by marine pilots during the demanding process of docking large vessels in severe weather conditions sheds light on the human factors involved in maritime navigation. Additionally, a study focusing on the impact of added mass on loading flexible dolphins underscores the importance of considering structural dynamics in mooring infrastructure design. The research on berth allocation and the quay crane allocation problem offers valuable insights into optimising port logistics and resource utilisation. Despite the rapid technological advancements in the maritime sector, the enduring relevance of paper navigational charts is emphasised in a study highlighting that traditional charts will not be entirely replaced despite the mandatory implementation of ECDIS in 2018. Traditional maritime practices are further explored in a study investigating the use of magnetic compasses on ships and the determination of residual deviation, showcasing the continued importance of fundamental navigation tools. Other notable papers in the collection address diverse topics such as the influence of sloping banks on manoeuvring large vessels, the design of port approaches using simulation-based methods, the utilisation of ECDIS for passage planning with a focus on the zone of confidence impact on cross-track limit determination, the performance of satellite-based correction services for precise hydrographic surveying, collision avoidance modelling employing fuzzy methodology, and an approach to accident reconstruction based on real-world cases. This diverse array of research contributes significantly to enhancing our understanding of critical maritime operations, safety protocols, and technological advancements, ultimately shaping the maritime industry's future. The keywords connecting all 15 papers are the following: ship handling; docking and mooring; precise navigation systems; ECDIS; hydrographic measurements; GNSS and AIS interference; bank effects and squat effects; anti-collision; port resource allocation; maritime safety; accident reconstruction.

Author Contributions

This Special Issue begins conceptually with an article addressing the complexities of docking increasingly larger ships in confined harbour basins (*Contribution 1*); this study focuses on technical solutions to enhance operational efficiency and safety. The contribution lies in providing innovative approaches to tackle the unique challenges of large vessels manoeuvring in restricted spaces, ultimately improving port operations and navigation safety. The use of laser systems for distance measurement and speed control of ships as well as for controlling angles of attack is well known in the tanker industry. Implementing the distance measurement system is relatively simple, as the tankers always berth in the same position, i.e., the ship's manifold, which is located in the middle part of the ship and is always aligned with the terminal's connecting arms. The middle part of the ship's hull is always on the wall side, which means that the lasers are directed towards the parallel part of the hull, as this simplifies all calculations and layout drawings. For container ships, however, the situation is more complicated: there is no exact fixed location for the ship's berthing point; if the measuring devices are on the shore, the distances between the shore and the final position of the ship are small (possibly below the expected minimum offset), and if the lidars are mounted on the STS crane, they are subject to movement, as they must be outside the bow and stern as the ship approaches, yet at the same time away from the ship's wings. The measurements must be very accurate, because with a 400 m long ship, even a small error in the angle of attack can lead to a completely wrong perception of the situation and a collision with the STS crane. As the ship's position in the longitudinal direction is very difficult or impossible to determine with lidar in the transverse direction, the authors initially proposed and developed the integration of lidar measurements with AIS data, a system tested in the Port of Koper. An additional docking setup was developed and implemented in the port of Świnoujście, where 3D LiDAR technology is utilised. Scanning the bow, stern, and sides of the ship provides vessel layout, position, and orientation without GNSS positioning and AIS dimensioning. By utilising advanced algorithms (Random Sample Consensus—RANSC) and real-time data processing, the information regarding the vessel's orientation, minimum distances to the ship's side, berth, and bow/stern ramp, as well as wind data, calculated velocities, and heading, is presented to the pilots through an onboard application. The designed application communicates with a server via an LTE internet connection, providing real-time updates and essential data for the pilots to make informed decisions during the docking process of the Ro-Pax vessel.

Another vessel-independent berthing assistant system was developed using reference points (*Contribution 14*), the approach suggested by marine pilots. The system defines a Berthing Support Area (BSA) within which safe berthing is provided, and this area contains reference points for perpendicular distance measurements at arbitrary positions. The system includes at least three reference points per ship length, evenly distributed along the quay to track a vessel's bow and stern (based on Lidar measurements), allowing for monitoring of the Rate of Turn (ROT) during the entire berthing process, even with steeper and uncommon berthing angles. In this case, instead of the individual LiDAR spot reflection echoes (which are handled in the sensors firmware), it calculates the perpendicular distance and approach speed from the quay to a vessel's hull based on a set of LiDAR spots. Based on the working principle of LiDAR, the first echo is used as the output distance. Changes in the distance are used to calculate a vessels' velocity and acceleration relative to the quay. Additionally, it provides data not only to the vessel's pilot but also to the tug masters, improving coordination between the vessel and tugboats while assisting the vessel when necessary, and providing warnings for high approach speeds or high rates of turn. The BSA was implemented at a berth in Wilhelmshaven, and a test campaign verified that the system meets pilot-derived requirements and complies with IMO Resolution A.915(22) [20].

This edition contains an additional article on the mooring of ships, in particular tankers, to flexible mooring dolphins (*Contribution 8*). This study raises the critical question of how additional mass during mooring affects a particular type of ship. The research aims to understand the dynamics and loads that flexible dolphins are subjected to as shock absorbers during the mooring process by investigating the hydrodynamic effects of the ship. The key achievements of the study include simplifying the mathematical model by approaching the zero-frequency limit and focusing on pure rocking motions in a three-dimensional study using the strip theory approach. The study also uses conformal mapping with Lewis mapping for the ship geometry to analyse the flow around the ship. The study definitively contributes to improving the safety and efficiency of mooring operations for different types and sizes of tankers by emphasising the importance of considering the effects of additional mass and provides a theoretical framework for estimating these effects using advanced hydrodynamic modelling techniques, thus improving the overall understanding of ship–anchorage interactions. The author concludes that the values of added mass are very high and must always be taken into account in the load analysis of flexible dolphins, which is particularly important when increasingly large tankers call at existing terminals whose dolphins are already severely deteriorated. The berthing forces must be known and, if necessary, controlled by the docking system described above.

The exact positioning of the ship is undoubtedly important, but so is the accuracy of the hydrographic measurements. *Contribution 2* deals with the accuracy and performance of satellite-based correction services, particularly Trimble PP-RTX technology and GNSS virtual reference stations for hydrographic surveying. This study aims to assess how these techniques meet the minimum standards for hydrographic surveys set by the International Hydrographic Organization (IHO). A three-hour session in Sharm Obhur used the hydrographic vessel KAU to achieve this goal. A variety of equipment was used during the data collection, including the Trimble SPS855 GNSS receiver at the base station, the Kongsberg EM 712 multibeam echosounder for bathymetry data collection, the Valeport sound velocity profiler for sound velocity measurements, and the Applanix POS MV for position and orientation solutions. Real-time PP-RTX corrections were used during the session. In addition, virtual GNSS reference station (VRS) data were generated using the Kingdom of Saudi Arabia's Continuous Operation Reference Station (KSA-CORS) network. The study results include evidence that both the PP-RTX and VRS methods meet the IHO minimum standards for hydrographic survey tasks, with THU and TVU values within the acceptable range at a 95% confidence level. Statistical analysis of the differences in the seabed surface underpins the accuracy and reliability of these methods for hydrographic survey applications.

The next paper (*Contribution 3*) is a review study, which ties in with the previously processed bathymetric data and discusses electronic maps, the ECDIS system, and navigation. In particular, the effects of the zone of confidence (ZOC) on the determination of the safety parameter Cross Track Limit (XTL) in electronic chart displays and information systems (ECDIS) for voyage planning are examined. Maritime safety can be improved by understanding how the quality of chart survey data affects the determination of XTL values, which play a crucial role in determining the safe distance between a vessel and potential hazards. The review involved evaluating the navigational procedures of different shipping companies to identify variations in XTL determination practices and the potential risks associated with subjective XTL settings. Through a detailed analysis of the XTL determination practices of different shipping companies, the research aims to define a method for determining the XTL that is consistent with predefined criteria and improves the situational awareness of navigators. The results reveal significant differences in XTL reporting practices, emphasise the importance of accurate XTL values to avoid navigational risks, and provide insights that should lead to improvements in the training process for interpreting safety parameters on ECDIS screens. The methodology used in the study includes a thorough review of the existing literature on ECDIS safety parameters and voyage planning, an analysis of groundings related to the underutilisation of XTL, and a survey of shipping professionals to collect data on the practice of XTL determination. The authors suggest that the actual determination of XTL by experienced navigators provides material for future research to analyse the established routine among mariners. A detailed understanding of the interpretation of safety parameters by navigators has the potential to change and improve the training process.

The content of *Contribution 4* profiles the stress of officers on watch during simulated navigation tasks, focusing on collision avoidance manoeuvres. The goal of this research was to quantify stress and cognitive load using biometric markers such as pupil diameter, heart rate, electrodermal activity (EDA), and wrist acceleration. By collecting and analysing this physiological data, the researchers aimed to understand how high cognitive load and stress can affect the decision-making and performance of shipmasters, potentially leading to navigational errors and accidents at sea. This study highlights the importance of developing a mathematically based model that uses machine learning algorithms to accurately identify and quantify cognitive workload. The results of this study provide insights into participants' cognitive processes and the psychophysical states of the participants and lay the foundations for future large-scale experiments to further validate the results and develop practical applications to improve safety at sea. The methods used in the study included a combination of quantitative and qualitative approaches supported by machine learning algorithms to analyse the biometric data collected during the simulation.

Contribution 5 examines the cybersecurity landscape in the maritime sector, particularly on the vulnerabilities and risks associated with satellite navigation systems such as AIS and GNSS, especially GPS, as one of the most commonly used constellations in maritime transportation. This paper aims to raise awareness of the potential cyber threats to the maritime industry and provide insights into practical research on an AIS spoofing event, such as that analysed near the island of Elba. The paper's accomplishments include identifying the main cyber challenges in maritime shipping, emphasising the need for improved cybersecurity measures, and formulating recommendations to improve the security of shipping systems. The authors conducted a systematic literature search for documented guidelines to achieve these goals. Databases such as Scopus, Web of Science, and Google Scholar were searched for relevant documents on cyber and cyber security in shipping. This article suggests that manufacturers can improve the cybersecurity of their products in the maritime industry by taking several important steps. First, it emphasises the importance of changing attitudes toward cybersecurity and devoting more attention and resources to this critical aspect of maritime operations. Manufacturers are advised to focus on improving the security features of their ship systems to avoid vulnerabilities that could be exploited by cybercriminals. In addition, this article recommends conducting training and refresher courses for ship crews to ensure they are equipped to deal effectively with cyber security challenges. Contingency plans for worst-case scenarios should be in place and alternative modes of operation should be considered to maintain safe and reliable operations in an insecure cyber environment. In addition to this, the article highlights the need for manufacturers to work with national defence forces to explore alternative positioning, navigation and timing (PNT) systems that can complement GPS or serve as a backup to reduce reliance on a single vulnerable system.

The main objective of *Contribution 6* was to investigate the impact of the introduction of electronic chart display and information systems (ECDIS) on the skills and preferences of navigators, particularly in comparison to traditional paper nautical charts (PNC), with a focus on improving navigational safety in shipping. The researchers analysed navigators' opinions on ECDIS and PNC, identified potential challenges in transitioning from PNC to ECDIS, and suggested ways to improve cybersecurity and safety in shipping. Navigators' acceptance and understanding of ECDIS is crucial in ensuring safe navigation practices. If navigators positively evaluate and master ECDIS, this can lead to better situational awareness, better decision making and more efficient route planning, ultimately contributing to safer navigation. On the other hand, errors, misinterpretation of data, and potential safety risks can occur during navigation if navigators have a negative opinion or are not sufficiently trained in the use of ECDIS. The research used a combination of methods, including international questionnaires to gather feedback from officers and other stakeholders and a human–machine interface (HMI) survey. These tools provided valuable insights into seafarers' experiences, preferences and challenges in using ECDIS and ultimately contributed to developing training processes, curriculum improvements, and practical recommendations for the maritime industry.

It is well known that ship collisions are still one of the main causes of total losses and loss of life at sea. In this context, a multi-parametric ship collision avoidance decision model using fuzzy logic is proposed in *Contribution 7* to improve collision avoidance planning in maritime navigation. This paper presents an improved decision model that uses fuzzy logic to calculate heading changes based on four parameters, improving the current tool for trial manoeuvres in ARPA radar systems. The advantage of fuzzy logic lies in generating decisions based on imprecise data that mathematical notation cannot describe because the data are expressed in language. Fuzzy logic imitates the human way of thinking, which can solve complex tasks that can also be subject to major uncertainties. The model is designed to consider both the avoidance direction and the minimum course change according to the COLREG rules, thus ensuring safe overtaking distances and efficient manoeuvring sequences. The authors used a database of correct solutions from manual radar recordings to develop the fuzzy parameters and rules, presenting a novel approach to collision avoidance decision-making. The study applies methods based on fuzzy logic to enable complex decision-making in uncertain and dynamic maritime environments, ultimately contributing to improved navigation practices and reduced collision risks at sea.

This time, with *Contribution 9*, the fuzzy approach was presented to enhance logistics at the quay, presenting a fuzzy optimisation model that deals with the berth assignment problem (BAP) and the quay crane assignment problem (QCAP) in multiple quays, considering the uncertainty of ship arrival times. Quay utilisation can be improved by efficiently allocating berths and cranes to arriving ships, considering imprecise arrival times represented by triangular fuzzy numbers. The main outcome of the research is developing a robust berthing plan that accounts for both early and late arrivals while efficiently allocating resources such as quay cranes. The model was implemented using the CPLEX solver, a powerful optimisation tool that allowed researchers to quickly obtain optimal solutions for minor instances. However, the behaviour was undefined for 'medium' instances, and for 'large' instances, no solutions were found within the given processing time. This suggests the model may have scalability issues with more vessels and complex scenarios, leading to computational challenges and potentially longer processing times.

In modern electronic navigation, which can easily be disturbed and distorted, knowledge of traditional navigation methods and aids has regained importance. The main purpose of *Contribution 10* is to analyse the cyber challenges faced by maritime navigation, focusing on the importance of the magnetic compass as an aid despite the proliferation of electronic compasses and satellite systems. This paper aims to show that the magnetic compass needs to be regularly adjusted and compensated to ensure its reliability under different conditions. The researchers conducted a study that included practical compensations of the magnetic compass at sea and a survey of navigation officers to gain insights into the current use and requirements for proper use. Methods used in the study included closed-ended questionnaires with predefined single-choice answers to categorise respondents' profiles and collect data on their knowledge and practices related to magnetic compass deviation verification and compensation. The survey was distributed online to 320 navigation officers, with 123 responses collected and verified through face-to-face interviews or email correspondence with a select 10% of respondents. Using this methodology, the paper provides insights and recommendations to improve the compensation and use of magnetic compasses in maritime navigation to address the cyber challenges associated with modern navigation technologies.

Another aspect of this Special Issue concerns the design of ports and waterways. The main objective of *Contribution 11* is to demonstrate the effectiveness of real-time simulation methods for ship manoeuvres in planning port approaches, focusing on the modernised port of Ustka in Poland. This study aimed to determine the optimal approach and breakwater solution for a general cargo ship, considering various parameters such as ship dimensions, propulsion systems, waterway safety parameters and navigational conditions. By using simulation methods, the study successfully determines the safety conditions for port operations, including manoeuvring procedures when entering, berthing, casting off, and turning, for different types of ships. The main achievement of the work is that it demonstrates how simulation methods can streamline the design process, reduce time-consuming testing and improve the overall efficiency of port infrastructure development. This study introduces a two-stage methodology, including the innovative "soft-bank" approach, to speed up the research while maintaining the accuracy of the results. The authors used simulation software and a limited-task simulator to conduct multiple simulation experiments under different hydrometeorological conditions to ensure the statistical variability and reliability of results. By following a systematic approach and considering factors such as ship size limitations and port infrastructure constraints, this study provides valuable insights for optimising port-planning strategies and improving navigational safety in port operations.

This Special Issue's theme is also directly related to shipping and handling many ships on canals and in other restricted areas. The highly publicised Ever Given disaster, which brought one of the most important flows of goods in maritime traffic to a standstill for a whole week, also occurred during this period. In this paper, the researchers (*Contribution 12*) analysed the effects of embankments on the trajectory of container ships. They proposed minimum distances that could reduce the forces acting on the embankments, ultimately reducing the risk of ships running aground and increasing the safety of shipping traffic. This study focuses on the influence of sloping banks, addressing a critical aspect of navigation in narrow waterways such as the Suez Canal, where bank effects can significantly influence vessel behaviour. Using simulation tools (Full Mission Navigation Simulator) and methods, the study assessed the impact of bank effects on vessel trajectories and suggested specific distances that could help minimise these effects, which in turn contributes to improved safety measures in navigation. Specifically, a container ship model was used, representing a 347 m long and 43 m wide ship with a draught of 14 m. The fairway was inclined at a ratio of 1:4 to simulate the banks of the Suez Canal, with the canal depths set to 1.2 and 1.5 times the ship's draught. The simulations focused on one side of the fairway, as the ship only approached one side of the canal. The weather, sea conditions, and sea current were set to zero to isolate the effects of the sloping banks. The distance from the shore (d) was varied from 50 m to 10 m at 10 m intervals, and three different speeds were tested: 10, 7.5, and 5 knots. During the simulations, the steering was set to networked autopilot mode with track control. The rocking force caused by different shore types at different vessel speeds and distances from the shore was analysed, and it was found that the rocking force increased with higher vessel speed and steeper shore inclinations. It was confirmed that the maximum rocking force occurred at a smaller distance from sloping shores compared to vertical shores, indicating the importance of the distance under the keel in mitigating the effects of shores on the ship's trajectory.

The last original article in this Special Issue deals with accident analysis. In *Contribution 13*, a specific incident on a city ferry in the port of Świnoujście is analysed using a modified causal model. This paper develops guidelines for improving the safety of ferry crossing and evaluating the accident analysis model used. The study's achievements include uncovering a communication problem between stakeholders that complicates safety improvement models and providing insights into broader issues related to organisational management and design flaws in complex socio-technical systems such as maritime transport. Various calculations were conducted in the research, including determining anchor holding force coefficients, calculating anchor holding forces using classical and modified relationships, analysing load forces of chains by ships, assessing anchor stability through friction coefficients, estimating aerodynamic and hydrodynamic forces from wind and current, evaluating emergency engine thrust, and analysing forces on ships with increasing chain length and wind speed to recommend optimal anchoring practices. Using causal models enables a comprehensive understanding of the factors contributing to accidents, including human error, organisational management problems, and design flaws. By addressing these causes through improved accident investigation techniques, the research provides valuable insights that should lead to improved safety measures in maritime transportation systems. Implementing the findings and recommendations from this study can help prevent similar accidents in the future, ultimately leading to a safer and more secure maritime environment.

The last article (*Contribution 15*) belongs to the category of review articles that provides a comprehensive overview of the factors that influence the manoeuvrability of ships in difficult conditions, such as shallow and narrow waters. The authors investigated prediction methods and model development strategies that can improve the understanding and prediction of vessel performance in these environments. This goal was achieved through a thorough literature review and analysis of relevant studies from Scopus, WOS, and Sci Direct databases. The authors followed a systematic approach according to PRISMA guidelines, which included steps such as identification of research studies, screening, assessment of eligibility and detailed analysis of selected studies. Various tools and methods were used, including bibliometric analysis, visualisation of keyword networks and assessment of research studies based on credibility, substantive relevance, and research validity. The review discusses various prediction methods and model development strategies, starting with the captive model testing method, in which tests are conducted in controlled environments such as towing tanks to measure the rocking force and yaw moment, which provide insight into the hydrodynamic forces acting on the ship. It also explains the computational fluid dynamics (CFD) method of simulating the flow around the hull, rudder, and propeller, which allows for the prediction of hydrodynamic forces and ship behaviour under different conditions. Finally, it describes system identification techniques that analyse data from ship trials or model tests to determine the mathematical models that best represent the manoeuvring behaviour of the ship. By fitting mathematical models to experimental data, researchers can predict the ship's response to various steering inputs and environmental conditions to better understand the ship's manoeuvrability. These methods contribute to improving ship design, operational procedures, and safety measures in difficult waters. In addition, the authors emphasise the importance of bridge simulator training as a mandatory program for seafarer training to improve the understanding of ship manoeuvring, both for crewed ships and maritime autonomous surface ships (MASS).

4. Outlook Research

Climate change and the impact of ships on global warming are important issues today. Engine manufacturers and ship owners are working diligently to introduce clean fuels. In this transition phase, reducing propulsion power is an effective solution [21]. To support these efforts, the IMO has introduced the Energy Efficiency Index (IEEX) for existing ships, which aims to reduce pollution through the shaft power limiter (SHaPoLi), or engine power through a limiter (EPL) [22]. The EPL solution is relatively simple and inexpensive and causes minimal disruption to vessel operations; it is a mechanical stopper for limiting the fuel index. The IMO has even stipulated that the power limitation system can be overridden if necessary to ensure the safety of the ship and crew. This override must, however, trigger an alarm and be recorded in the vessel's Onboard Management Manual (OMM). Furthermore, the vessel's flag state (or its representative) and the relevant port authority must be notified immediately. However, in the real world, some EPLs cannot be easily removed (mechanically controlled engines), as seen in some installation photographs

(Figure 1), raising concerns about whether skippers have sufficient power when needed. New ships are already fitted with engines of significantly lower power, larger propellers that turn more slowly to make propulsion more efficient, and smaller rudders.

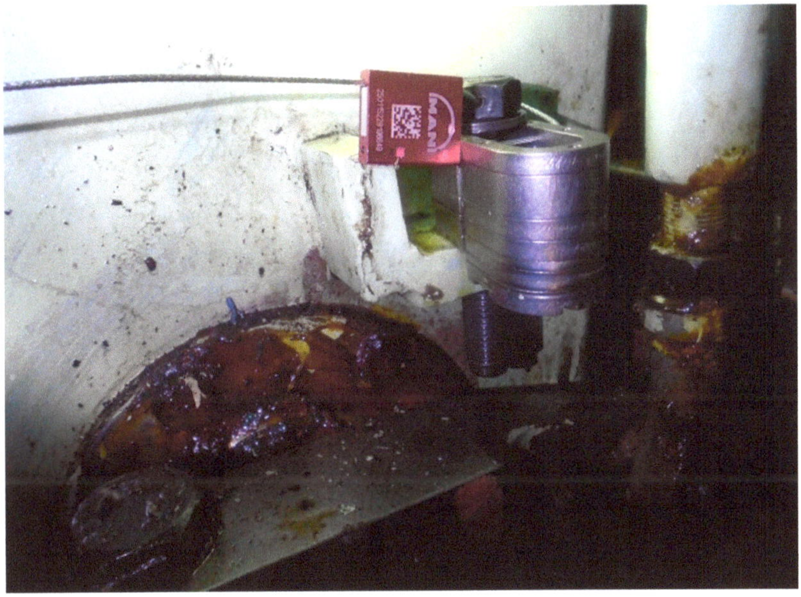

Figure 1. Engine power limiter of a mechanically driven type engine (changing a set of governor's fuel index).

Pilots in the Suez and elsewhere have complained of insufficient jet speed at the rudder, highlighting the need for a power boost (kick) when entering zones of bank effect. This scenario requires further research into manoeuvring in difficult conditions, such as crosswinds, currents, and narrow winding channels, especially for less manoeuvrable ships.

Figure 2 illustrates the standard maximum continuous rating (SMCR) of engines installed on container ships up to 10,000 TEU (green polygon), along with projections for larger mother ships during the short-term development of container vessels up to 14,500 TEU (orange), and eventually up to 18,000 TEU (red) [22]. It also shows that modern vessels exceeding 20,000 TEU (such as the Ever Given) are powered by engines with less than 60,000 kW, which is comparable to the engine size of ships built 10 years earlier, whose capacity (expressed in TEU) was approximately three times smaller, as indicated by the dashed arrow.

The following equation gives an estimate of the speed of the efflux velocity V_o as a function of the applied engine power (P), the diameter of the propeller D_p and the density of the sea [23]. The lower the propulsion power and the larger the propeller, the lower the velocity of the jet inflowing to the rudder, which leads to a ship's reduced manoeuvrability:

$$V_o = 1.48\left(\frac{P}{\rho_w D_\mathrm{p}^2}\right)^{\frac{1}{3}} \tag{1}$$

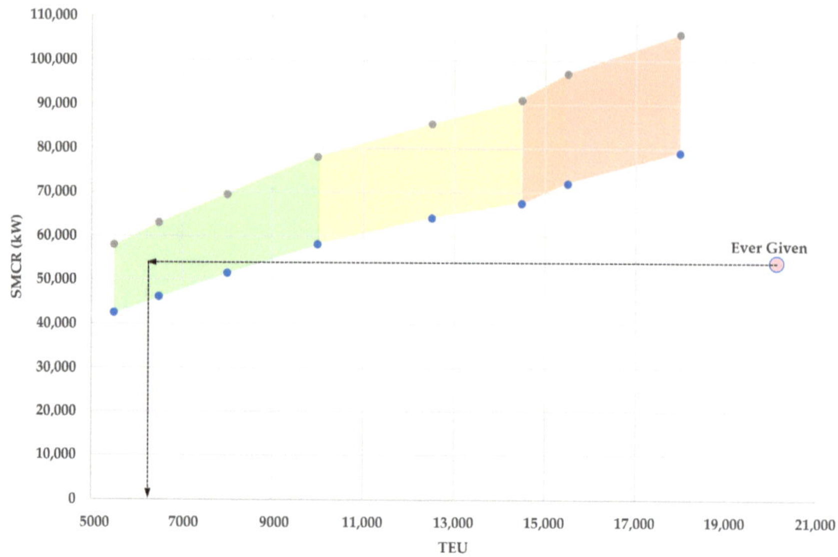

Figure 2. Main engine propulsion power vs. vessel size expressed in TEU.

Maritime simulators are ideal for training in crisis procedures and general ship management. With recent power failures and accidents involving large ships, especially those with high-voltage systems, more investment is needed in fully integrated simulator training, including simulators for ship handling, engineering, and possibly VTS (Vessel Traffic Services).

Research into introducing new electronic charts and modernising ECDIS is also highly topical. The transition from existing ECDIS systems to the new S-100 ECDIS, as called for in Resolution MSC.530(106) [24], risks incompatibilities, technical upgrades, data interoperability issues, training needs, and compliance issues. Ensuring seamless compatibility, technical readiness, data consistency, adequate training, and regulatory compliance is critical to mitigating these risks and successfully transitioning to the updated electronic chart display and information systems.

The slow uptake of e-navigation services should also be mentioned. Technological solutions (VDES) for the transmission of extended data volumes are already available and software solutions both on the side of VTS centres and on the side of ships' ECDIS systems allow the implementation of maritime traffic management (STM), which undoubtedly contributes to greater safety at sea and more efficient operation of shipping companies. The communication system is also more resistant to intentional interference, a daily occurrence with AIS systems today.

Maritime safety is also highly questionable concerning illegal oil transhipment; the dark fleet operates virtually without restrictions, and jamming and spoofing of GNSS signals is commonplace and poses a major security threat. The maritime pilots in Suez are virtually without GNSS support, solutions are being sought, and a highly accurate ground-based timing system should be considered to support technological applications and provide services when space-based signals are unavailable.

Finally, the recent attacks by Yemen's Ansarallah, or Houthi, fighters should be mentioned, as they have significantly impaired maritime safety, especially in the Red Sea. Since November last year (2023), the Houthis have intensified their campaign against commercial shipping, attacking dozens of vessels believed to be either bound for Israel or in any way associated with Israel. These actions have caused significant disruption to global trade routes with major shipping companies [25]. The Houthis attack ships using a variety of methods, including rockets, drones, mines, and boats loaded with explosives. Researchers can contribute some insights and innovations to improve security, such as developing AI tools for detection, exploring satellite-based surveillance, and developing risk assessment models, but without diplomatic efforts to resolve the conflict in Palestine/Israel coupled with international support to rebuild and stabilise the region, the incentives for such attacks will only increase.

Funding: The publication of the paper is supported by the research project (L7-1847; Developing a sustainable model for the growth of the "green port") and the research group (P2-0394; Modelling and simulations in traffic and maritime engineering) at the Faculty of Maritime Studies and Transport, financed by the Slovenian National Research Agency.

Acknowledgments: This Special Issue would not have been possible without the contributions of the authors and reviewers. We would like to congratulate all the authors. I would also like to take this opportunity to express our sincere thanks to all the reviewers. Finally, I would like to thank Capt. Migel Mehlmauer from the Slovenian Maritime Administration for the help with AIS data assessment.

Conflicts of Interest: The author declares no conflicts of interest.

List of Contributions:

1. Perkovič, M.; Gucma, L.; Bilewski, M.; Muczynski, B.; Dimc, F.; Luin, B.; Vidmar, P.; Lorenčič, V.; Batista, M. Laser-Based Aid Systems for Berthing and Docking. *J. Mar. Sci. Eng.* **2020**, *8*, 346. https://doi.org/10.3390/jmse8050346.
2. Elsobeiey, M.E. Accuracy Assessment of Satellite-Based Correction Service and Virtual GNSS Reference Station for Hydrographic Surveying. *J. Mar. Sci. Eng.* **2020**, *8*, 542. https://doi.org/10.3390/jmse8070542.
3. Kristić, M.; Žuškin, S.; Brčić, D.; Valčić, S. Zone of Confidence Impact on Cross Track Limit Determination in ECDIS Passage Planning. *J. Mar. Sci. Eng.* **2020**, *8*, 566. https://doi.org/10.3390/jmse8080566.
4. Žagar, D.; Svetina, M.; Košir, A.; Dimc, F. Human Factor in Navigation: Overview of Cognitive Load Measurement during Simulated Navigational Tasks. *J. Mar. Sci. Eng.* **2020**, *8*, 775. https://doi.org/10.3390/jmse8100775.
5. Androjna, A.; Brcko, T.; Pavic, I.; Greidanus, H. Assessing Cyber Challenges of Maritime Navigation. *J. Mar. Sci. Eng.* **2020**, *8*, 776. https://doi.org/10.3390/jmse8100776.
6. Car, M.; Brčić, D.; Žuškin, S.; Svilicic, B. The Navigator's Aspect of PNC before and after ECDIS Implementation: Facts and Potential Implications towards Navigation Safety Improvement. *J. Mar. Sci. Eng.* **2020**, *8*, 842. https://doi.org/10.3390/jmse8110842.
7. Brcko, T.; Androjna, A.; Srše, J.; Boć, R. Vessel Multi-Parametric Collision Avoidance Decision Model: Fuzzy Approach. *J. Mar. Sci. Eng.* **2021**, *9*, 49. https://doi.org/10.3390/jmse9010049.
8. Grm, A. Ships Added Mass Effect on a Flexible Mooring Dolphin in Berthing Manoeuvre. *J. Mar. Sci. Eng.* **2021**, *9*, 108. https://doi.org/10.3390/jmse9020108.
9. Lujan, E.; Vergara, E.; Rodriguez-Melquiades, J.; Jiménez-Carrión, M.; Sabino-Escobar, C.; Gutierrez, F. A Fuzzy Optimization Model for the Berth Allocation Problem and Quay Crane Allocation Problem (BAP + QCAP) with *n* Quays. *J. Mar. Sci. Eng.* **2021**, *9*, 152. https://doi.org/10.3390/jmse9020152.
10. Androjna, A.; Belev, B.; Pavic, I.; Perkovič, M. Determining Residual Deviation and Analysis of the Current Use of the Magnetic Compass. *J. Mar. Sci. Eng.* **2021**, *9*, 204. https://doi.org/10.3390/jmse9020204.
11. Łazuga, K.; Quý, N.M.; Gucma, L. Cost-Effective Design of Port Approaches Using Simulation Methods Based on the Example of a Modernized Port in the Ustka. *J. Mar. Sci. Eng.* **2021**, *9*, 211. https://doi.org/10.3390/jmse9020211.
12. Baric, M.; Mohovic, R.; Mohovic, D.; Pavic, V. The Simulation of Sloped Bank Effect Influence on Container Ship Trajectory. *J. Mar. Sci. Eng.* **2021**, *9*, 1283. https://doi.org/10.3390/jmse9111283.
13. Gucma, L.; Androjna, A.; Łazuga, K.; Vidmar, P.; Perkovič, M. Reconstructing Maritime Incidents and Accidents Using Causal Models for Safety Improvement: Based on a Case Study. *J. Mar. Sci. Eng.* **2021**, *9*, 1414. https://doi.org/10.3390/jmse9121414.
14. Mentjes, J.; Wiards, H.; Feuerstack, S. Berthing Assistant System Using Reference Points. *J. Mar. Sci. Eng.* **2022**, *10*, 385. https://doi.org/10.3390/jmse10030385.

15. Maljković, M.; Pavić, I.; Meštrović, T.; Perkovič, M. Ship Maneuvering in Shallow and Narrow Waters: Predictive Methods and Model Development Review. *J. Mar. Sci. Eng.* **2024**, *12*, 1450. https://doi.org/10.3390/jmse12081450.

References

1. Jungen, H.; Specht, P.; Ovens, J.; Lemper, B. The Rise of Ultra Large Container Vessels: Implications for Seaport Systems and Environmental Considerations. In *Dynamics in Logistics: Twenty-Five Years of Interdisciplinary Logistics Research in Bremen, Germany*; Freitag, M., Kotzab, H., Megow, N., Eds.; Springer International Publishing: Cham, Switzerland, 2021; pp. 249–275.
2. Paulauskas, V.; Paulauskas, D. Dependence of Ships Turning at Port Turning Basins on Clearance under the Ship's Keel. *Sustainability* **2024**, *16*, 2819. [CrossRef]
3. Degrieck, A.; Uyttersprot, B.; Sutulo, S.; Guedes Soares, C.; Van Hoydonck, W.; Vantorre, M.; Lataire, E. Hydrodynamic ship–ship and ship–bank interaction: A comparative numerical study. *Ocean Eng.* **2021**, *230*, 108970. [CrossRef]
4. Perkovic, M.; Gucma, M.; Luin, B.; Gucma, L.; Brcko, T. Accommodating larger container vessels using an integrated laser system for approach and berthing. *Microprocess. Microsyst.* **2017**, *52*, 106–116. [CrossRef]
5. Fan, S.; Yang, Z.; Wang, J.; Marsland, J. Shipping accident analysis in restricted waters: Lesson from the Suez Canal blockage in 2021. *Ocean Eng.* **2022**, *266*, 113119. [CrossRef]
6. Yu, Y.; Ahn, Y.J.; Lee, C.H. Using FRAM for Causal Analysis of Marine Risks in the Motor Vessel Milano Bridge Accident: Identifying Potential Solutions. *Appl. Sci.* **2023**, *13*, 8764. [CrossRef]
7. Zhang, Q.; Zhang, L.; Sun, A.; Meng, X.; Zhao, D.; Hancock, C. GNSS Carrier-Phase Multipath Modeling and Correction: A Review and Prospect of Data Processing Methods. *Remote Sens.* **2024**, *16*, 189. [CrossRef]
8. Ito, K.; Tsuyuzaki, T.; Yuasa, D.; Choi, Y.; Kim, Y.B. Development of a Portable Interface System Sharing the Positioning and Heading Information to Support a Berthing Vessel. *J. Mar. Sci. Eng.* **2022**, *10*, 1637. [CrossRef]
9. Androjna, A.; Perkovič, M. Impact of Spoofing of Navigation Systems on Maritime Situational Awareness. *Trans. Marit. Sci.* **2021**, *10*, 361–373. [CrossRef]
10. Androjna, A.; Pavić, I.; Gucma, L.; Vidmar, P.; Perkovič, M. AIS Data Manipulation in the Illicit Global Oil Trade. *J. Mar. Sci. Eng.* **2024**, *12*, 6. [CrossRef]
11. Potamos, G.; Stavrou, E.; Stavrou, S. Enhancing Maritime Cybersecurity through Operational Technology Sensor Data Fusion: A Comprehensive Survey and Analysis. *Sensors* **2024**, *24*, 3458. [CrossRef] [PubMed]
12. Dimakopoulou, A.; Rantos, K. Comprehensive Analysis of Maritime Cybersecurity Landscape Based on the NIST CSF v2.0. *J. Mar. Sci. Eng.* **2024**, *12*, 919. [CrossRef]
13. Li, P.; Wang, Y.; Yang, Z. Risk assessment of maritime autonomous surface ships collisions using an FTA-FBN model. *Ocean Eng.* **2024**, *309*, 118444. [CrossRef]
14. Namgung, H.; Kim, J.S. Collision Risk Inference System for Maritime Autonomous Surface Ships Using COLREGs Rules Compliant Collision Avoidance. *IEEE Access* **2021**, *9*, 7823–7835. [CrossRef]
15. Ni, S.; Wang, N.; Li, W.; Liu, Z.; Liu, S.; Fang, S.; Zhang, T. A deterministic collision avoidance decision-making system for multi-MASS encounter situation. *Ocean Eng.* **2022**, *266*, 113087. [CrossRef]
16. Namgung, H. Local route planning for collision avoidance of maritime autonomous surface ships in compliance with colregs rules. *Sustainability* **2022**, *14*, 198. [CrossRef]
17. Vagale, A.; Oucheikh, R.; Bye, R.T.; Osen, O.L.; Fossen, T.I. Path planning and collision avoidance for autonomous surface vehicles I: A review. *J. Mar. Sci. Technol.* **2021**, *26*, 1292–1306. [CrossRef]
18. Hebbar, A.A.; Schröder-Hinrichs, J.-U.; Yildiz, S. Vessel Traffic Management in the Era of Maritime Autonomous Surface Ships and Digitalization: Experiences in European Waters. In *Area-Based Management of Shipping: Canadian and Comparative Perspectives*; Chircop, A., Goerlandt, F., Pelot, R., Aporta, C., Eds.; Springer Nature Switzerland: Cham, Switzerland, 2024; pp. 185–205.
19. Guo, W.; Zhang, X.; Wang, J.; Feng, H.; Tengecha, N.A. Traffic Organization Service for Maritime Autonomous Surface Ships (MASS) with Different Degrees of Autonomy. *J. Mar. Sci. Eng.* **2022**, *10*, 1889. [CrossRef]
20. IMO A 22/Res.915. Revised Maritime Policy and Requirements for a Future Global Navigation Satellite System (GNSS). January 2002. Available online: https://wwwcdn.imo.org/localresources/en/KnowledgeCentre/IndexofIMOResolutions/AssemblyDocuments/A.915(22).pdf (accessed on 3 September 2024).
21. Gospić, I.; Martić, I.; Degiuli, N.; Farkas, A. Energetic and Ecological Effects of the Slow Steaming Application and Gasification of Container Ships. *J. Mar. Sci. Eng.* **2022**, *10*, 703. [CrossRef]
22. Perkovič, M.; Batista, M.; Luin, B. Ship handling challenges when vessels are outgrowing ports. In Proceedings of the Port Management & Navigation Seminar (Trelleborg), Dubai, United Arab Emirates, 12–13 December 2023; p. 64. Available online: https://www.researchgate.net/publication/376456166_Ship_Handling_Challenges_When_Vessels_are_Outgrowing_Ports (accessed on 3 September 2024).
23. Castells-Sanabra, M.; Mujal-Colilles, A.; LLull, T.; Moncunill, J.; Martínez de Osés, F.X.; Gironella, X. Alternative Manoeuvres to Reduce Ship Scour. *J. Navig.* **2021**, *74*, 125–142. [CrossRef]

24. IMO MEPC.335(76). 2021 Guidelines on the Shaft/Engine Power Limitation System to Comply with the EEXI Requirements and Use of a Power Reserve. June 2021. Available online: https://wwwcdn.imo.org/localresources/en/KnowledgeCentre/IndexofIMOResolutions/MEPCDocuments/MEPC.335(76).pdf (accessed on 3 September 2024).
25. Perkovič, M.; Gucma, L.; Feuerstack, S. Maritime Security and Risk Assessments. *J. Mar. Sci. Eng.* **2024**, *12*, 988. [CrossRef]

Journal of
Marine Science and Engineering

MDPI

Article

Laser-Based Aid Systems for Berthing and Docking

Marko Perković [1,*], Lucjan Gucma [2], Mateusz Bilewski [2], Bartosz Muczynski [2], Franc Dimc [1], Blaž Luin [1], Peter Vidmar [1], Vivien Lorenčič [1] and Milan Batista [1]

1 Faculty of Maritime Studies and Transport, University of Ljubljana, 6320 Portorož, Slovenia; franc.dimc@fpp.uni-lj.si (F.D.); blaz.luin@fpp.uni-lj.si (B.L.); peter.vidmar@fpp.uni-lj.si (P.V.); vivien.lorencic@fpp.uni-lj.si (V.L.); milan.batista@fpp.uni-lj.si (M.B.)

2 Faculty of navigation, Maritime University of Szczecin, 70-500 Szczecin, Poland; l.gucma@am.szczecin.pl (L.G.); m.bilewski@am.szczecin.pl (M.B.); b.muczynski@am.szczecin.pl (B.M.)

* Correspondence: Marko.Perkovic@fpp.uni-lj.si

Received: 21 April 2020; Accepted: 9 May 2020; Published: 12 May 2020

Abstract: The berthing of an ultra large ship is always a difficult issue and becomes yet more complex when vessels must be handled in restricted manoeuvring areas of limited depth, exposed to a forceful crosswind, or manoeuvring in a strong current, or all three. The final approaching manoeuvre and precise positioning is particularly demanding at container terminals where many STS cranes are located along the quay, seriously limiting margin for error in the process of mooring a ship, especially when the cranes are located nearby a bridge wing or at the very edge of the pier. In order to avoid collisions, the final manoeuvre (side-push) must be fully controlled; the ship's orientation must be parallel with the quay while maintaining the minimum lateral approaching velocity without significantly shifting the vessel longitudinally. The mooring of a Ro-Ro vessel is occasionally even more challenging: a precise docking manoeuvre is normally executed without any towing assistance. In this paper low cost laser-based berthing and docking systems developed for the ports of Koper and Swinousce are presented and several berthing manoeuvres are analysed and compared with the most commonly used GNSS-based navigational aid system portable pilot units (PPU).

Keywords: container ship; precise positioning; portable pilot unit; gyro error; container terminal; berthing and docking; LiDAR scanner; STS crane

---

1. Introduction

Every port represents its own set of challenges to provide safe navigable waterways along with properly designed harbour approach channels [1]; adequate basin area available for manoeuvring and wharf approaching [2]; and guaranteeing completely safe berthing for particular ships for the duration of their stay [3]. Adverse weather can render even a well-designed mooring facility unsafe, but as long as the ship can safely leave the port or is carefully monitored while moored, such a port may still be considered safe. Ordinary good navigation and seamanship will not render a port unsafe [4]. This paper focuses on near shore vessel positioning required for safe berthing or docking, as well as monitoring vessels moored alongside. In berthing and unberthing situational awareness in the surrounding area is essential for the pilots and masters in order to take appropriate actions especially at the final manoeuvring stage where the speed of approach, distance, and angle of attack must be precisely determined. At the end, every vessel meets the berthing wharf, transferring a significant amount of energy, mostly depending on the ship's speed and its displacement tonnage [5]. It is obvious that the approaching speed must be controlled to avoid damaging the ship's hull, fenders, and other elements of berthing infrastructure.

According to the British Standard [6], risk reduction measures can consist of equipment installed on board the ship or at the berth. Such equipment could be either installed at shore with display units visible from the ship's bridge, or Portable Pilot Units (PPU) used on board the ship [7]. Various levels of awareness can be obtained from a PPU, which is currently used in many ports. It is a computer-based system that a pilot brings onboard a vessel to use as a decision-support tool for navigation, docking procedures or lock entry [8]. The unit is interfaced with either vessel positioning and heading sensors, or it may have its own advanced sensors providing centimetre position accuracy (less than 2 cm RMS when L1/L2 RTK GNSS service is available) and heading precision up to 0.01 deg. An advanced system is commonly used at LNG terminals and when other sensitive operations like the handling of very large vessels, narrow passages, side-by-side mooring and locking are involved in the process. PPU systems are definitely useful and highly recommended for pilots when handling vessels in the "ballistic" manoeuvring phase and during the final "side-pushing" or "positioning" phase when the elements of global satellite-aided positioning system (GNSS) are not obstructed and when the pilot has enough time to properly set-up the advanced PPU system; i.e., fill in receiving antenna positions (in body fixed coordinate system) and identifying the ships gyro heading offset, if possible.

At some locations, such as the port of Koper, the approaching and berthing pilotage phase is extremely short, so there often is not enough time to set-up an advanced PPU system (pilots sometimes execute more than 10 manoeuvres in their 12 h shift). Even so the GNSS satellite signal may be obstructed or redirected in certain navigational areas, thus consequently ordinary PPU based on ship's Automatic Identification System (AIS) does not always provide reliable data. Precise vessel position, transversal speed, and vessel-approaching angle are of great importance for safely berthing large container vessels particularly when inadequate fenders await and ship-to-shore (STS) cranes are near. The greatest pressure currently on ports is "Scale Enlargement," a euphemism for the incessantly increasing size of vessels along with the extreme pressure put on ports to accommodate this phenomenon. The rapid increase in ship sizes forces ports into a state of virtually constant adjustment. Container vessels are the best example of ships that grow faster than ports [9,10]. As early as 2004 an article warned of berthing challenges that would arise with ever bigger container ships; two particular challenges would be that they become more difficult to handle, and, worse, it would be difficult, at times impossible, to see the side of the quay once up close [11].

In recent years the cargo throughput in the port of Koper has averaged an annual growth of 8% with container growth as much as 16% yearly without an increase in the number of vessel calls. Management of this throughput was made possible by dredging activities and pier extension, allowing for the reception of larger vessels. Figure 1a depicts the evolution of the size of container ships calling at Koper in the last decade. The size doubles whether expressing deadweight tonnage (DWT) or Gross Tonnage (GT). When one of the larger vessels is alongside even minor yawing or list can cause contact with STS cranes, with catastrophic results. The extremely small gap between ship and crane is evident from Figure 1b.

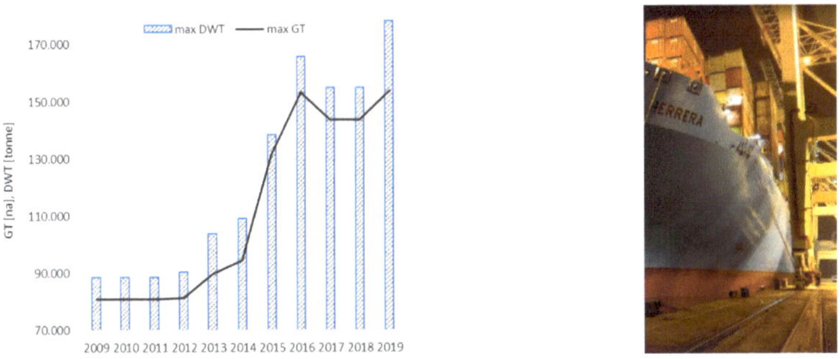

(a) (b)

Figure 1. Evolution of the size of container ships calling at Koper (**a**) and ship-to-shore (STS) crane close proximity (**b**).

The largest cranes, necessary for the largest vessels, must be fitted on the extant rails, which were designed, originally, to provide a safe distance between the vessels and cranes. This is not speculative, as one can see from the disastrous events of Figure 2 and Table 1, all occurring in 2019. Each of these collisions had a major impact on the port, aside from the already significant cost of replacing the damaged cranes. The port, at least near the incident, closes, the replacement can take up to a year, and of course operators may find an alternative port in the meantime.

(a) (b) (c)

Figure 2. Container ship/crane collisions (**a**) with bow; (**b**) stern and (**c**) ships' bridge wing.

Table 1. Container ship/crane accidents in 2019.

Date and Location	Vessel Name and Capacity/Size	Accident Description
2019/01/28, Vancouver,	C/S "Ever Summit", 78612 dwt, 7024 TEU, 300 × 42.9 m	Berthing: while a ship was pushed alongside the dock the stern struck the crane located at the edge of the terminal; the force of impact knocked the crane rails off the quay, after which the crane collapsed onto the ship.
2019/02/03, Port of Shuwaikh, Kuwait,	C/S "Belgian Express", 25775 dwt, 1794 TEU, 179.8 × 27.6 m	Departure: while leaving the berth the vessel collided with the gantry crane.
2019/06/07, Hai Pong, Container Terminal, Vietnam	C/S "Nagoya Express", 103646 dwt, 8749 TEU, 335 × 42 m	Departure: while unmooring, the vessel collided with a gantry crane that did not collapse but was critically damaged.
2019/07/14, Terminal Petikemas Semarang in Indonesia	C/S "Soul of Luck" 21519 dwt, 1642 TEU, 168.1 × 27.2 m	Berthing: a feeder vessel approached at a high angle and hit the gantry crane with its bow - the crane collapsed.
2019/12/09, Antwerp, DP World Terminal	C/S "APL Mexico City", 115024 dwt, 9326 TEU, 328.2 × 45.2 m	Mooring breakaway: vessel drifted as her mooring broke due to strong wind, unsuccessful towing assistance, vessel's bridge wing hit the STS crane (unmanned), which collapsed and destroyed the quay.

During the last decade, the development of ship sizes of Ro-Ro/Ro-Pax vessels and the scale enlargement of product tankers have also been problematic. As the vessel sizes and amount of traffics increases, safe and efficient berthing becomes more challenging worldwide. Among the many incidents in recent years, in the port of Koper in 2017 a Ro-Ro vessel, having moored then broken away, crashed into the bulk terminal, knocking over a crane, seriously damaging the pier (100 meters had to be replaced).

The worst of this type of accident occurred just as this paper was being finalized (on 6 April 2020 in Busan port). Due to an uncontrollable high-speed, the container ship "Milano Bridge" collided with an entire strip of container cranes on the dock. Only one of the six gantry cranes remained functional after the accident. As possible causes, the investigators cited not only speed, but also a shorter reaction time after turning, insufficient power from tugs due to the high forward speed and lack of effect of the bow thrusters. The crew had not taken into account the fact that the ship was empty and was sailing with a partially submerged rudder, nor had they sufficiently considered the local weather conditions and, on top of all that, the pilot panicked as the ship approached the dock. Given the extraordinary cost of the cranes, the economic impact requires no elaboration.

2. Berthing Aid System, Development and Some Applications

2.1. An Historic Overview

The need for berthing support systems first emerged in the oil and gas industry. With increasing size and complexity of tankers, the owners had to pay more attention to potential risk factors when planning to handle larger ships, preventing possible damage to both jetties and ships, and, of course, preventing environmental pollution (along with insurance and demurrage). In the early 1970s the first mooring load monitoring system was developed in Norway, based on strain-gauged load measuring pins installed into the quick release hooks to measure the tension in the mooring line [12]. The first real docking and monitoring system integrated with sonar-based speed of approach measurement, environmental and meteorological sensors, was designed in 1990 [12]. At almost the same time developments in the measurement of speed of approach were tending to favour radar-based applications-ultrasonic systems had a problem with surrounding water and fouling of the ships' hull. The radar also had some difficulties [11], not being a focused sensing device, backscattering data must be filtered and any changes in surroundings like a moved or moving crane would require additional filtering, a software upgrade. Finally, with the development of eye-safe lasers more accurate and reliable berthing and docking measurements in all weather conditions were made possible. Comparison of ultrasound, radar and laser diode ranging systems using a dedicated scale is presented in Table 2, considering: range, power, dependability of weather, beam size, method of measurement, refresh rate, price and lifetime. Ultrasound can be eliminated, and both radar and laser ranging solutions are capable of being used in berthing and docking systems. A laser-based ranging system is not subject to weather; the most significant problem theoretically is reflectivity of the measured object (the vessel in this case). With radar a microwave is reflected by the flat surfaces and the method of measurement used here–frequency modulated continuous wave (FMCW)–provides a very reliable solution with the same antenna for received and transmitted signals. The pulse system used in the lasers is resistant to external interferences and usually is coded. The FMCW method used in radars is very rugged and not affected by interference. Because of the range factor, radar is not used to their full potential; lasers have better flexibility. Another consideration is that radar is approximately 5 times more expensive than laser.

Table 2. Comparison of methods.

Parameter	Ranging Method		
	Laser Diode	**Radar**	**Ultrasounds**
Range	to 300 m	to 2 km	to 10 m
Emission power	10 mW (laser)	10 mW (96 GHz)	1W (sound)
Beam	Very narrow (<2 deg)	Narrow (<5 deg)	Wide (>5 deg)
Weather dependability	Low	Very Low	Significant
Price	Average	Very High	Low
Method of measurement	FMCW	Pulse	Continuous-interrupted
Refresh rate	0.1 kHz	1 kHz	0.001 kHz
Lifetime	Average	Average	High

Another advancement is the integration of the ranging system with others, like positioning, environmental, and PPU. Among others a non-commercial alternative has been provided by a laser measurement system integrated with a high precision positioning pilot navigation docking system (PNDS), which was proposed for the docking of LNG vessels [13]. A low-cost integrated laser ranging and berthing system integrated with meteorological and oceanographical data (MetOcean) was developed for the safe passage through the narrow winding channel and the final berthing of large vessels calling at the container terminal in the port of Koper [14] (the precise descriptions apply to the methods used at the port of Koper; others may, for instance, place the LiDARs in different strategic locations). Installing ranging devices on moving STS cranes is more complicated compared to jetty installation, where sensors are at fixed positions. When a container vessel is berthing, STS cranes are tied down to the wharf structure (wharf anchorage) at s known position, so the exact LiDAR positions are always known. LiDARs and processing units are located on the STS (ship to shore) crane, just above the wheels or about at 2 m height (Figure 3). For the sake of accuracy, while the vessel is approaching, the STS crane (one with LDS units) shall be located wherever the vessel's outline is parallel. The parallel body of the vessel hull can be found at the middle part and can be as much as 60% of the vessel length. The equipment used for measurements is a combination of different sensors and modules that are remotely controlled by pilots through specific user interfaces.

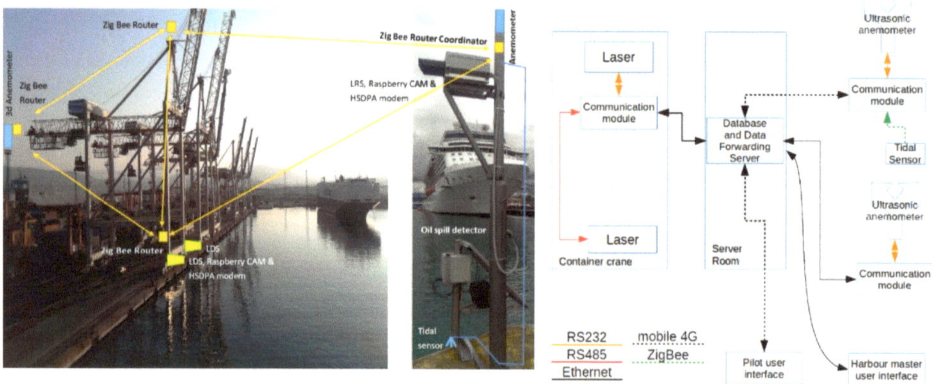

(a) (b)

Figure 3. Laser ranging and berthing system (**a**) port of Koper installation, (**b**) communication protocol.

The list of equipment is shown in Table 3. In order to extend laser diode lifetimes, devices are powered on demand. The laser distance meters in the docking system provide 2 cm accuracy for distances up to 300 m, while the ranging sensor provides 1.5 cm accuracy up to 400 m. All anemometers are able to measure wind velocities in three directions between 0 and 65 m/s, which is sufficient in the area of interest. Measurement can be provided at a sample rate of 20 Hz. The resolution of the ultrasonic tidal sensor is 1 cm.

Table 3. Comparison of methods.

Device	Model	Application
Laser Distance Meter (diode based)	Riegl LD90	Ranging sensor
Laser Distance Meter (diode based)	Jenoptik LDM301	Docking system
ZigBee to SIO module	Digi XBEE-PRO XBP24	For remote tidal sensor data transm.
Utrasonic Distance Meter	MaxBotix XL-MaxSonar WR1	For tide measurement
Ultrasonic 3D anemometer	Gill WindMaster PRO	Wind velocity measurement
Ultrasonic 2D anemometer	Gill WindObserver 65	Wind velocity measurement

2.2. Some Applications

The berthing aid or docking system measures the precise distance from which lateral speed and orientation of an approaching vessel is calculated in real time. Even though this system can monitor vessel movement only when it is allied with the jetty (i.e., when both laser beams are reflected from the parallel part of vessel hull called parallel body), it still provides invaluable information to the captain or pilot when berthing a ship. Additionally, such a system can be used as a vessel drift warning system, identifying uncontrolled vessel movements caused by passing vessels that cause ship-to-ship interaction, or when a ship is moved by strong winds or currents. Furthermore, the entire docking operation is thus recorded, and the data can be used to evaluate potential incidents.

In order to measure the longitudinal position of an approaching vessel, the laser system must be integrated with the vessel's AIS system, providing GNSS data and the ships' heading information. When precise front-end docking is required (Ro-Ro), a differential GNSS system which provides centimetre accuracy can be used along with 3D laser scanning devices.

Data gained from the berthing aid system are additionally applied for the development of new port design recommendations and standards; i.e., researchers have been investigating different methods of fender design. Sakakibara and Kubo [15] used a docking sonar system and found actual berthing conditions that were not in line with those used for design purposes. They have presented a measurement of pressure on pneumatic fenders; Metzger et al. [16] conducted a similar study based on using ultrasonic distance sensors calculating vessel kinematics and analysing pressure on fenders. Yamse et al. [17] analysed more than 2500 instances of berthing velocities in various conditions (14 terminals in 6 countries) and found that their results did not fit the Brolsma curve [18] used as a standard for designing fenders based on vessel velocity, displacement tonnage, and navigational conditions. One of the most important discoveries [17] is that winds blowing under 10 m/s do not contribute to the increase in berthing velocity. At winds under 10 m/s the berthing velocity depends solely on the ship and its port assistance. They found that there is no need to divide ports and berthing velocities into five groups; Rather they have grouped berthing velocities in types A and B, based on operator experience and environmental conditions (Table 4). Ueda [19] statistically analysed laser-based measured berthing velocities-obtained at different container terminals located in Japan and elsewhere in East Asia. Large vessel berthing impact was analysed by Hein [20] where vessel kinematics was measured with radar-based systems, while Roubus et al. [21] and Kirbiš et al. [22] used a laser-based system. Berthing velocities are relatively low in the port of Koper, as fairly indicated in the study. Berthing velocities are significantly lower even for group A, where excellent operators are handling vessels in temperate MetOcean conditions.

Table 4. Berthing velocities.

Berthing Velocity		**Port of Koper Study [22]** *Large Container Ships*		**Various Ports [17]**	
				Group A	*Group B*
		Angle of Attack (deg)	**Berthing Velocity (cm/s)**	**Berthing Velocity (cm/s)**	
	Average	0.48	3.05	3.9–5.4	6.7–9.3
95% (2δ)	Confidence value	1.74	7.21	7.7–11.3	14.1–16.9
99% (3δ)	Confidence value	2.26	8.96	9.4–16.0	19.3–23.8

In the scope of unmanned navigation, which is still solving a variety of technical problems, largely to do with automation [23], the laser ranging and berthing system is one of the core subsystems of autonomous surface vessels required for collision avoidance [24,25].

3. Performance: Comparison Between the Pilot Navigation System and LiDAR

3.1. System Structures and Measurements

Pilot navigation systems are sometimes seen as alternatives to the pier-mounted laser and radar systems. Suppliers claim superior accuracy in terms of positioning, heading and rate of turn (ROT) measurements. The systems, however, mostly rely on non-RTK capable GNSS receivers that are vulnerable to disturbances of satellite signals. It was assumed that even under normal conditions, the reliability of such systems could be compromised in some cases, especially if the geometry of the structures in the area is such that it could cause multipath signal travel. Therefore, a comparison of different systems' performances for berthing applications was carried out. In order to carry out such a comparison, AIS logs from the pilot navigation systems and from ships were decoded using special purpose Python scripts in conjunction with Libais library. The event timestamps are from the pilot navigation systems that have time synchronization through the GNSS. For laser measurements, NTP time synchronization on logging computers was used in order to obtain correct measurement time.

When the ship's fixed GNSS position data was analysed, AIVDO/AIVDM designated messages were used; And when the pilot navigation system was used as a positioning source, the supplemental GPGGA message was used to obtain positions and custom HEHDT and HEROT vendor specific messages were used for heading and rate of turn.

All measurements were resampled and aligned for comparison using the Pandas Python processing library. In Figure 4a an overview of measurement equipment positions is shown. Several LIDARs were evaluated, from high frequency low end model (HS) with ±50 mm accuracy to higher precision crane-mounted (JEL, JER) ±20 mm accuracy, and also high-precision (HP) with ±1 mm accuracy that was used as a reference to compare all other sensors. Two Jenoptik LDS301 units were mounted on a crane at a fixed distance and used to determine the orientation relative to the quay and fenders. During the testing, an additional LiDAR was used (Riegel), (parameters in Table 5). Distance *d* was measured, unknown variable in the test case. In order to compare the accuracy of pilot navigation systems with LIDARs, it was necessary to calculate the position of the ship's edge from the AIS and GNSS data. AIS and GNSS positions are in the form of geographical latitude and longitude converted into a local coordinate system that converted positions into distance units (meters). The ship's GNSS position is shown as the point $P_g(x_g,y_g)$. The vector $\vec{O_s}$ is a unit vector representing ship orientation and is calculated from the heading data.

$$\vec{O_s} = \left(\cos\left(\pi/2 - \psi_g\right), \sin\left(\pi/2 - \psi_g\right)\right) \tag{1}$$

Vector, perpendicular to the ship orientation is

$$\vec{O_p} = \left(\cos\left(-\psi_g\right), \sin\left(-\psi_g\right)\right) \tag{2}$$

They are used to obtain points P_1 and P_2 on the ship's edge facing the pier

$$\vec{P_1} = \vec{Pg} + h\vec{O_p} \tag{3}$$

and P_2 a translation of s meters along $\vec{O_s}$ direction is calculated

$$\vec{P_2} = \vec{P_1} + s\vec{O_s} \tag{4}$$

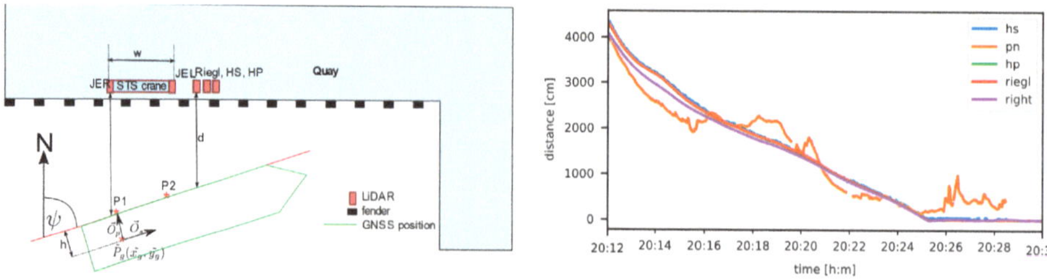

(**a**) (**b**)

Figure 4. Measurement equipment setup and ship data (**a**), measured distance *d* versus time (**b**).

Table 5. LIDAR parameters.

Laser Rangefinder Model	Abbr.	Range (m)	Accuracy (mm)	Sampling Frequency (Output) (Hz)	Wavelength (nm)
Riegl LD90	Riegl	150 1000 *	±15	1	905
Jenoptik LDM301	JEL JER	300 3000 *	±20	100	905
Foresight FST-B200	HS	200	±50	200	905
Foresight FST-100M	HP	100	±1	10	635

* (Retro-Reflective targets).

Once the points $P_1(x1,y1)$ and $P_2(x2,y2)$ are obtained, equations for the line defining the ship's edge are obtained. The distance d is obtained by calculating the distance between the obtained line and the LiDAR positions perpendicular to the pier.

In Figure 5a, it can be seen that as the vessel is tugged closer to the pier that measured distances between different sensors are minimized. However, the pilot navigation system data is apparently showing unrealistic changes in positions due to suspected multipath problems, at least from positions recorded in the log. The errors when the vessel position was supposed to be constant (after 20:26) were about 5 meters. However, a comparison between the pilot navigation system and ship's own GNSS receiver reveals that the ship's receiver performs significantly worse, which is shown in Figure 5a.

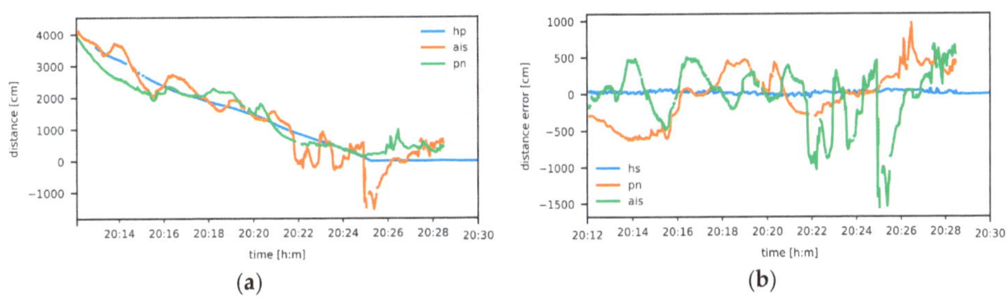

Figure 5. Comparison of LIDAR reference, ship and pilot navigation system (**a**) and distance errors of ship GPS, pilot navigation system and worst LIDAR (**b**).

To determine errors shown in Figure 5b, the differences between the range obtained from ship GNSS, pilot navigation system, the worst and best LiDAR as references are shown. Even the worst, low-cost LiDAR is orders of magnitude more precise than common pilot navigation systems, especially in the areas affected by the multipath disturbances. However, with the pilot navigation system, position error was mostly within 5 m while the ship GNSS yielded far worse results, even though it has antennas placed at much more favourable locations than portable pilot systems. In order to illustrate the quantification of the measurement results presented in Figure 5, Root Mean Square Error (RMSE) and maximum absolute errors recorded are shown in Table 6. The results indicate significant errors in the GNSS-based systems in the vicinity of the overhead cranes and the strong benefit of using even low-end LIDAR (relative errors calculated at 5 m). The results suggest that it would be beneficial to augment the GNSS data using local LiDAR measurements in order to obtain more accurate position during the docking procedures.

Table 6. Comparison of positioning errors.

Sensor	RMSE (cm)	Max Error (cm)	Mean Relative Error (%)	Max Relative Error (%)
HS LiDAR	28	73	2	14
Pilot Navigation	355	969	25	189
AIS	406	1553	31	307

3.2. Fusion of AIS, PPU and LiDAR Measurements

Fusion of the AIS position data and local measurements taken with LiDARs or radars mounted in the dock would be beneficial for pilots during docking procedures. This way LiDAR precision and reliability would be combined with AIS data, which provides longitudinal positioning along the pier (Figure 6). The calculation is done using the UTM coordinate system. Location of lasers onshore is given by points (x_{01},y_{01}) and (x_{02}, y_{02}). Coordinates of the vessel's port side parallel body are given by

$$x_i = x_{0i} + d_i \sin\alpha\,,\ y_i = y_{0i} - d_i \cos\alpha,\ i = 1,2 \tag{5}$$

where d_i is offset of the LiDARs and the angle a represents the orientation of the pier

$$\alpha = \arctan\frac{y_{02} - y_{01}}{x_{02} - x_{01}} \tag{6}$$

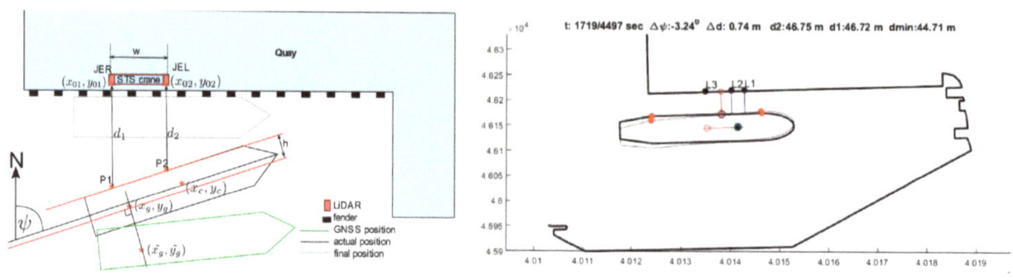

Figure 6. Fusion of GNSS and LiDAR measurements, model (**a**), final positioning (**b**).

Coordinates of the GNSS antenna on the ship, which is where AIS position data is obtained from, are (x_a, h_a) in the ship's relative coordinate system, which is the right side coordinate system with its origin in ship's centre, where x axis is directed from stern to bow. The location of the AIS transmitter from the left side parallel body is

$$h = \frac{B}{2} + h_a \tag{7}$$

Approximate GNSS coordinates transmitted by the ship over the AIS are $\left(\widetilde{x_g}, \widetilde{y_g}\right)$ and heading is $\widetilde{\psi_g}$. The obtained approximate ship coordinates will be used in the following calculation. The equation defining the starboard side parallel body edge of the vessel is

$$y = kx + n \tag{8}$$

where coefficient k and intersection n are calculated from points determined by the LiDAR measurements.

$$k = \frac{y_2 - y_1}{x_2 - y_1} \text{ and } n = y_1 - kx_1. \tag{9}$$

The precise GNSS antenna line is obtained from LiDAR-measured ship edge Equation (8) by translating a distance h towards the centre, obtaining the line

$$y = kx + n - h\sqrt{1+k^2} \tag{10}$$

Since precise GNSS antenna line is obtained from LiDAR measurements, longitudinal position of the vessel must be determined from the AIS data by projecting the GNSS coordinates (x_g, y_g) on the line specified by the Equation (10), obtaining

$$\begin{aligned} x_g &= \frac{\widetilde{x_g}+k\left(\widetilde{y_g}-n+h\sqrt{1+k^2}\right)}{1+k^2} \text{ and} \\ y_g &= \frac{k\widetilde{x_g}+n+k^2\widetilde{y_g}-h\sqrt{1+k^2}}{1+k^2}. \end{aligned} \tag{11}$$

This way we obtain the projected coordinates on the line specified by quay-mounted LIDARs, where x_g and y_g are relative coordinates from a reference point on the quay. They are calculated from approximate values of the corresponding coordinates obtained from the GNSS receiver $\widetilde{x_g}$ and $\widetilde{y_g}$. The k and n are the projection line coefficient and intersection and h is the offset of the ship GNSS antenna position from the side edge. The process is illustrated in the Figure 6. The ship heading is obtained by

$$\psi = \frac{\pi}{2} - \arctan(k) \tag{12}$$

The centre of the ship is therefore at the following point

$$\begin{aligned} x_c &= x_g - x_a\cos\psi + h_a\sin\psi, \\ y_c &= y_g - x_a\sin\psi - h_a\cos\psi. \end{aligned} \tag{13}$$

This way the ship's position and orientation are obtained by merging the accurate LiDAR measurements of ship to pier distances and longitudinal position from GNSS obtained AIS data.

Figure 6b depicts the real interface of data fusion. The actual berthing of Maersk Herrera: The bold line is the outline of the vessel measured by lasers; The thin line is that obtained by the PPU. The PPU takes the position from the ship's GNNS while the orientation of the ship is additionally filtered by the PPU (parameters in Table 7). As the data suggests, the position of the ship is quite good, but the orientation is insufficiently prepared for berthing. Extensive analysis of berthing led us to discover that aside from multipath disturbances, the greatest problem is the tendency for gyro error to be greater than the maximum IMO standards of what should be accepted. The orientation is attributable to gyro error; and for a ship that size, such an orientation error could mean a deviation of up to ten meters from where the pilot and captain think the bow of the vessel is.

Table 7. Automatic Identification System (AIS) and global satellite-aided positioning system (GNSS) receiver features.

GNSS Data Source	Abbr.	Position/Orientation Accuracy	Type
Ship's GNSS	AIS	variable	depending on the vessel age and type
SafePilot CAT ROT	PN	0.1 deg heading	pilot navigation system
SafePilot CAT I	PN1	0.7 m RMS	pilot navigation system

3.3. *PPU: Limitation of the GNSS and Ship's GYRO*

Marine pilots in the Port of Koper have observed sudden jumps of the vessel or tug in certain areas, on their monitoring equipment. The effect may well be related to the presence at least of two types of reflective objects in the surroundings: port cranes and the surface of the sea [26]. In order to apply the variance model, a strong correlation must exist between the satellite elevation angle and the GNSS signal quality. But variance models become inefficient when observations are affected by multipath, signal diffraction and receiver characteristics. For measurements collected under non-ideal observational conditions, direct signal quality measures such as signal-to-noise ratio (SNR) based variance models can be more appropriate for assessing the quality of GNSS observations. Furthermore, pseudorange multipath error mitigation methods exist (e.g., [27]), but they are not yet widely applied in the commercial maritime fleet.

The behaviour of the navigation solution and consequently its accuracy is subjected to a combination of diffuse forward scattering and fluctuations of very low frequency [28]. Secondary path signals with longer propagation time distort the amplitude and phase of the direct-path signals. As a consequence, the code ranges are more susceptible to multipath than carrier phases. Since the signals received from lower-elevated satellites are more prone to multipath (elevation-dependent weighting concept), observing the elevations of the satellites in view may contribute to a realistic model. But modelling of the propagation situation, e.g., [29] (p. 156), in the case of the port of Koper is based on navigational solution reports from vessels' AIS data. The crucial knowledge regarding which satellites were actually included in the determination of navigational solutions when anomalous behaviour occurs is missing. In the absence of signal quality monitoring, but assuming that no interfering in-band signals were affecting the navigation solution, a decision was made that spatial distribution of anomalous changes of navigational solutions would be mapped from the reports, and not from the GNSS satellite-receiver tracks.

A collection of AIS Class A messages, contains data: time, latitude, longitude, heading, Rate of Turn, Course over Ground, Speed over Ground. Data were preliminarily analysed without flags for the differential correction status (corrected, uncorrected), Position Accuracy (high <=, low > 10 m) and RAIM (in use, not in use).

The following figure, Figure 7a,b present the reported positions of two vessels. Green dots represent positions with speed corresponding to SOG, while blue lines represent anomalous events of rate of position exceeding SOG by more than twice.

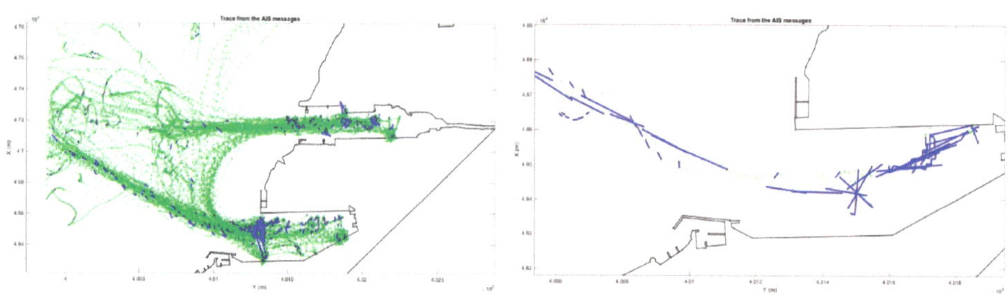

(a) (b)

Figure 7. Anomalous parts of the tug's Zeus track (**a**) and container ship Murat K track (**b**) as observed in AIS messages from September 1-24 2019, Port of Koper, coordinate system D96 (UTM).

The gyrocompass is an indispensable navigational instrument which should determine the direction of the ship's heading in relation to geographic (true) north. Various systems on board ship are under the control of the Master Gyro; the gyrocompass system transmits information to the Steering Console, RADAR, ECDIS, AIS, GNSS, VDR, GMDSS equipment, and other equipment or systems as required, such as PPU. The accuracy of predicted position can be significantly affected upon the occurrence of gyrocompass errors, especially during berthing manoeuvres. According to performance standards IMO A.424(XI), gyro error should not exceed 0.75 × secant latitude and the root mean square value of the differences between individual heading indications, and the mean should be less than 0.25° × secant latitude. Gyro settling error can be estimated from the orientation of a moored ship alongside a quay. A long-term study involving the collection of AIS data from the container terminal, where larger ships are assumed to be constantly moored parallel with the quay, provided 56 berthing events at berth 7 at port of Koper for our evaluation. The average observed absolute gyrocompass error equalled 1.4°, with a standard deviation of 1.1°, significantly higher than IMO standards. There is no reason to believe that this is not generally the case involving the gyrocompass.

4. Advanced Integrated LiDAR Based Docking System Specifically for Ro-Pax

4.1. Layout

To extend the capabilities of a positioning system based on laser distance measurements, an extended system based on laser scanners has been proposed. A test version of such a system has been implemented at a terminal ferry in the port of Świnoujście, Poland. It consists of two 3D LiDAR sensors located in such a way that bow/stern and the sides of the ship can be scanned at any time during the berthing/unberthing manoeuvre. LiDARs specification is given in Table 8 and the full architecture of the system is presented in Figure 8 and consists of laser scanners, two CCTV cameras, server, AIS receiver, anemometer, power system and data network.

Table 8. LiDARs specification (outdoor use).

Laser Class	1 (IEC 60825-1:2014, EN 60825-1:2014)
Aperture angle	Horizontal 85° (working range with 4 measuring planes, 25° extension of working range with 2 measuring planes to a total of 110°), vertical 3.2°
Scanning frequency	12.5 Hz ... 50 Hz
Angular resolution	0.125°, 0.25°, 0.5°
Working range	0.5 m ... 300 m
Scanning range at 10% remission	50 m
Amount of evaluated echoes	3
Detectable object shape	Almost any
Systematic error	± 300 mm
Statistical error	100 mm
Integrated application	Field evaluation
Number of field sets	16 fields
Simultaneous evaluation cases	16
Ambient operating temperature	−40 °C + 70 °C

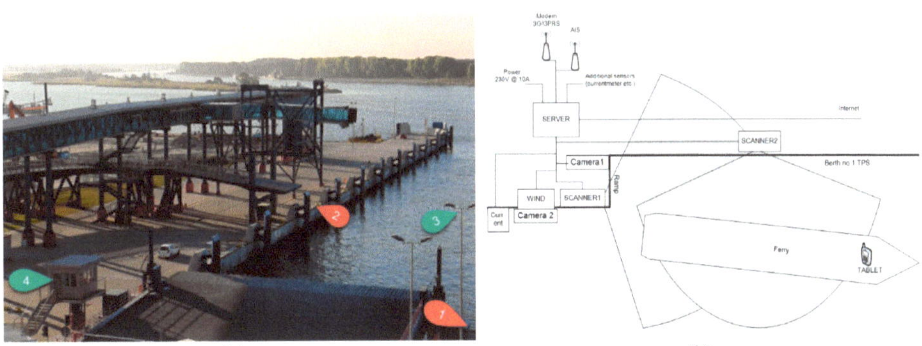

(**a**) (**b**)

Figure 8. Ferry post no 1 in Świnoujście port and the location of the system devices: 1, 2—laser scanners, 3—anemometer and camera, 4—camera, AIS receiver, server (**a**), photo taken by C.Aszkiełowicz. System diagram (**b**).

The data from the system are provided by on-board tablets which communicate with a server hosted application via LTE internet connection. Such an approach makes it simple to update and extend the application remotely, without the need to access the end users' devices. The application provides information about the ship's outline in relation to the berth, together with minimum distances: ship's side, berth and ship's bow/stern ramp. It also provides wind data, calculated velocities (longitudinal and transversals) and calculated heading.

4.2. Positioning, Velocity Determination and Ship Orientation Algorithm

Since it is impossible to establish the exact geometry of the hull and the actual draft of a ship using only two LiDARs, the system has been adapted for two specific ferries that use the berth on a daily basis. For each ship a family of outline shapes has been prepared, taking into consideration expected changes in draft. The positioning algorithm, after reading data points from both LiDARs, performs the following operations (Figure 9a):

1) Removes points identified as berth points (based on (x,y) coordinates)
2) Removes isolated points, i.e., points that do not have any neighbouring points in a 5 m radius
3) Approximates ship's side-line using the RANSAC algorithm
4) Removes outliers
5) Estimates ship's side using the Least Square Method (LMS)
6) Removes false points, e.g., points from seagulls flying low, inside the scanner's range
7) Finds the best estimation for ship's stern/bow
8) Returns best fit of ship's position.

The position of the ship's stern/bow is estimated by minimizing the sum of squares of distances between points and the ship's outline. This operation is repeated no less than 100 times (until the result does not improve by more than 0.00001). In Figure 10, the ship's sideline is illustrated, along with the points that were used to determine it. The points that are detected by the individual LiDAR are drawn in corresponding colours, in red and blue.

The RANSAC algorithm is used specifically to estimate the best line that fits points acquired by both LiDARs by minimizing a monotonically increasing function of the absolute values of the signed error for each point:

$$e_M(d;\Theta) = \frac{\Theta_1 x_1 + \Theta_2 x_2 + \Theta_3}{\sqrt{\Theta_1^2 + \Theta_2^2}} \tag{14}$$

The exact implementation of the RANSAC algorithm, based on [30], performs the following steps:

1) Selects two points randomly.
2) Solves in regard to the line parameters.
3) Determines how many points from the set of all points fit with a tolerance of 0.15 m.
4) If less than 70 percent of all points exceeds the assumed tolerance, repeats steps 1–3 but not more than 1000 times.
5) Otherwise, terminate the algorithm.

After the RANSAC algorithm is terminated a set of points located not more than 0.2/0.5 m (depending on the given ship) is selected. This set is used to re-estimate the line using the Least Square Method. The ship's state obtained from RANSAC is base position and course vector (x_i, y_i, K_i); it is transformed in every time step into velocities (v_x, v_y, ω). In this case the filtering and tracking itself is performed by algorithm based on a very simple and robust exponential smoothing by Holt method with Pegels' multiplicative trend and additional damping (Figure 9b) [31]. The applied algorithm allows the estimation of the location of the ship's gravity centre and course in the next time step (x_{i+1}, y_{i+1}, K_{i+1}), the ship's velocities in 3 degrees of freedom (v_x, v_y, ω), and the future (predicted) ship parameters given up to 3 min (ship tendency). The smoothing is done in separate steps for all three parameters (v_x, v_y, ω) according to the following formula:

$$\begin{aligned} X_t(m) &= S_t R_t^{\varphi} \\ S_t &= \propto X_t + (1-\propto)(S_{t-1} R_{t-1}^{\varphi}) \\ R_t &= \gamma\left(\frac{S_t}{S_{t-1}}\right) + (1-\gamma) R_{t-1}^{\varphi} \end{aligned} \tag{15}$$

where:
α, γ—smoothing coefficient,
ϕ—dampening coefficient,
$X_t(m)$—estimated parameter.

The values of smoothing coefficients (α, γ, ϕ) have been estimated based on the real time simulation trials for Ro-Pax ferries where minimum values of mean square errors were found for all estimated values.

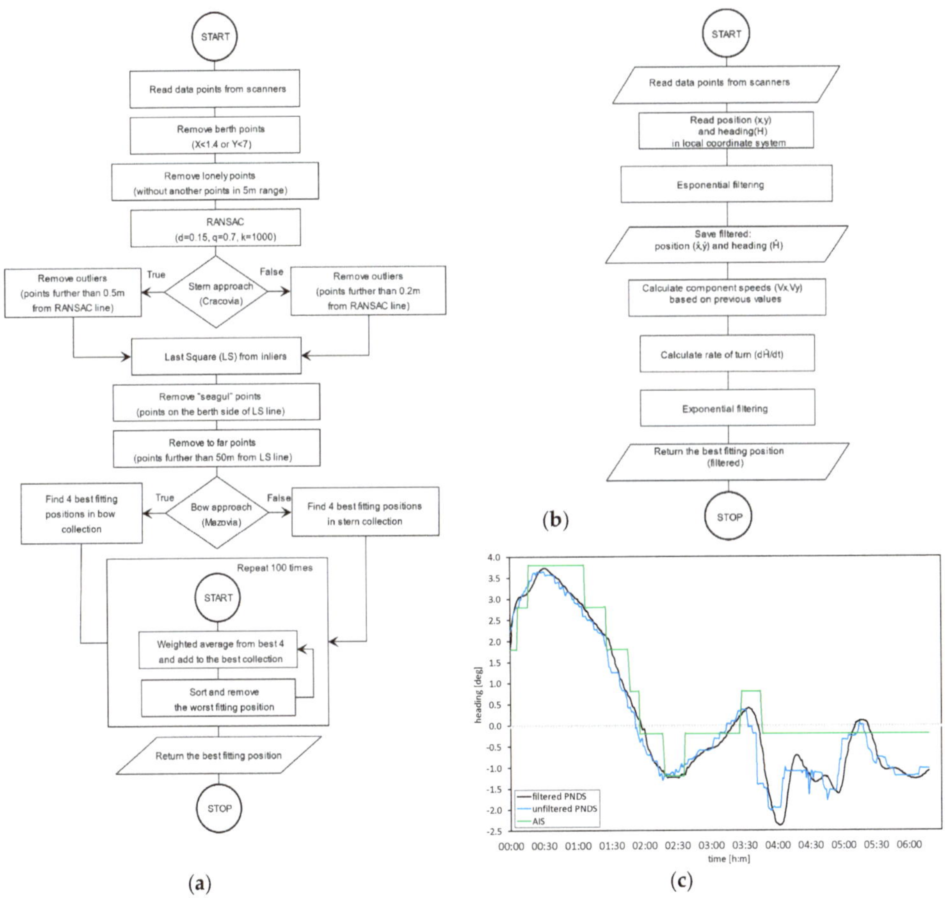

Figure 9. Algorithm for ship's position estimation (**a**); estimation and smoothing algorithm for course and velocities (**b**) with processed ship's true heading (**c**).

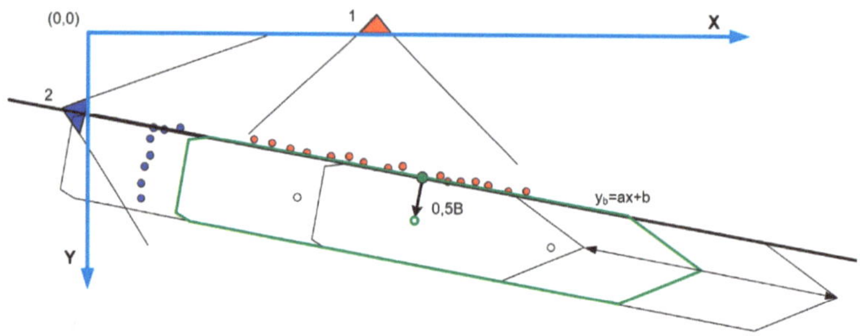

Figure 10. Estimation of ship's position from LIDARs measurements.

4.3. Verification of the System's Performance

During the stage of system development, the verification of system performance and algorithm accuracy was based on the comparison between the position calculated by the system and the position reported by the AIS system. A system of cameras was utilized as well to visually evaluate the stages of the manoeuvre and ship's dynamics. All comparisons were made using the developer version of the application. The first stage of the verification was based on a comparison between the position reported by the AIS system (GNSS position) and the position calculated by the PNDS algorithm. Due to the number of AIS messages, equal to around 6 messages per minute, only positions at the time of a new message were compared.

The point for which the AIS position is reported was evaluated statistically by measuring and analysing the position of the ferry when it was moored. For both ships it was evaluated as the point located on the longitudinal symmetry axis, 153 m from the aft perpendiculars for the Cracovia ferry and 140 m from the aft perpendicular for the Mazovia ferry. The observed position deviation for the estimated AIS points is equal to 0.82 ± 0.40 m. An example of values reported by the AIS and calculated by PNDS system are shown by Figure 11 (Ais R-heading, in relation to the north, Scan R-heading in relation to the berth; check both green and blue frames). Taking into consideration the lack of detailed information about the location of CCRP on both ships, delays during transmission, and the reception of AIS messages and the general accuracy of the GNSS system, the observed position accuracy may be acknowledged as very high and in accordance with PNDS system assumptions.

4.4. Verification of the System Under Real Conditions

The PNDS system was installed on two ferries. In both cases a tablet docking station was installed in accordance with the recommendations and requirements of the captains. Each captain was fully informed about the functionality and restrictions of the system and was trained to use it. Two developers took part in a first voyage after the installation to check the performance of the system and answer any question the captains might have. Because of differences in cargo operations the ferries approach their berth in different ways.

The Mazovia ferry moves to the southern part of the berth, enters the turning circle located next to the Basen Bałtycki, rotates and approaches the berth moving north, parallel to the berth. The general average approach is shown in Figure 12a. The Cracovia ferry reduces velocity to 0–1 knot and using thrusters approaches parallel to the berth. At the last stage, the ship corrects location along the berth (Figure 12b).

Longitudinal speed, lateral speed, and angular speed are calculated directly form changes of position and rotation of the ship's outline in time. The application of kinematic equations and an exponential filter gives the basis for an assumption that the accuracy of speed indication directly depends on how often and how accurate the system calculates course and position. Independently from the above assumption, a set of comparisons during manoeuvres was conducted.

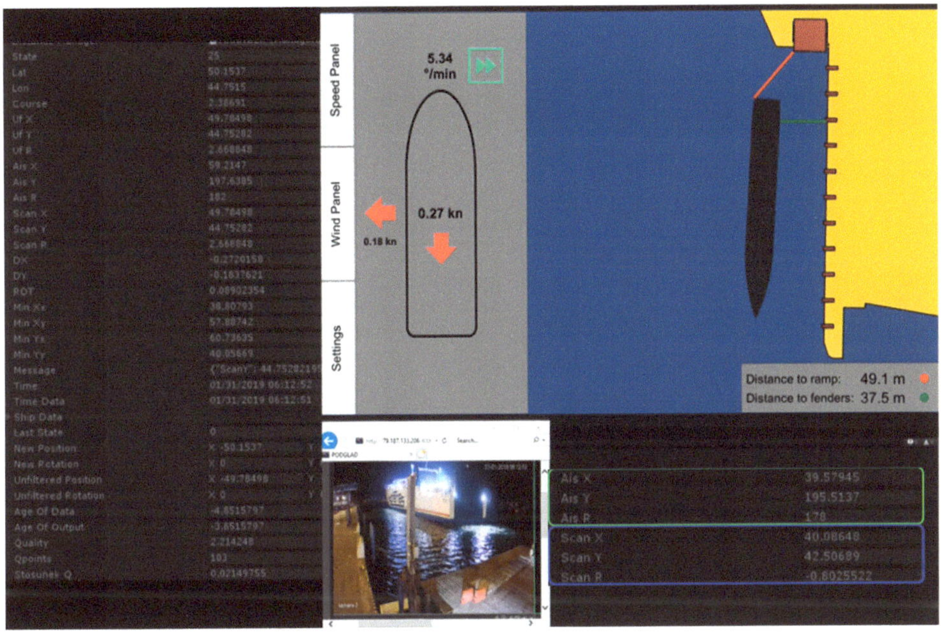

Figure 11. Test application prepared for the purpose of verification of pilot navigation docking system (PNDS) system and comparison of position reported AIS and calculated by PNDS system (green and blue frames).

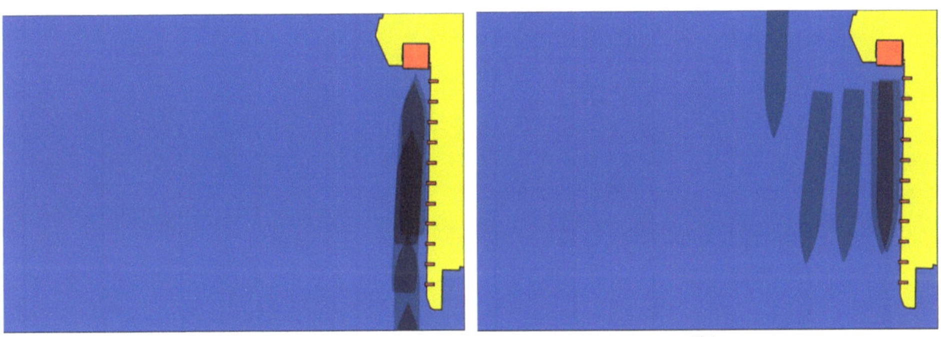

(**a**) (**b**)

Figure 12. Mazovia ferry longitudinal approach (**a**), Cracovia ferry in transversal approach (**b**).

All comparisons included recording speed values from on board equipment (like log and gyrocompass) and values from the PNDS system. Because the PNDS system can calculate the speed only when the ship is in range of both laser scanners, the time when the comparison could be made was only a few minutes during each manoeuvre. In 72% of randomly chosen moments linear velocities shown by on-board equipment did not differ by a more than 15%. In the case of angular speed, the average difference was ± 1.5 degree/min. It was noted during those test that the on board equipment that is used during manoeuvres displays longitudinal and lateral speeds with an accuracy of 0.1 knots (Figure 13a), while the PNDS systems displays those values with an accuracy of 0.01 knots (Figure 13b). Thus, the PNDS system provides more accurate data.

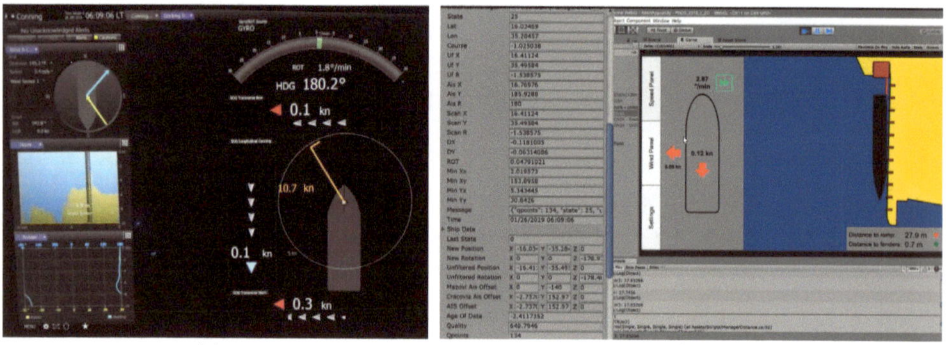

(a) (b)

Figure 13. Comparison of speed and tendency of the Cracovia ferry during berth approaching: data available through ship's conning station (**a**), and PNDS (**b**).

Values of distances calculated by the system were verified with the use of a standard, handheld laser distance meter when the ships were moored. Due to the opened bow door on the Mazovia ferry it was not possible to measure an exact distance between the ramp and the ship's outline (Figure 14a). For the Cracovia ferry the calculated distance from the ramp to the stern differed by a margin of 15 cm when compared to manual measurements. Such a difference is acceptable as per system assumptions. Figure 14b,c shows ship's side measurements.

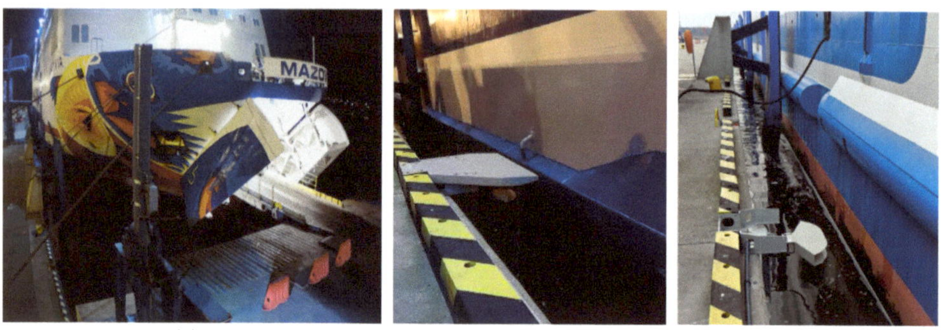

(a) (b) (c)

Figure 14. Location of the m/f Mazovia bow in relation to the ramp during cargo operations on Ferry post no 1 (**a**). Location of the m/f Mazovia side in relation to the berth. Laser scanner, shore fender and ship's bumper are visible. The system reports correct distance between the ship and fenders (**b**). Location of the m/f Cracovia side in relations to the berth. Laser scanner, shore fender and ship's bumper are visible. The system reports incorrect distance between the ship and fenders (**c**).

Regarding the distance between the berth and sides of the ferries, the maximum observed difference was over 90 centimetres. Two main factors responsible for such a difference were identified:

(1) Documentation of both ferries does not represent the accurate shape of the external hull-just the general layout and the shape of a theoretical outline at deck levels. This data was used to create hypothetical ship outlines that are used for the system during calculations.

(2) On both ferries the external hull has an extruded bumper throughout its length. This bumper is the point of direct contact between the ship and the fender. Without the information about current ship's draft and the shape of the external hull it is not possible to evaluate whether the laser scanner measures the distance to the side of ship or to the bumper.

This can lead to a situation where the ship is in contact with the fender, but the laser scanner measures the distance to the ship's side, so the system reports the distance of around 0.5 m. Because of this, the system does not report exact distances between 0 and 0.5 m Instead it shows that the distance is less than 0.5 m.

5. Conclusions

Experts in maritime commerce rightly lament the dangerous speed of growth of vessels, particularly in the container business. However, in one instance a great deal of expense and an increasing threat has been solved at an extremely low cost. The laser-based berthing and docking system is the safest means of ensuring the integrity of the berthing and docking process (the system installed at the port of Koper has been operating without failure or degradation for five years at the moment of this writing). PPU is perhaps good enough when obstructions like cranes do not exist, but where they do, their accuracy is not sufficiently reliable. Where PPU appears good enough, a laser-based system is an inexpensive means of even safer berthing and if nothing else should be made available for purposes of redundancy. A novel way of fusing GNSS positioning data with LIDAR measurements has been proposed. The benefit of such a system is the combination of ship GNSS longitudinal position with its lesser accuracy with the high-precision lateral positioning, thus minimizing the risk of collision.

The loss of a crane, which we have been using as our worst case scenario, is catastrophic for a port even without death and injury—the expense is immense, and the time to repair the area of a fallen crane and the replacement of it is considerable. There have been numerous ship-to-crane collisions resulting in damage measured in millions of Euros, and a lidar-based system is significantly less than 1 percent of likely damage in the case of a single accident. The only limitation in regard to the system is the impossibility of spreading the lasers far enough apart at, for instance, container terminals to be absolutely certain of the orientation of the ship to the quay.

For the needs of Ro-Ros, which require the most precise docking, an advanced integrated docking system was developed based on 3D scanners combined with the PPU system, finally providing a working solution to the problem of the need for clear outlines of vessels both laterally and longitudinally as they approach the pier.

Author Contributions: Conceptualization, M.P., L.G.; methodology, L.G., M.P., B.L.; data collection: B.L., M.P., L.G., V.L.; software, M.B. (Milan Batista), B.L., B.M., M.B. (Mateusz Bilewski); validation, M.P., B.L., P.V., F.D. All authors have read and agreed to the published version of the manuscript.

Funding: This research received no external funding.

Acknowledgments: The authors acknowledge the Port of Koper, the Slovenian Maritime Administration, marine pilot Matjaž Felicjan, and Blaž Kirbiš.

Conflicts of Interest: The authors declare no conflict of interest.

List of Acronyms

AIS	Automatic Identification System
AIVDO/AIVDM	NMEA AIS sentence format, where; AI–Mobile AIS station; VDO–AIS VHF Data-link Own-vessel report; VDM-AIS VHF Data-link Message
C/S	Container Ship
CCRP	Consistent Common Reference Point
CCTV	Closed-Circuit Television Camera
DWT	Deadweight Tonnage
ECDIS	Electronic Chart Display and Information System
FMCW	Frequency Modulated Continuous Wave
GMDSS	Global Maritime Distress and Safety System
GNSS	Global Navigation Satellite System
GT	Gross Tonnage
HEHDT/HEROT	Vendor specific messages where; HDT–True Heading Data; ROT–Rate of Turn
IMO	International Maritime Organization

LDS	Laser Docking Sensor or System
LiDAR	Light Detection and Ranging
LMS	Least Square Method
LNG	Liquefied Natural Gas
LTE	Long-Term Evolution
MetOcean	Meteorology and Oceanography
NMEA	National Marine Electronics Association; tandard for data interchange in marine navigation systems
NTP	Network Time Protocol
PNDS	Pilot Navigation and Docking System
PPU/PN	Portable Pilot Unit/Pilot Navigation Sistem
RADAR	Radio Detecting And Ranging
RAIM	Receiver autonomous integrity monitoring
RANSAC	Random sample consensus
Ro-Pax	Roll-On-Roll-Off-Passenger-ship/ferry
Ro-Ro	Roll-On/Roll-Off
ROT	Rete of Turn
RTK	Real-Time Kinematic
SNR	Signal-to-Noise Ratio
SOG	Speed Over Ground
STS	Ship to Shore
TEU	Twenty-Foot Equivalent Units
UTM	Universal Transverse Mercator
VDR	Voyage Data Recorder

References

1. PIANC. *Harbour ApproacH Channels Design Guidelines, PIANC Report n 121*; The World association for Waterborne Transport infrastructure: Brussles, Belgium, 2014.
2. ROM 3.1-99. *Recommendations for Maritime Works, Design of the Maritime Configuration of Ports, Approach Channels and Harbour Basins*; Puertos del Estado: Madrid, Spain, 2007.
3. Kharchanka, A. The Meaning of a Good Safe Port and Berth in a Modern Shipping World. Ph.D. Thesis, University of Groningen, Groningen, The Netherlands, 2014. Available online: https://www.rug.nl/research/portal/files/32037387/Complete_thesis.pdf (accessed on 7 January 2020).
4. Rabeaux, J. Safe Ports and Berths in a Nutshell, Defence Guides DCG-SPB-GBR-19-V1, West of England Insurance Services. 2016. Available online: https://www.westpandi.com/getattachment/bda5d2f7-8283-43e1-abbb-9727e7c7a0cb/safe_ports_and_berths_4pp_v2_lr.pdf (accessed on 20 April 2020).
5. Carbonari, S.; Gara, F.; Antolloni, G.; Gara, F.; Lorenzoni, C.; Mancinelli, A. A performance-based approach for the design of coupled dolphin-fender berthing structures. *Marine Struct.* **2019**, *64*, 78–91. [CrossRef]
6. BS 6349-4:2014. *Maritime Works—Part 4: Code of Practice for Design of Fendering and Mooring Systems*; BSI Standards Limited: London, UK, 2014.
7. Selwyn, P. Better docking for both pilots and terminals. *J. Ports Termin.* **2015**, *66*, 1–2.
8. APA. Portable Pilots' Units: A Best Practices Summary, American Pilots' Association. 2016. Available online: http://www.americanpilots.org/PPUBestPracticesAPA_%2027Oct2016.pdf (accessed on 7 January 2020).
9. Perkovic, M.; Brcko, T.; Luin, B.; Vidmar, P. Ship Handling Challenges When Vessels are Outgrowing Ports. In Proceedings of the IMLA-INSLC 19 Proceedings, South African Maritime Training Academy, Western Cape, South Africa, 5–7 September 2016.
10. Merk, O. *Container Ship Size and Port Relocation, Discussion Paper*; International Transport Forum: Paris, France, 2018.
11. Integrating the Laser—Docking Made Easier with Advanced Systems Engineering. Available online: https://www.motorship.com/news101/industry-news/integrating-the-laser-docking-made-easier-with-advanced-systems-engineering (accessed on 20 April 2020).
12. Thomas, S. An integrated approach to vessel berthing. *J. Ports Termin.* **2012**, *27*, 1–2.

13. Gucma, L.; Bak, A.; Jankowski, S.; Zalewski, P.; Perkovic, M. Laser docking system integrated with Pilot Navigation Support System, background to high precision, fast and reliable vessel docking. In Proceedings of the 17th Saint Petersburg International Conference on Integrated Navigation Systems, St. Petersburg, Russia, 31 May–2 June 2010.
14. Perkovic, M.; Gucma, M.; Luin, B.; Gucma, L.; Brcko, T. Accommodating larger container vessels using an integrated laser system for approach and berthing. *Micropocess Microsist* **2017**, *52*, 106–116. [CrossRef]
15. Sakakibara, S.; Kubo, M. Ship berthing and mooring monitoring system by pneumatic-type fenders. *Ocean Eng.* **2007**, *34*, 1174–1181. [CrossRef]
16. Metzger, A.T.; Hutchinson, J.; Kwiatkowski, J. Measurement of marine vessel berthing parameters. *Mar. Struct.* **2014**, *39*, 350–372. [CrossRef]
17. Yamse, S.; Ueda, S.; Okada, T.; Arai, A.; Shimizu, K. Chracteristics of Measured Berthing Velocity and the Application for Fender Design of Berthing Ship. In Proceedings of the 33rd PIANC World Congress, San Francisco, CA, USA, 1–5 June 2014.
18. Rankine, B. *Berthing Velocities and Brolsma's Curves. Project Report*; Marine Consulting Engineers: London, UK, 2010.
19. Ueda, S. Observation of Berthing Velocity (Japan and East Asia) and Statistical Analysis (Overall ports). In Proceedings of the 33rd PIANC World Congress, San Francisco, CA, USA, 1–5 June 2014.
20. Hein, C. Berthing velocity of large container ships. In Proceedings of the 33rd PIANC World Congress, San Francisco, CA, USA, 1–5 June 2014.
21. Roubos, A.; Groenewegen, L.; Peters, D.J. Berthing velocity of large seagoing vessels in the port of Rotterdam. *Mar. Struct.* **2017**, *51*, 202–219. [CrossRef]
22. Kirbiš, B.; Perkovič, M.; Luin, B. Port of Koper Fighting the Growth of the Ships and Trying to Accommodate Large Container Vessels Using Laser-Ranging System. Available online: https://www.researchgate.net/publication/327816531_Port_of_Koper_Fighting_The_Growth_of_Ships_and_Trying_to_Accommodate_Large_Container_Vessels_Using_a_Laser-Ranging_System (accessed on 11 May 2020).
23. Autonomous Shipping Concepts. 2020. Available online: https://www.norclub.no/blog/autonomous-shipping-concepts/ (accessed on 7 January 2020).
24. Van Cappelle, L.E.; Chen, L.; Negenborn, R.R. Survey on Short-Term Technology Developments and Readiness Levels for Autonomous Shipping. In *Computational Logistics*; Cerulli, R., Raiconi, A., Voß, S., Eds.; Springer: Berlin/Heidelberg, Germany, 2018.
25. Spange, J. Autonomous Docking for Marine Vessels Using a Lidar and Proximity Sensors. Ph.D. Thesis, Norwegian University of Science and Technology, Trondheim, Norway, 2016.
26. Rumora, I.; Sikirica, N.; Filjar, R. An Experimental Identification of Multipath Effect in GPS Positioning Error. *TransNav Int. J. Mar. Navig. Saf. Sea Transp.* **2018**, *12*, 29–32. [CrossRef]
27. Pirsiavash, A.; Broumandan, A.; Lachapelle, G.; O'Keefe, K. GNSS Code Multipath Mitigation by Cascading Measurement Monitoring Techniques. *Sensors* **2018**, *18*, 32. [CrossRef] [PubMed]
28. Mekik, Ç. An investigation on multipath errors in real time kinematic GPS method. *Sci. Res. Essays* **2010**, *5*, 2186–2200.
29. Hofmann-Wellenhof, B.; Lichtenegger, H.; Wasle, E. *GNSS—Global Navigation Satellite System*; Springer: Berlin/Heidelberg, Germany, 2008.
30. Zhang, X.; Xu, W.; Dong, C.; Dolan, J.M. Efficient L-shape fitting for vehicle detection using laser scanners. In Proceedings of the IEEE Intelligent Vehicles Symposium (IV), Los Angeles, CA, USA, 11–14 June 2017.
31. Taylor, J.W. Exponential Smoothing with a Damped Multiplicative Trend. *Int. J. Forecast.* **2003**, *19*, 715–725. [CrossRef]

Journal of
Marine Science and Engineering

MDPI

Article

Accuracy Assessment of Satellite-Based Correction Service and Virtual GNSS Reference Station for Hydrographic Surveying

Mohamed Elsayed Elsobeiey

Department of Hydrographic Surveying, Faculty of Maritime Studies, King Abdulaziz University, Jeddah 21589, Saudi Arabia; melsobeiey@kau.edu.sa

Received: 15 June 2020; Accepted: 16 July 2020; Published: 20 July 2020

Abstract: The aim of this paper is to assess the performance of satellite-based correction service, Trimble PP-RTX, and Virtual Reference Stations (VRS) for bathymetry determination, and check how far these techniques meet the minimum standards of the International Hydrography Organization (IHO) for hydrographic surveys. To this end, a three-hour duration session was conducted at Sharm Obhur using KAU-Hydrography 1 vessel. This session includes Global Navigation Satellite System (GNSS) data at the base station using Trimble SPS855 GNSS receiver, multibeam records using Kongsberg EM 712 multibeam echo sounder, sound velocity profile using Valeport's sound velocity profiler, Applanix POS MV measurements, and real-time PP-RTX corrections. Moreover, the VRS GNSS data was generated using Kingdom of Saudi Arabia Continuous Operation Reference Station network (KSA-CORS). It is shown that the Total Horizontal Uncertainty (THU) and Total Vertical Uncertainty (TVU) of the PP-RTX technique are 5.50 cm and 5.90 cm, respectively, which meets the IHO minimum standards for all survey orders at 95% confidence level. The THU and TVU of the VRS technique, on the other hand, are 5.75 cm and 7.05 cm at 95% confidence level, respectively. These values meet the IHO standards for all survey orders as well. Statistical analysis of the seabed surface differences showed a −0.07 cm average difference between the PP-RTX seabed surface and the reference seabed surface with a standard deviation of 3.60 cm. However, the average difference between the VRS-based seabed surface and the reference seabed surface is −0.03 cm and a standard deviation of 3.61 cm.

Keywords: IHO; virtual reference station; bathymetry; hydrographic surveying

1. Introduction

Safe marine navigation requires an accurate bathymetry determination. Typically, differential carrier-phased-based Global Navigation Satellite System (GNSS) techniques are used in high-accuracy surveying applications. These techniques inherit their high accuracy from the fact that both the GNSS base and rover receivers are close and share the same errors and biases [1]. The shorter the baseline is, the more there is similarity of errors and biases at both stations. As such, the effects of orbital errors, ionospheric, and tropospheric errors are significantly reduced by forming differenced observables [2]. However, as the baseline length increases, the errors at the reference and the rover receivers become less correlated and they would not cancel out sufficiently through differencing [3]. This leads to unsuccessful fixing for the ambiguity parameters, which in turn deteriorates the positioning accuracy [4]. Therefore, it is very important to have short baselines, by using reference base station close to the rover. To overcome the baseline length limitation, both global commercial satellite correction services, e.g., Trimble CenterPoint RTX (Real Time eXtended) and Virtual reference station (VRS) represent good alternatives.

1.1. Trimble PP-RTX Technique

Trimble CenterPoint RTX technology utilizes real time GNSS data from globally distributed network of tracking stations to generate Trimble RTX corrections. Such corrections include precise satellite orbits, satellite clock corrections, and observation biases for any location on the earth at a rate of 1 Hz [5]. The corrections are then delivered to subscribers via a set of geostationary satellites or by the internet [6,7]. The mathematical models for corrections generation are out of the scope of this paper and may be found in detailed elsewhere [8]. According to a previous study [9], the horizontal accuracy of Trimble RTX service ranges from 2.00–50.00 cm root mean square (RMS), depending on the subscription level, and the vertical accuracy is 5.00 cm. However, an initialization time of 1–20 min is required depending on the subscription level.

PP-RTX is a high accuracy post processed RTX-aided inertial processing method that utilizes the precise corrections derived from Trimble RTX. This technique has been developed and implemented in Position and Orientation System Post-Processing Package Mobile Mapping Suite (POSPac MMS) to enable cm level positioning for mobile mapping without the need for a reference station. Figure 1 shows the PP-RTX implementation in POSPac MMS [10].

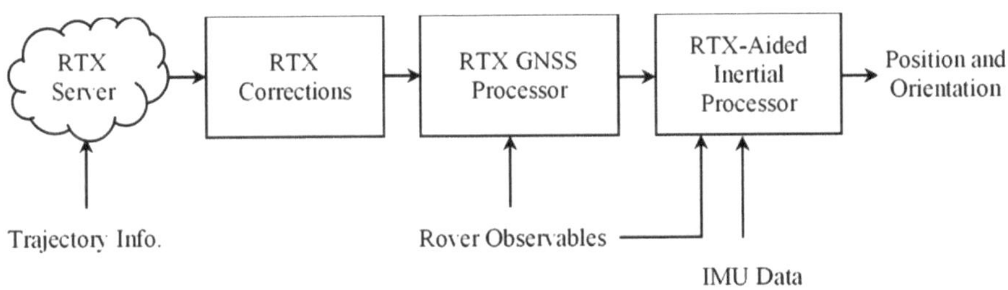

Figure 1. PP-RTX implementation in POSPac MMS [10].

The PP-RTX is available 1 h after data collection and horizontal and vertical accuracies are less than 3.00 cm and 6.00 cm, respectively. To reach full accuracy, mission duration of about 30 min is required for standard PP-RTX [10].

1.2. VRS Principles

Virtual reference station (VRS) is one of main important solutions for high precision real-time kinematic (RTK) applications [11]. The idea is to utilize real observations from an existing network of multiple reference stations to generate observations at a specific location of a nonexisting station, i.e., a virtual station [2]. In this case, VRS data will be used as if they were collected from a normal local reference station [12]. Hence, neither special data format nor software changes in the rover receiver are required to use the VRS approach [13].

In general, the data from at least three reference stations surrounding the VRS is used to calculate the measurements at the VRS. Figure 2 shows an example of three reference stations (R_1, R_2, and R_3) with VRS denoted as V and rover receiver indicated as r. Coordinates of the reference stations are known and fixed. However, the position of the VRS is assumed as the approximate position of the rover receiver to assure that short baselines are formed.

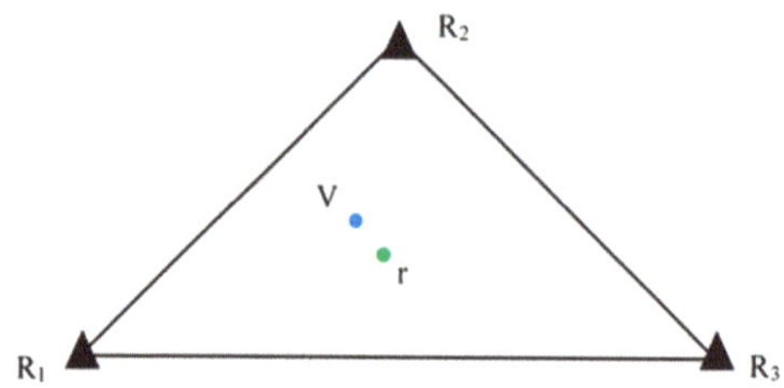

Figure 2. Network of reference stations (R_1, R_2, R_3), virtual station (V), and rover receiver (r).

The general form of the mathematical models of the carrier-phase observables can be written as follows [14]:

$$\Phi(t) = \rho(t) + cdt_r^s(t) + \lambda N_r^s + \Delta_r^s(t) \quad (1)$$

where, $\Phi(t)$ is the carrier-phase measurements at specific time t, scaled to distance (m); $dt_r^s = [dt_r - dt^s]$ represents the difference between receiver clock error dt_r and satellite clock error dt^s; λ is the carrier-phase wavelength; N_r^s is the carrier-phase ambiguity integer number; c is the speed of light in vacuum (m/sec); ρ is the true geometric distance between satellite antenna phase center and receiver antenna phase center at reception time (m); $\Delta_r^s = \Delta^{Trop} + \Delta^{Iono} + \Delta^{Orbit}$ is the summation of the slant tropospheric delay Δ^{Trop}, ionospheric delay Δ^{Iono} and orbital error Δ^{Orbit}.

Since the reference stations coordinates are known, baselines of the reference network can be solved and the carrier-phase ambiguities can be determined. The results of reference network processing include the residual error for all reference stations and at each processed epoch $\Delta_r^s(R_1,t), \Delta_r^s(R_2,t)$ and $\Delta_r^s(R_3,t)$. Denoting R for reference station and V for virtual reference station, Equation (1) can be written for both reference and virtual reference stations, as follows:

$$\Phi(R,t) = \rho(R,t) + cdt_r^s(R,t) + \lambda N_r^s(R) + \Delta_r^s(R,t) \quad (2)$$

$$\Phi(V,t) = \rho(V,t) + cdt_r^s(V,t) + \lambda N_r^s(V) + \Delta_r^s(V,t) \quad (3)$$

Differencing Equations (2) and (3) leads to:

$$\Phi(V,t) - \Phi(R,t) = \rho(V,t) - \rho(R,t) + c[dt_r^s(V,t) - dt_r^s(R,t)] + \lambda[N_r^s(V) - N_r^s(R)] + \Delta_r^s(V,t) - \Delta_r^s(R,t) \quad (4)$$

The observations at reference stations $\Phi(R,t)$ are measured, i.e., known. The true geometric range between satellite antenna phase center and both reference and virtual antenna phase center $\left(\rho(R,t) \text{ and } \rho(V,t)\right)$ are known, because the coordinates of both stations are known. The receiver and satellite clock difference $(dt_r^s(V,t)$ and $dt_r^s(R,t))$ can be eliminated by differencing. The ambiguity terms $(N_r^s(V)$ and $N_r^s(R))$ and errors at reference stations $\Delta_r^s(R,t)$ are resolved by network processing solution. Thus, Equation (4) can be written as:

$$\Phi(V,t) = \Phi(R,t) + \rho(V,t) - \rho(R,t) + \Delta_r^s(V,t) \quad (5)$$

From Equation (5), we can notice that the only unknown to estimate the observations at VRS is the error term $\Delta_r^s(V,t)$. Many interpolation techniques can be used to compute the corrections at the VRS location from the errors estimated at the reference stations. Linear combination model (LCM) was proposed previously [15] to model the orbital error Δ^{Orbit}, ionospheric delay Δ^{Iono}, slant tropospheric delay Δ^{Trop}, and to significantly reduce the effect of multipath and observation noise. The results showed a 100% success rate of carrier phase ambiguity resolution for every epoch. The distance-based linear interpolation method (DIM) was introduced by researchers [16] to estimate the ionospheric correction at a rover station, based on its distances from a network of reference stations. Further improvement to DIM was introduced by other researchers [17], taking into account the spatial correlation of regional differential ionosphere delays using differential distance and elevation parameters, defined on a single-layer ionosphere shell at an altitude of 350 km [18]. Linear interpolation method (LIM) was developed by other investigators [19,20] to produce a regional ionospheric correction model epoch-by-epoch and satellite-by-satellite, using dual-frequency phase observations from at least three GPS reference stations. In addition to ionospheric error, researchers [21] extended LIM method to estimate distance-dependent tropospheric and orbital errors at rover station using network of reference stations. To consider the spatial correlation of the combined corrections across network of reference stations, researchers [22] introduced low-order surface model (LSM). The coefficients of LSM can be estimated using least squares adjustment of the data collected at the reference stations. In addition to all the previously mentioned methods, least squares collocation method (LCM) can be used to interpolate distance-dependent errors at rover stations, using such errors at reference stations [23]. The performance of all models discussed in this section is similar [18].

KSA Reference Network

Kingdom of Saudi Arabia continuous operation reference station network (KSA-CORS) includes more than 200 GNSS stations. All stations are occupied with high end geodetic GNSS receivers. High rate GNSS data can be obtained from the web site of the general commission for survey (GCS) https://ksacors.gcs.gov.sa. KSA-CORS is used to generate 1 Hz GNSS data at a virtual location in the study area, which is denoted as VBase station (the VRS is chosen at the same location as the physical base station to maintain the same satellite geometry). Figure 3 shows the KSA-CORS network and the study area location is shown at the red dot.

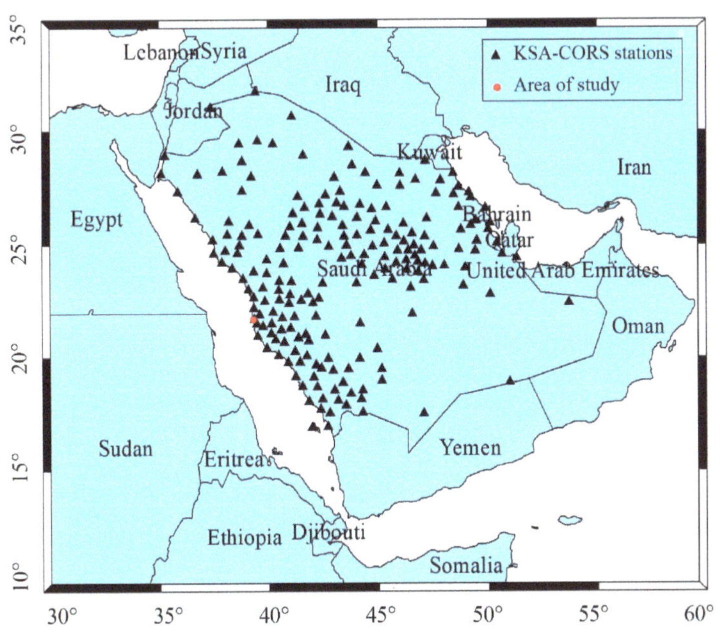

Figure 3. Saudi Arabia continuous operation reference station network (KSA-CORS).

The main objective of this paper is to evaluate the performance of satellite-based correction service, e.g., Trimble PP-RTX, and VRS techniques for bathymetry determination, and assess how far these techniques meet the minimum IHO standards for hydrographic surveys.

2. Field Test

KAU-Hydrography 1 vessel, Figure 4, was used to carry out a hydrographic surveying session of 3 h duration at Sharm Obhur where the Faculty of Maritime Studies (FMS) is located. The base station was setup on the rooftop of FMS main building using Trimble SPS855 GNSS receiver. Figure 5 shows the surveying lines and the base station location during the field test. The distance between the base station and the vessel was within 2.0 km. Kongsberg EM 712 multibeam echo sounder was used for bathymetry data collection. Valeport's sound velocity profiler (SVP) was used to measure und velocity, temperature, and pressure through water layers. To obtain a robust and accurate position and orientation solution, Applanix POS MV was used to blend GNSS data with angular rate and acceleration data from an IMU and heading from GNSS Azimuth Measurement System (GAMS).

Figure 4. KAU-hydrography 1 vessel.

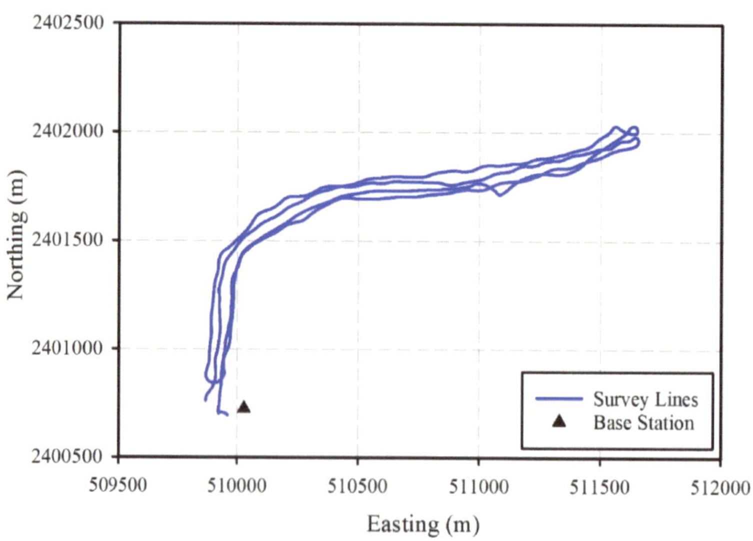

Figure 5. Survey lines and base station location.

3. Data Processing and Results

Applanix POSPac MMS software was used to process GNSS data from the real base station and POS MV data to generate the smoothed best estimate of trajectory (SBET) file. The SBET generated using the real base station is used as a reference in this study. A second SBET file was generated using the PP-RTX corrections, while a third SBET was generated using the VBase GNSS data. Both PP-RTX and VBase trajectories were compared with the reference trajectory. Figure 6 shows the easting, northing, and up difference between Base and PP-RTX trajectories. Additionally, Figure 7 shows the differences between the Base and VBase trajectories.

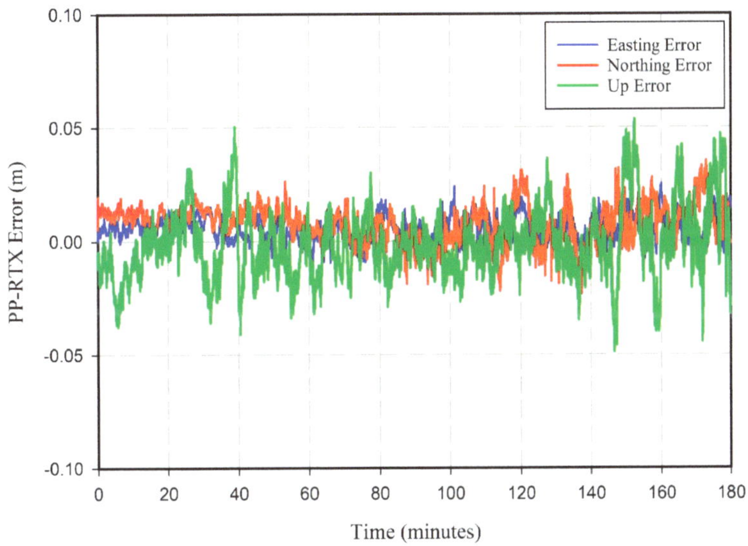

Figure 6. Easting, northing, and up errors of the PP-RTX trajectory.

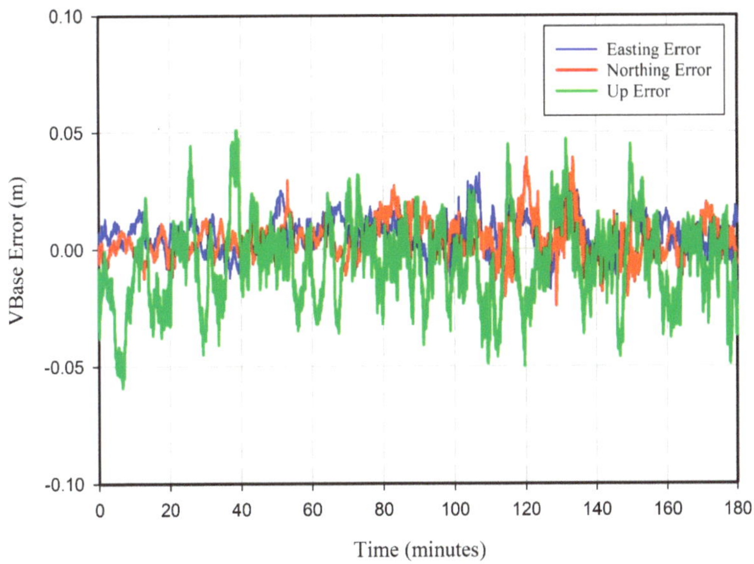

Figure 7. Easting, northing, and up errors of the VBase trajectory.

To evaluate whether the PP-RTX and VBase solutions satisfy the IHO's hydrographic surveys minimum standards, both the total horizontal uncertainty (THU) and the total vertical uncertainty (TVU) of both solutions were computed at 95% confidence level, as follows [24]:

$$THU_{95\%}^{2D} = 2.44\sqrt{\frac{\sum_{i=1}^{n}\left(\hat{N}_{Base} - \hat{N}\begin{array}{c}PP-RTX\\VBase\end{array}\right)_i^2 + \left(\hat{E}_{Base} - \hat{E}\begin{array}{c}PP-RTX\\VBase\end{array}\right)_i^2}{n}} \quad (6)$$

$$TVU_{95\%}^{1D} = 1.96\sqrt{\frac{\sum_{i=1}^{n}\left(\hat{U}_{Base} - \hat{U}\begin{array}{c}PP-RTX\\VBase\end{array}\right)_i^2}{n}} \quad (7)$$

where $THU_{95\%}^{2D}$ represents the total 2D horizontal uncertainty of northing and easting position error at 95% confidence level; $\hat{N}_{Base}$, $\hat{E}_{Base}$ are the northing and easting coordinates of the real Base-based trajectory(the reference solution), respectively; $\hat{N}_{PP-RTX}$, $\hat{E}_{PP-RTX}$ are the northing and easting coordinates of the PP-RTX-based trajectory; $\hat{N}_{VBase}$, $\hat{E}_{VBase}$ are the northing and easting coordinates of the VBase-based trajectory; n is the total number of epochs; $TVU_{95\%}^{1D}$ represents the total 1D vertical uncertainty of the Up component at 95% confidence level.

Typically, the expected accuracy (RMS) using single base station is 0.8 mm + 1 PPM (part per million) for the horizontal component and 15 mm + 1 PPM for the vertical component [10]. That means that for a 2.0-km baseline, at one sigma level, 1.0 cm and 1.7 cm RMS is expected for horizonal and vertical components, respectively. Such accuracy must be considered and added to Equations (6) and (7) to compute THU and TVU at 95% confidence level. Table 1 summarizes the THU and TVU values at 95% confidence level estimated for both the PP-RTX and the VBase solutions.

Table 1. Total horizontal uncertainty (THU) and total vertical uncertainty (TVU) of the PP-RTX and VBase solutions estimated at 95% confidence level.

	PP-RTX	VBase
THU (cm)	5.50	5.75
TVU (cm)	5.90	7.05

Table 1 shows that both PP-RTX and VBase systems deliver comparable accuracies. Tables 2 and 3, on the other hand, show the IHO minimum standards for hydrographic surveys. The values of THU and TVU in these tables are estimated as follows [25]:

$$THU = const. + \%\text{ of depth} \quad (8)$$

$$TVU = \pm\sqrt{a^2 + (b \times d)^2} \quad (9)$$

where a represents that portion of the uncertainty that does not vary with depth; b is a coefficient which represents that portion of the uncertainty that varies with depth; and d is the depth. The depth values used in Tables 2 and 3 are 40 m for special order survey and 100 m for other survey orders.

Table 2. International Hydrography Organization (IHO) minimum standards for hydrographic surveys (THU) [25].

Survey Order	Special	1a	1b	2
Constant [m]	2	5	5	20
Varible [% of depth]	0	5	5	10
THU (m)	2	10	10	30

Table 3. IHO Minimum standards for hydrographic surveys (TVU) [25].

Survey Order	Special	1a	1b	2
Constant (a) [m]	0.25	0.50	0.50	1.00
Varible (b) [% of depth]	0.75	1.30	1.30	2.30
TVU (m)	0.39	1.39	1.39	2.51

Comparing THU and TVU of both techniques from Table 1 with the minimum IHO standards in Tables 2 and 3, it is clear that both PP-RTX and VBase techniques meet the IHO minimum standards for all survey orders at 95% confidence level. To further investigate the difference between the surface generated using the PP-RTX technique and VBase technique, Caris HIPS and SIPS 11.00 was used to process the multibeam data and generate three gridded surfaces at a resolution of 0.50 m. Figure 8 shows the bathymetry of the survey area estimated using the first SBET file (the reference surface).

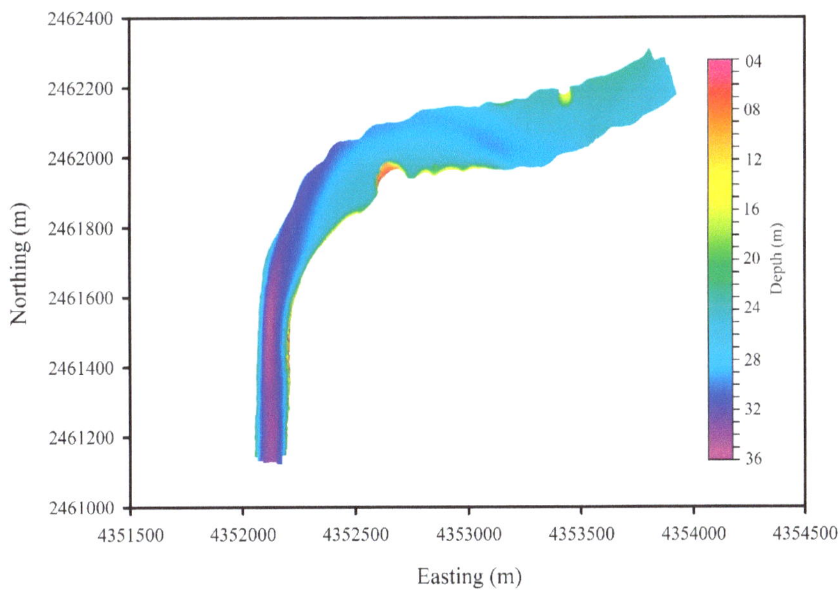

Figure 8. Bathymetry of the survey area using real base station.

The main difference between the three surfaces is the source of the SBET file estimated earlier. Thus, at the end, we have the reference surface, the PP-RTX-based surface, and the VBase-based surface. Both PP-RTX surface and VBase surface are then compared with the reference surface using Caris HIPS and SIPS. Figure 9 shows the differences between the PP-RTX surface and the reference surface. Figure 10, on the other hand, shows the difference between the VBase surface and the reference surface.

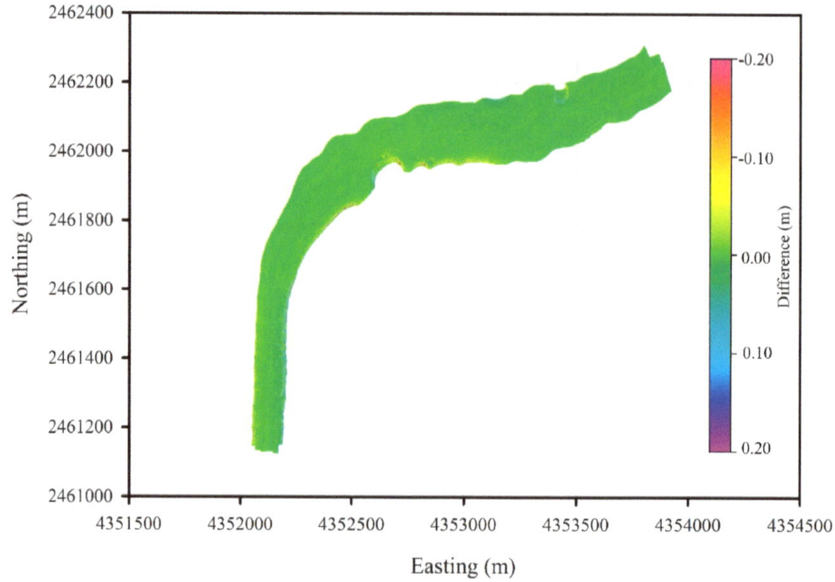

Figure 9. Difference between the PP-RTX-based surface and the reference surface.

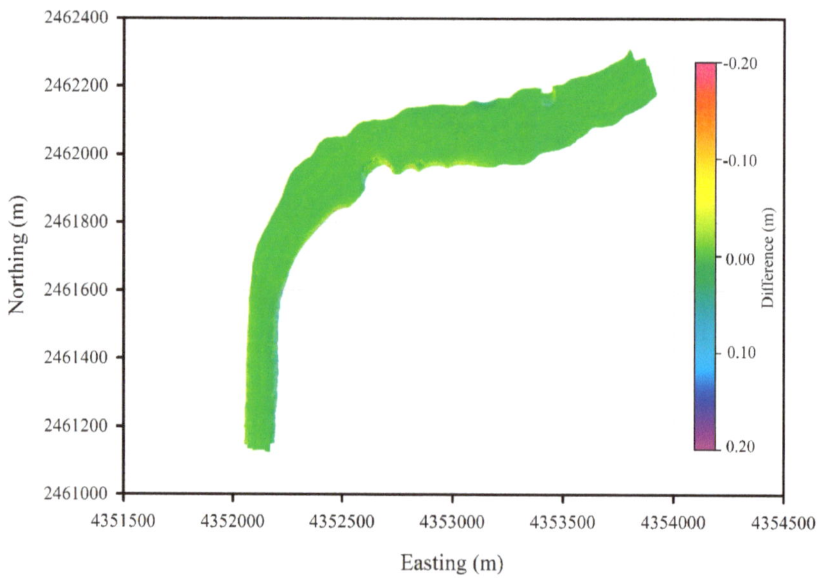

Figure 10. Difference between the VBase-based surface and the reference surface.

Figures 9 and 10 show that the main differences between the PP-RTX-based surface, the VBase surface, and the reference surface are at the channel edges beams. This is because of the slope is higher at channel edges and any horizontal shift in position will cause a significant change in depth. However, these differences are not significant and meet the IHO special order hydrographic survey standards. Table 4 summarizes the statistical analysis of the PP-RTX and VBase surface differences with the reference surface. Moreover, Figure 11 illustrates the statistical results of the PP-RTX and VBase surface differences.

Table 4. Statistical analysis of surface differences.

	PP-RTX	VBase
Minimum (m)	−3.82	−4.76
Maximum (m)	4.56	4.57
Mean (cm)	−0.07	−0.03
Standard Deviation (cm)	3.60	3.61
Total Count	1,375,305	1,375,213

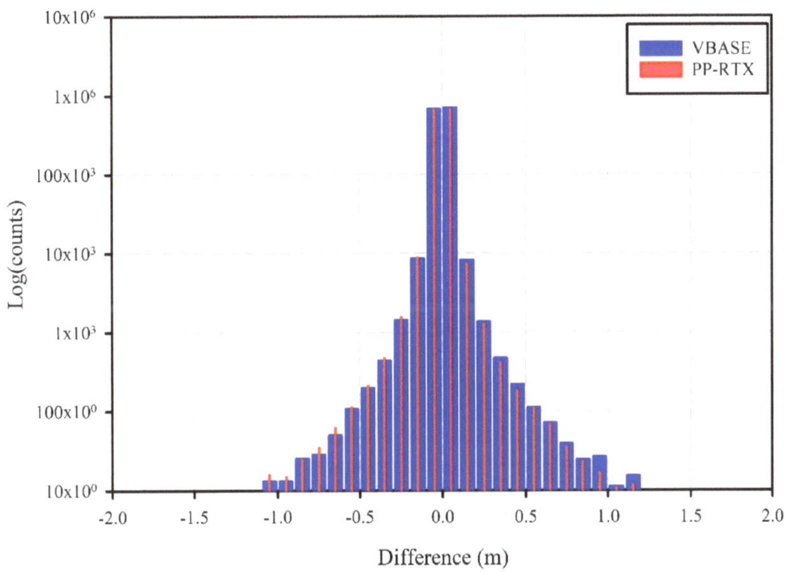

Figure 11. Histogram of the PP-RTX and VBase surface difference.

Table 4, illustrating a statistical analysis of the seabed surface differences, shows a −0.07 cm average difference between the PP-RTX seabed surface and the reference seabed surface with a standard deviation of 3.60 cm. However, the average difference between the VBase seabed surface and the reference seabed surface is −0.03 cm and a standard deviation of 3.61 cm. It is clear from Table 4 and Figure 11 that both PP-RTX and VBase techniques could provide very accurate comparable results without the need for an existing real base station in the survey area.

4. Conclusions

This study aimed to evaluate the performance of a satellite-based correction service, e.g., Trimble PP-RTX, and a virtual GNSS reference station for bathymetry determination and assess how far these techniques meet the minimum IHO standards for hydrographic surveys. A three-hour duration session was conducted at Sharm Obhur using KAU-Hydrography 1 vessel. This session included GNSS data at a base station using Trimble SPS855 GNSS receiver, multibeam records using Kongsberg EM 712 multibeam echo sounder, sound velocity profile using Valeport's sound velocity profiler, Applanix POS MV measurements, and real-time PP-RTX corrections.

Our results showed that THU and TVU of the PP-RTX technique are 5.50 cm and 5.90 cm at 95% confidence level, respectively. The THU and TVU of the VBase technique, on the other hand, are 5.75 cm and 7.05 cm at 95% confidence level, respectively. These results mean that both the PP-RTX and VBase techniques meet the IHO minimum standards of all hydrographic survey orders. Statistical analysis of the seabed surface differences showed a −0.07 cm average difference between the PP-RTX seabed surface and the reference seabed surface with a standard deviation of 3.6 cm. However, the average difference between the VBase seabed surface and the reference seabed surface is −0.03 cm and a standard deviation of 3.61 cm. From this study, we can conclude that both PP-RTX and VBase techniques meet the IHO standards for different hydrographic survey orders without the need for an existing real base station in the survey area.

Funding: This research was funded by Deanship of Scientific Research (DSR), King Abdulaziz University, Jeddah, grant number G-562-980-1441 and the APC was funded by DSR, King Abdulaziz University, Jeddah.

Acknowledgments: This work was funded by the Deanship of Scientific Research (DSR), King Abdulaziz University, Jeddah under grant no. (G-562-980-1441). The author, therefore, acknowledges with thanks DSR technical and financial support. The author acknowledges Trimble for providing access for PP-RTX corrections. The GNSS data used for VRS generation was obtained from the general commission for survey (GCS).

Conflicts of Interest: The author declares no conflict of interest.

References

1. El-Rabbany, A. *Introduction to GPS: The Global Positioning System*, 2nd ed.; Artech House: Boston, MA, USA, 2006; p. 230.
2. Hofmann-Wellenhof, B.; Lichtenegger, H.; Wasle, E. *GNSS—Global Navigation Satellite Systems: GPS, GLONASS, Galileo, and More*; Springer: Vienna, Austria; New York, NY, USA, 2008; p. 516.
3. Luo, X. *GPS Stochastic Modelling: Signal Quality Measures and ARMA Processes*; Springer Science & Business Media: Berlin, Germany, 2013; p. 331.
4. El-Rabbany, A. An Autonomous GPS Carrier-Phased-Based System for Precision Navigation. In Proceedings of the IEEE 2006 Intelligent Transportation Systems Conference (ITSC '06), Toronto, ON, Canada, 17–20 September 2006; pp. 783–787.
5. Hutton, J.; Gopaul, N.; Zhang, X.; Wang, J.; Menon, V.; Rieck, D.; Kipka, A.; Pastor, F. Centimeter-level, robust gnss-aided inertial post-processing for mobile mapping without local reference stations. *ISPRS Int. Arch. Photogramm. Remote Sens. Spat. Inform. Sci.* **2016**, *XLI-B3*, 819–826. [CrossRef]
6. Zhang, F.; Brandl, M.; Chen, X.; Drescher, R.; Glocker, M.; Landau, H.; Leandro, R.; Nitschke, M.; Salazar, D.; Weinbach, U. Trimble CenterPoint RTX—A First Study on Supporting Galileo. In Proceedings of the European Navigation Conference, Vienna, Austria, 23–25 April 2013.
7. Alkan, R.M. Cm-level high accurate point positioning with satellite-based GNSS correction service in dynamic applications. *J. Spat. Sci.* **2019**, 1–9. [CrossRef]
8. Leandro, R.; Landau, H.; Nitschke, M.; Glocker, M.; Seeger, S.; Chen, X.; Deking, A.; BenTahar, M.; Zhang, F.; Ferguson, K.; et al. RTX Positioning: The Next Generation of Cm-Accurate Real-Time GNSS Positioning. In Proceedings of the 24th International Technical Meeting of the Satellite Division of The Institute of Navigation (ION GNSS 2011), Portland, OR, USA, 20–23 September 2011; pp. 1460–1475.
9. Trimble. Trimble RTX Frequently Asked Questions. Available online: https://positioningservices.trimble.com/wp-content/uploads/2019/02/Trimble-RTX-FAQ-2020-Brochure.pdf (accessed on 5 June 2020).
10. Feld, C. What is the PP-RTX service and how to use it in POSPac. Applanix PP-RTX Technical Note. Available online: https://support.geocue.com/wp-content/uploads/2020/04/Applanix_TechNote_PPRTX.pdf (accessed on 5 June 2020).
11. Hu, G.R.; Khoo, H.S.; Goh, P.C.; Law, C.L. Development and assessment of GPS virtual reference stations for RTK positioning. *J. Geod.* **2003**, *77*, 292–302. [CrossRef]
12. Vollath, U.; Buecherl, A.; Landau, H.; Pagels, C.; Wagner, B. Multi-Base RTK Positioning Using Virtual Reference Stations. In Proceedings of the 13th International Technical Meeting of the Satellite Division of the Institute of Navigation (ION GPS 2000), Salt Lake City, UT, USA, 19–22 September 2000; pp. 123–131.
13. Jin, S. *Global Navigation Satellite Systems: Signal, Theory and Applications*; Intech: Rijeka, Croatia, 2012; p. 426.
14. Kleusberg, A.; Teunissen, P.J.G. *GPS for Geodesy*, 2nd ed; completely rev. and extended ed.; Springer: Berlin, Germany; New York, NY, USA, 1998; p. 650.
15. Han, S.; Rizos, C. *Instantaneous Ambiguity Resolution for Medium-Range GPS Kinematic Positioning Using Multiple Reference Stations*; Springer: Berlin/Heidelberg, Germany, 1998; pp. 283–288.

16. Gao, Y. Carrier phase based regional area differential GPS for decimeter-level positioning and navigation. In Proceedings of the 10th International Tech Meeting Satellite Division Inst Navigation, Kansas City, MO, USA, 16–19 September 1997; pp. 1305–1313.
17. Gao, Y.; Li, Z. Ionosphere effect and modeling for regional area differential GPS network. In Proceedings of the 11th International Technical Meeting of the Satellite Division of The Institute of Navigation (ION GPS 1998), Nashville, TN, USA, 15–18 September 1998; pp. 91–98.
18. Dai, L.; Han, S.; Wang, J.; Rizos, C. A study on GPS/GLONASS multiple reference station techniques for precise real-time carrier phase-based positioning. In Proceedings of the 14th International Technical Meeting of the Satellite Division of The Institute of Navigation (ION GPS 2001), Salt Lake City, UT, USA, 11–14 September 2001; pp. 392–403.
19. Wanninger, L. Improved ambiguity resolution by regional differential modelling of the ionosphere. In Proceedings of the 8th International Technical Meeting of the Satellite Division of The Institute of Navigation (ION GPS 1995), Palm Springs, CA, USA, 12–15 September 1995; pp. 55–62.
20. Wanninger, L. Enhancing differential GPS using regional ionospheric error models. *Bull. Geod.* **1995**, *69*, 283–291. [CrossRef]
21. Wübbena, G.; Bagge, A.; Seeber, G.; Böder, V.; Hankemeier, P. Reducing distance dependent errors for real-time precise DGPS applications by establishing reference station networks. In Proceedings of the 9th International Technical Meeting of the Satellite Division of The Institute of Navigation (ION GPS 1996), Kansas City, MO, USA, 17–20 September 1996; pp. 1845–1852.
22. Fotopoulos, G. Parameterization of Carrier Phase Corrections Based on a Regional Network of Reference Stations. Accepted for publication. In Proceedings of the 13th International Technical Meeting of the Satellite Division of the Institute of Navigation (ION GPS-00), Salt Lake City, UT, USA, 19–22 September 2000.
23. Raquet, J.F. *Development of a Method for Kinematic GPS Carrier-Phase Ambiguity Resolution Using Multiple Reference Receivers*; UCGE Rep 20116, University of Calgary: Calgary, AB, Canada. Available online: http://hdl.handle.net/1880/25880:1998 (accessed on 10 May 2020).
24. Elsobeiey, M. Performance Analysis of Low-Cost Single-Frequency GPS Receivers in Hydrographic Surveying. *Int. Arch. Photogramm. Remote Sens. Spat. Inf. Sci.* **2017**, *XLII-4/W5*, 67–71. [CrossRef]
25. International Hydrographic Bureau (IHO). *IHO Standards for Hydrographic Surveys, Special Publication No. 44*, 5th ed; February 2008; International Hydrographic Bureau: Monaco, 2008. Available online: https://iho.int/uploads/user/pubs/standards/s-44/S-44_5E.pdf (accessed on 15 May 2020).

Journal of
*Marine Science
and Engineering*

MDPI

Review

Zone of Confidence Impact on Cross Track Limit Determination in ECDIS Passage Planning

Miho Kristić [1,*], Srđan Žuškin [2,*], David Brčić [2] and Sanjin Valčić [2]

[1] Nautical Department, University of Dubrovnik, Ćira Carića 4, 20000 Dubrovnik, Croatia
[2] Faculty of Maritime Studies, University of Rijeka, Studentska ulica 2, 51000 Rijeka, Croatia; brcic@pfri.hr (D.B.); svalcic@pfri.hr (S.V.)
* Correspondence: miho.kristic@unidu.hr (M.K.); szuskin@pfri.hr (S.Ž.); Tel.: +385-98-938-7737 (M.K.)

Received: 26 June 2020; Accepted: 25 July 2020; Published: 27 July 2020

Abstract: The technology breakthrough that Electronic Chart Display and Information System (ECDIS) has brought to modern navigation has the capability to improve the safety of navigation. This could be achieved only when the capabilities of the system are known by an end-user. Cross Track Limit (XTL) is an ECDIS safety parameter, set by the navigator, which enhances the navigational task automation in the function of workload reduction. Determination of factors affecting the value of XTL safety parameter, with special consideration to chart data reliability, is elaborated in this paper. Chart data reliability depends on the quality of chart survey data, which in many cases are outdated and unreliable. Analysis of past research on this subject is used to define the factors affecting XTL. Practices of different shipping companies with regards to XTL are analyzed and compared in order to confirm if there is a uniform practice between them. Nevertheless, shipping companies have a different or no practice of obtaining XTL, which allows the navigator to define safety parameters by a subjective opinion. In this paper, method of XTL determination for a specific vessel is suggested, considering previously defined factors. Finally, crucial influence of survey data to the safety of navigation is presented in this study.

Keywords: Electronic Chart Display and Information System; Zone of Confidence; passage planning; route validation; safety parameters

---

1. Introduction

The implementation of ECDIS on board ships has brought changes in navigational procedures that have particularly affected navigator's routines with regards to preparing and executing a voyage. Preparation and execution of voyage are of vital importance for the safety of life at sea, navigational safety, and pollution prevention [1]. As a result of the abovementioned changes, there are still some problems that could pose a threat to navigational ventures, including proper handling with the ECDIS system, previous knowledge, and interpretation. New tools require additional level of knowledge and understanding [2,3]. The determination of parameters defining safety margins in the system is proposed in this paper. The Cross Track Limit (XTL) could be described as the minimum safety corridor along the navigational route which is defined by end user. Meanwhile, the Cross Track Distance (XTD) usually represents the XTL value or individual max cross track distance for each route leg. Furthermore, Cross Track Error (XTE) alarm will be triggered when the vessel deviation from the route plan is larger than the set XTL value. However, the XTL value and determination methods are still not regulated, which could mislead the navigator. Simultaneously, the XTL calculation method is rarely prescribed from companies' internal navigational procedures in accordance with the International Safety Management Code (ISM Code). Meanwhile, in case of missing navigational procedures, the value will be determined by the navigator. Theoretical knowledge of safety parameters and their adjustment are essential for navigation safety. In order to properly and adequately set the XTL parameter, the user must be able to know its value, as well as the affecting factors. When these prerequisites are satisfied, the user can create the route and validate it.

The aim of the research is to define the effect of chart accuracy and other factors on the determination of the XTL value and suggest XTL determination method. The research is based on analyses of previous studies and recommendations on factors affecting value of the XTL. The influence of navigational chart accuracy represented by Category of Zones of Confidence (CATZOC) which is assigned to the geographical areas to indicate whether data meets the minimum set for criteria is expressed by Zone of Confidence (ZOC) value which is one of the main factors for positional and depth accuracy of the chart survey. The ZOC value directly reflects the depth accuracy for safety depth or UKC determination (vertical value) and position accuracy (horizontal value) for XTL determination. These two parameters are significant ECDIS minimum safety settings which need to be proper set by end-user. Furthermore, navigational procedures from a few different international shipping companies were evaluated in order to identify procedures' similarities. Research of several shipping companies' ECDIS navigational procedures reveals that, although describing same procedures, there are still notable differences in practice [4]. In this paper different approaches to the XTL determination were identified, showing significant differences, which can potentially lead a vessel towards danger, causing marine accidents, such as grounding or pollution. The XTL is primarily safety tool in the ECDIS system. Its role to keep a vessel on safe distance from navigational dangers depends on navigator's knowledge given the proper setting of safety parameters. Although ECDIS offers great safety tools not limited solely to the XTL, if not properly used or understood it could become danger to safety of navigation.

The previous research chapter refers to general ECDIS features, providing relevant information for the research, including previous research for Zone of Confidence (ZOC) impact together with Cross Track Limit Determination. Furthermore, in the next chapter the research methodology is based on general consideration of Passage Planning with respect of Cross Track Limit parameter together with discussion of XTL determination among the different shipping companies. The results have been shown significant differences in XTL determination. Due to findings, the aim of the research is to define method of XTL determination for one of the significant ECDIS safety parameter which is suggested according to the pre-defined criteria. In the conclusion chapter, summarized findings are presented in order to raise situational awareness and detailed understanding of navigators' interpretation of safety parameters together with educational process improvement.

2. Previous Research

Setting of safety parameters is not a novelty in a long history of navigation. Modern day's navigation, which ECDIS is a part of, includes large number of pre-set safety parameters. These parameters tend to increase safety of navigation and automatize some tasks. It is elaborated by several authors that correct use of ECDIS can only be achieved if the end-user understands safety settings and alarms, but also correctly interprets information on the ECDIS display [2,3]. Safety settings that are set by the user himself have great significance for the navigational safety [5]. These settings must be defined for each leg, and includes safety contour, safety depth, shallow contour, deep contour, XTL, and look-ahead settings [6]. Setting of the XTL is part of basic passage planning settings by the end-user on ECDIS system. It has a major impact on safety of navigation, as it defines safety distance between the vessel and potential hazards. Furthermore, it is important to find the proper value determination which will consider all the significant factors. Small or insufficient value of XTL will cause navigational hazards which are close to the route to pass undetected; meanwhile, if oversized it will create a huge number of alarms [7]. Relative terms are not acceptable in safe navigation; therefore, the XTL parameter must be exact.

Well investigated and analyzed grounding of m/t Ovit in September 2013 among other factors was caused by poor passage planning and no usage of the XTL safety parameter at all [8,9]. Official investigation of grounding of vessel "Nova Cura" on 20th April 2016 on her voyage from Eregli on the

Black Sea (Turkey) to Aliaga (Turkey) reveals interesting facts. Vessel run aground on shallows that were wrongly shown on Electronic Navigational Charts (ENC) by 0.2 nautical miles northerly of actual position. Due to unfamiliarity with the CATZOC concept, the crew were not aware that significant errors could be present on the ENC. As a result, a poorly set XTL value leads to the vessel grounding [10]. Another investigation of the incident highlighted importance of adequate XTL setting and revealed serious deficiencies in passage planning and execution [11]. The ro-ro ferry Commodore Clipper's route was planned too close to a charted hazard, at the area that is well known by lower quality of survey data. Unfortunately, the quality of survey was not considered during passage planning process and was a contributing factor to this accident. Theoretical knowledge for ECDIS minimum safety parameters setting at the beginning relies on educational standards from the International Maritime Organization (IMO) model Courses [12]. These standards are cornerstone for setting ECDIS safety awareness and reduce the vessel incidents caused from poor ECDIS handling.

Chart accuracy is one of the most relevant factors affecting safety of navigation. At the time before the ECDIS system, navigators were using source diagram data on the Paper Navigational Chart (PNC) that was providing survey data accuracy. Unfortunately, this data was mainly showing survey age without chart quality information (Figure 1).

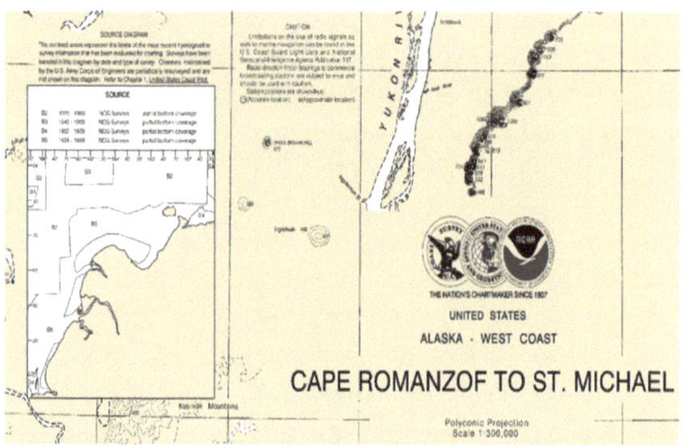

Figure 1. Source diagram data on Paper Navigational Charts (PNC) [13].

From the source diagram data user could only assume data quality depending on the year of the survey, so navigation officers and masters were mostly passing on safe distance from the obstructions, which depended on many factors and their experience. Chart data strongly depends on used survey hydrographic technique. Older survey data could have significant errors. According to Mariners' guide to accuracy of ENC (IHO S-67), survey vessels were able to use satellite-based navigation system for survey purpose only from 1990's, with an accuracy in the range from 2 to 20 m. It was shore-based electronic positioning that was used by survey vessels from late 1940's until 1990's with accuracy from 20 to 100 m. Before this, the accuracy was even worse, as survey vessels used prominent marks on the shore and sextant for measuring horizontal angles in order to find a position, so accuracy of survey was about 50–500 m. If the survey was conducted offshore, it could be based on celestial navigation only, so position accuracy was no better than 1 to 2 nautical miles, and sometimes even worse [14].

Figure 2 shows that even nowadays on the ENC chart is visible that the CATZOC for the area of Cape Romanzof is still unassessed category like on the Paper Navigational Charts on Figure 1.

Figure 2. Category of Zones of Confidence (CATZOC) on Electronic Navigational Charts (ENC) chart [13].

Navigational charts contain mixture of data from different surveys throughout many years with different methods that are connected to form a single chart. Details and interpretation of data quality was varied between hydrographic offices, so International Hydrographic Organization (IHO) developed a new international system, the ZOC system, that will be used by all countries within S-57 Electronic Navigational Charts [14].

Today, Electronic Navigational Charts use the ZOC system to explain the survey quality [15]. The survey quality is based on typical survey characteristics together with used technique of sounding. The system consists of six categories, starting from most reliable to unassessed, graphically presented on electronic navigational chart with respective symbol (Table 1). These CATZOC categories are shown in Table 1 together with graphically presentation on ENC, position accuracy and typical survey characteristics. The 'M_QUAL' quality information layer within ENC contains CATZOC as a mandatory attribute, as well as other optional 'M_QUAL' attributes, such as Positional Accuracy (POSACC) and Sounding Accuracy (SOUACC).

Table 1. Position accuracy within the CATZOC [14].

ZOC	Graphically Presentation on ENC	Position Accuracy	Typical Survey Characteristics
A1	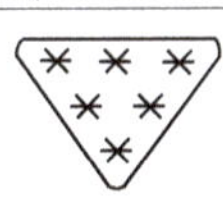	±5m + 5% depth	Controlled systematic survey high position and depth accuracy achieved using Differential Global Positioning System (DGPS) or a minimum three high quality Lines Of Position (LOP) and a multibeam, channel or mechanical sweep system.
A2	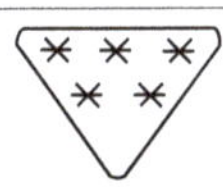	±20 m	Controlled systematic survey achieving position and depth accuracy less than *ZOC* A1 and using a modern survey echosounder and a sonar or mechanical sweep system.
B	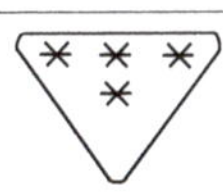	±50 m	Controlled systematic survey achieving similar depth but lesser position than *ZOC* A2 and using a modern survey echosounder but no sonar or mechanical sweep system.
C	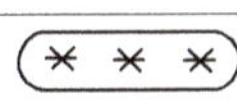	±500 m	Low accuracy survey or data collected on an opportunity basis such as soundings on passage.
D		worse than ZOC C	Poor quality data or data that cannot be quality assessed due to lack of information.
U		Unassessed—The quality of the bathymetric data has yet to be assessed.	

Additionally, attributes Survey date—start (SURSTA), Survey date—end (SUREND) and Technique of sounding measurement (TECSOU) are used to indicate the dates of the survey and the technique of sounding [15]. Furthermore, the use of Satellite Derived Bathymetry (SDB) technology is found mostly accurate to derive depths from remote sensed (RS) data to address crucial aspect of sounding in function of navigation safety [16].

Position accuracy of navigation hazards as reefs, shallow water, etc., affected by previous bottom surveys is shown in Table 1. It means that underwater hazards can be closer to vessel's route planning than we assume. Nowadays, navigators must be used to rely on navigation charts with the fact that data on charts are sometimes unreliable. Illustration of such an uncertainty is shown in Table 2, which contain analysis of 14 million square kilometers of coastal ENC.

Table 2. Distribution of CATZOC in coastal ENC [14].

Category	Area Percentage of English Channel	Area Percentage of Singapore & Malacca Strait	Area Percentage of World's Coastal ENC	Confidence
A1	3.6%	1.4%	0.7%	Very Good
A2	9.4%	0.2%	1.0%	Very Good
B	62.9%	2.5%	30.5%	Good
C	21.3%	76.2%	21.8%	Fair
D	2.8%	1.1%	20.5%	Low
U	0.0%	18.5%	25.4%	Low

A disturbing fact is that 45.9% of world's coastal ENC has a low confidence of bathymetric data, and less than 2% of coastal waters have very good level of confidence. Situation beyond coastal waters is even worse, with most ocean waters having CATZOC C, D or U [15]. In addition, almost all Electronic Navigational Charts are provided with chart uncertainty depiction of the used chart data [17]. Furthermore, a new study in Sopot also shows that the bathymetric data in ENCs are outdated with significant depth and position discrepancies from the actual results of bathymetric measurements carried out [18]. Furthermore, a significant gap has been investigating from The Canadian Hydrographic Service (CHS) in the Canadian Arctic by using remote sensing techniques (Satellite-Derived Bathymetry (SDB)) to support hydrographic applications [19,20]. In addition, CHS currently has 32% of the Northern Marine Transportation Corridors (NMTC) and only 6% of Canadian Arctic navigational water adequately surveyed to the modern standards [21,22]. According to the study in US waters, only 44% has been surveyed, but only 18% with proper new sounding method without discrepancies in bathymetry. Furthermore, for detailed survey remaining area in US waters by using nowadays method to the modern standards would take 12 million linear nautical miles of survey or approximately 177 years of a single platform by running constantly at typical surveyed speed [23]. Despite that, the automated identification of discrepancies between ENCs and surveyed area could be identified by using of an automatic algorithm for fast gap determination [24].

Meanwhile, the hydrographic survey frequency is mainly influenced by the country government financial means due to complex and costly procedures [25]. Despite the significant coast, the survey frequency is also very important in the area where seabed is constantly changing. Recently, the General Bathymetric Chart of the Oceans (GEBCO) released these Multi-Beam Echo-sounder Systems (MBES) datasets, preprocessed and processed with Computer Aided Resource Information System (CARIS). The steep slopes and the rough sand or mud seafloor possibly affect the integrity of the acquired raw data in this chosen method for the area of Arabian Gulf [26].

Detailed knowledge of bathymetry is increasingly important for navigation safety, given the ever-decreasing safety margins for both surface and underwater operations [27]. Furthermore, the importance of CATZOC data for XTL value determination is recognized by other sources [14,28]. According to the American Practical Navigator, during passage planning process the navigator should check the quality of bathymetric data using CATZOC in order to assist safe distance determination from navigational hazards. The XTL of each leg during the voyage can be adjusted considering navigational areas from open sea to confined waters. The navigator also must consider other factors in order to plan for worst case scenario.

Among other factors, safety value for collision avoidance possibility should be considered as the additional external factor [29]. Furthermore, XTL determination should be considered according to the Master's decision [30]. Accordingly recommended relevant settings for Passage Planning and Route Monitoring on Very Large Crude Carrier (VLCC) are given in Table 3.

Table 3. Recommended Cross Track Limit XTL settings for Very Large Crude Carrier (VLCC) [30].

Navigational Area	Minimum XTL Settings	
	Port	Starboard
harbour and confined water	0.25 NM	0.25 NM
coastal navigation	0.5 NM	0.5 NM
open sea navigation	1 NM	1 NM

The defined division of the navigational areas (Confined, Coastal, and Open sea area) is used in this paper relating to the different company procedures. According to [31], the XTL is defined by the end-user to port and starboard side of the planned route; however, it can differ at each leg. Considering the frequency of changing the safety parameters during the voyage monitoring, XTL value resultant will change as well. This means that, on a long voyage, there will be several different XTL values throughout the voyage plan.

3. General Consideration of Passage Planning with Respect to Cross Track Limit Parameter

The navigational process can be divided into four components: appraisal, detailed voyage planning, voyage plan execution, and the vessel monitoring in the implementation plan [1]. During the voyage planning process, the navigator selects best route verifying ship's constraints against route constraints or hazards [32]. Navigational hazards inside safety corridor bounded by the XTL will be automatically detected during the route validation process. This process forms final part of passage planning process. The ECDIS system allows operator to easily input value of the XTL for each leg of the route during passage planning. Furthermore, it is possible to unsymmetrically adjust the XTL value, i.e., the value of port and starboard side of the XTL can be different (Figure 3). Voyage plan is set with different XTL determination value for each leg by using ECDIS model Japan Radio Company JAN 9201 in this study.

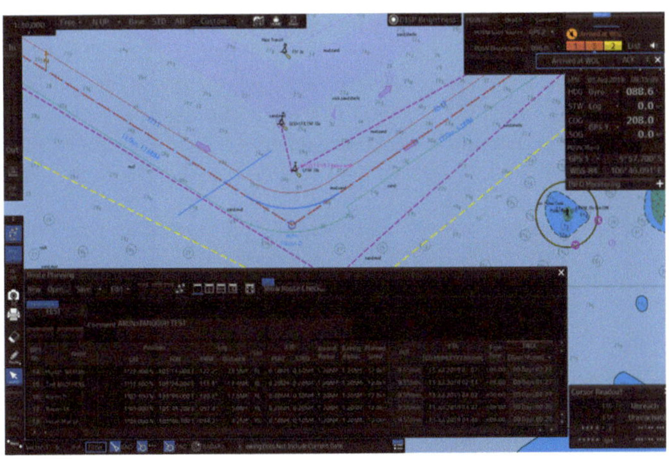

Figure 3. Electronic Chart Display and Information System (ECDIS) display with the part of planned route on model Japan Radio Company—JAN 9201.

Route validation is final part of route planning process. It is an automatic route scan, initiated by navigator at the end of passage planning. Dangers on both sides of route to the width of the XTL are scanned, and alarm is initiated when a danger is detected. Finally, at the end of the scanning process, the list of hazards will be generated. This will be used by the user to correct and modify a route to safely pass from departure to arrival point. Another importance of the XTL is active role during the voyage execution. In a case when a vessel exits pre-set XTL, Cross Track Error alarm will inform user for the necessity to return to the safe waters. The XTE alarm is the requirement of the International Maritime Organization (IMO) resolution, once when XTL is exceeded [33,34]. The IHO publication S-67 gives an example of minimum distance calculation from potential hazard. Furthermore, an example presents a situation where underwater reef is marked on navigational chart where ZOC is equal to A2, vessel's maximum breadth is 30 m, Global Navigation Satellite System (GNSS) positioning accuracy is 15 m, and vessel's orientation/motion increase additional 5 m. This example illustrates how to set the safe distance from underwater hazards where the vessel should avoid hazard at distance at least 55 m (Table 4).

Table 4. Minimum distance from hazard [14].

chart accuracy	20 m
half vessel's beam	15 m
GNSS accuracy	15 m
vessel's orientation/motion	5 m
total offset	55 m

This is an example of minimum distance from hazard, where grounding or other accident should not occur, but in this case, vessel will still pass extremely close to the danger. The significant factors for safe distance calculation in order to determine safer XTL value in function of increasing navigation safety and raising situational awareness are used in this paper.

4. Determination of the XTL Value Among the Different Shipping Companies

The potential problem lies in different XTL determination among the different shipping companies. Furthermore, some companies have defined minimal values of the XTL for different stages of voyage, while others provide safety factors to be considered. Unfortunately, some shipping companies do not provide any guidance to shipboard crew with regards to the XTL, nor any other safety parameter. In this paper, the analyzed shipowner companies vary from fleet size to the type of vessels.

Shipping company "A" is large shipping company having different type of vessels including but not limited to Very Large Crude Oil Carriers (VLCC) and Liquefied Natural Gas Carriers (LNGC). After Company "A" Ship Management system (SMS) navigational procedure analysis it was found that the company has defined minimum values of the XTL with regards to navigational area as in Table 5.

Table 5. XTL values with regards to navigational area according to Ship Management System (SMS) company "A".

Navigational Area	Minimum XTL Settings	
	Port	Starboard
within ports and confined water	0.03 NM–0.1 NM	0.03 NM–0.1 NM
coastal navigation	0.5 NM	0.5 NM
open sea navigation	1 NM	1 NM

Shipping company "B" has a large passenger's vessels fleet, and within its company SMS (Safety Management System) there is a guidance to define the XTL. According to the guidance, the XTL should be set to maximum allowable, considering the closest shore, grounding, isolated danger or any other navigational hazard. Shipping company "C" has a fleet of bulk carriers. It has no procedure whatsoever to calculate the XTL, nor mentions it in the company SMS. In this case the value of the XTL is decided by the master or officer of the watch (OOW) that oversees passage planning. It is solely depending on knowledge and experience of the operator but is not regulated by shipowner company's internal acts. It is important to draw a conclusion and identify advantages and disadvantages from the presented shipping companies' practices. Company "A" practice provide default minimum values which in some cases could be insufficient. It is not clear what factors were considered when creating minimum required values of the XTL. As factors to be considered are not known by the user, it is hard to adapt the XTL to different circumstances of the passage plan. The operator should set the maximum XTL with regards to prevailing situation but avoid creating numerous alarms. Company "B" in its navigational procedures explains factors that user should consider when deciding the XTL value. The value of CATZOC and vessel's position accuracy is not considered, so this practice could be described as incomplete. Finally, company "C" does not provide any guidance. This is an example of bad practice, as obviously company's procedures are not adapted to modern navigation procedures. Furthermore, it is not in line with ISM Code which requests from shipping companies to provide procedures and instructions that will ensure safety of vessel, crew and marine environment [35].

5. Determination of the XTL Value in the ECDIS System with Case Study

In order to correctly determine the XTL, identification of all influential factors is required. Recommendations from previous researches were considered while creating appropriate solution. Determination factors for minimum distance determination from navigational hazard were used in this paper, with addition of navigational area safety allowance. Navigational area safety allowance serves as a safety buffer, in order to allow vessel to pass on safe distance from hazard. Navigational area is defined through the company SMS navigational procedures. According to analyses various shipowner companies, differences has been notice in navigational area defining. For one of the companies, navigational areas are based on basically two navigational areas for safety settings determination in ECDIS system: open and coastal area; meanwhile, other companies established three navigational areas. The safety buffer value from company SMS is used directly from the currently navigational area according to vessel's position. Mostly used common division of three navigational areas on harbor/confined, coastal and open sea is used in order to reflect different navigational circumstances in disparate navigable waters. Accordingly, factors that must be considered are: CATZOC accuracy of bathymetric data, vessel's beam, own position accuracy, navigational area safety allowance and vessel's orientation safety allowance due to environmental effects and course change. When all above mentioned factors are considered, the following Equation (1) is created:

$$XTL = d_{zoc} + d_b + d_{pos} + d_{na} + d_{so}\ (\mathrm{m}) \tag{1}$$

where d_{zoc} = ZOC position accuracy; d_b = half vessel's beam; d_{pos} = own position accuracy; d_{na} = navigational area safety allowance; d_{so} = vessel's orientation safety allowance. The suggested equation allows definition of XTL for different circumstances. Therefore, the XTL values corresponding to different legs of voyage could be calculated and used during passage planning appraisal process and passage monitoring. In the following example, a vessel with length over all 280 m and breadth 43 m on specific route was considered for the case study. With regards to ZOC position accuracy, reference depths required for calculation of ZOC A1 position accuracy should be taken from navigational charts covering desired route. Furthermore, in this example used reference depths are 30 m (Harbor and confined waters), 50 m (Coastal navigation), and 200 m (Open Sea). Position accuracy for ZOC D is not exact, and according to ZOC system it is worse than zone C. Correspondingly XTL for harbor and confined waters in ZOC D was not calculated, as passing confined water on such accuracy is not advisable. Own vessel position accuracy obtained by GNSS is 15 m according to IHO S-67 sample. Accuracy of Global Positioning System (GPS) can be even better; as modern GPS receivers have been achieving horizontal accuracy of 3 m for 95% of the time [36]. Navigational area safety allowance considering different circumstances in each area are defined by master's experience or company SMS. According to the good guidance from chosen shipping companies in the study, safety value of navigational area safety allowance corresponds to 50 m for Harbor and confined waters, 0,5 NM (926 m) for Coastal navigation, and 1 NM (1852 m) for Open Sea area. Vessel's orientation safety allowance could be calculated by using modified formula for increment of ship's path due to vessel's turn [37]. Formula was modified to reflect the fact that XTL divide a path on two corridors, port and starboard side of the planned route. This value can be determined according to equation:

$$d_{so} = (LOA \times \sin \alpha)/2 \tag{2}$$

where LOA = Length Over All and α = drift angle. In the following case study, vessel's orientation safety allowance was calculated for 20° of drift, giving resultant of 48 m. Finally, based on all mentioned safety parameters, a following table as a case study is created for reference ship and used reference depths for different navigational areas (Table 6).

Table 6. Example of XTL determination (m).

Navigational Areas	ZOC					d_b	d_{pos}	d_{na}	d_{so}	XTL Value				
	A1	A2	B	C	D					A1	A2	B	C	D
harbor & confined waters	6.5	20	50	500	>500	21.5	15	50	48	141	154.5	184.5	634.5	N/A
coastal navigation	7.5	20	50	500	>500	21.5	15	926	48	1018	1031	1061	1511	>1511
open sea	15	20	50	500	>500	21.5	15	1852	48	1952	1957	1987	2437	>2437

According to suggested formula, XTL value is strongly affected by the position accuracy of bathymetric data. Results obtained for ZOC A1, A2 and B are very similar to values suggested by shipping company "A" and also in specific research [30] for Coastal/Open sea part. However, the results for ZOC C and D are well above comparable values in previous researches and ECDIS procedures by shipping companies. This is especially important considering previously noted accidents that happened while passing area well known by low confidence of survey data. Such differences of the XTL value in poorly surveyed area can make difference in safe passage of vessel or its grounding. Furthermore, the calculated XTL value in this research paper is acceptable from practical point of view. Consequently, its usage will not evoke numerous alarms and discourage navigator to use adequate setting.

Furthermore, precise XTL determination could be very significant for Vessel Traffic Services in maritime traffic monitoring for timely warning in case of impending danger. In [38], the decision support system is developed for route exchange. The exchange route segments should also take into consideration the suggested formula for XTL determination value for each vessel.

Safety of navigation must be priority in passage planning, considering all accessible information and identifying all hazards on navigational venture. Notably, value of the XTL is very dynamic, frequently changing as vessel is passing through more or less congested waters depending on reliability of the survey data. By recognizing that fact, in cases where there is more than one CATZOC category on the voyage leg, the one with the worst position accuracy should be chosen as relevant for the XTL determination. Other option is to split voyage leg in more segments by inserting additional waypoints at transition from one ZOC to another.

6. Conclusions

The implementation of new tools on board requires new skills and knowledge from the navigator. The XTL is designed as the basic safety tool in passage planning. Its role is to increase safety of navigation, by scanning the route during route validation process, and ensure choice of optimal route with regards to navigational hazards. That target can be achieved only when safety parameters are properly defined by the end user. Incorrectly defined parameters can lead to unwanted events such as grounding.

After having analyzed shipping companies' practice, the guidance for the XTL determination differs and are not precise. This allows different and possible incorrect interpretations by the navigator. Such a wrong interpretation of safety parameter setting had caused grounding accident, which was presented in chapter containing previous researches. The importance of CATZOC data study prior and during passage planning process is stressed as critical factor. Charts covering many parts of the world contain survey data that are old and, in some cases, unreliable. Errors caused by unreliable survey data are significant and must be considered. Using previous studies and common practice by shipping companies, the equation for the XTL determination in the ECDIS system for particular size of vessel is suggested in this paper. Integration of suggested equation in passage planning process has an ability to increase safety on navigation, as paramount target. Finally, the objective of this paper is to increase the awareness of ECDIS users with regards to position accuracy of navigational charts and its effect to navigational tools, especially the XTL. It is sometimes unexpected that information behind modern technology could be from 19th century. Passage planning is complex process where all information relevant to the voyage should be considered in order to perform safe voyage plan. Use of ZOC system provides additional information to the user and increase safety awareness during passage planning and execution of passage as well in order to avoid collision.

The actual setting of the XTL by experienced navigators represents material for future research, in order to analyze established routine among seafarers. A detailed understanding of navigators' interpretation of safety parameters, has a potential to change and improve educational process.

Author Contributions: Conceptualization, M.K. and S.Ž.; methodology, M.K., S.Ž. and D.B.; formal analysis, M.K. and S.V.; investigation, S.Ž. and D.B.; resources, M.K. and S.Ž.; data curation, D.B. and S.V.; writing—original draft preparation, M.K and S.Ž.; writing—review and editing, M.K., S.Ž., D.B. and S.V.; funding acquisition, D.B. All authors have read and agreed to the published version of the manuscript.

Funding: This research was funded by Croatian Science Foundation, grant number IP-2018-01-3739.

Acknowledgments: The authors would like to express deepest appreciation to all maritime professionals who participated in the survey and gave us the valuable opportunity to do this research.

Conflicts of Interest: The authors declare no conflict of interest.

References

1. International Maritime Organization. *Resolution A.893 (21), Guidelines for Voyage Planning*; IMO: London, UK, 2000.
2. Brčić, D.; Žuškin, S. Towards Paperless Vessels: A Master's Perspective. *J. Marit. Transp. Sci.* **2019**, *55*, 183–199. [CrossRef]
3. Brčić, D.; Kos, S.; Žuškin, S. Navigation with ECDIS: Choosing the Proper Secondary Positioning Source. *TransNav Int. J. Mar. Navig. Saf. Sea Transp.* **2015**, *9*, 317–326. [CrossRef]
4. Šakan, D.; Žuškin, S.; Brčić, D.; Valčić, S. Analysis of Primary Position Validation in ECDIS system. In *Advances in Marine Navigation and Safety of Sea Transportation, Proceedings of the 13th International Conference on Marine Navigation and Safety of Sea Transportation, Gdynia, Poland, 12–14 June 2019*; Weintrit, A., Neumann, T., Eds.; Taylor & Francis Group: London, UK, 2019; pp. 5–15. [CrossRef]

5. Žuškin, S.; Brčić, D.; Kos, S. Partial structural analysis of the ECDIS EHO research: The safety contour. In Proceedings of the 7th International Conference on Maritime Transport, Barcelona, Spain, 27–29 June 2016; Martinez de Oses, F.X., Castells Sanabra, M., Eds.; Universitat Politecnica De Catalunya—UPC: Barcelona, Spain, 2016; pp. 246–262.
6. United Kingdom Hydrographic Office. *Admiralty Guide to ECDIS Implementation, Policy and Procedures—NP232*, 2nd ed.; UKHO: London, UK, 2016; p. 71.
7. Witherby Seamenship International. *ECDIS Passage Planning and Watchkeeping*, 4th ed.; WSI: Glasgow, UK, 2018; p. 62.
8. Marine Accident Investigation Branch (MAIB). Report on the Investigation of the Grounding of m/t Ovit in the Dover Strait. Available online: www.maib.gov.uk (accessed on 10 December 2018).
9. Brčić, D.; Kos, S.; Žuškin, S. Partial structural analysis of the ECDIS EHO research: The handling part. In Proceedings of the 24th International Symposium on Electronics in Transport, Ljubljana, Slovenia, 29–30 March 2016; pp. 80–87.
10. Dutch Safety Board. Digital Navigation: Old Skills in New Technology. Available online: www.safetyboard.nl (accessed on 9 August 2019).
11. Marine Accident Investigation Branch (MAIB). Report on the Investigation of the Grounding and Flooding of the ro-ro Ferry Commodore Clipper in the Approaches to St Peter Port, Guernsey. Available online: www.maib.gov.uk (accessed on 20 July 2020).
12. Žuškin, S.; Brčić, D.; Šabalja, Đ. A contribution to improving the standards of ECDIS training. *Pomor. Sci. J. Mar. Res.* **2013**, *27*, 131–148.
13. Office of Coast Survey, NOAA. Available online: www.nauticalcharts.noaa.gov/updates/what-does-the-age-of-the-survey-mean-for-nautical-charts (accessed on 31 January 2020).
14. International Hydrographic Organization. *IHO Publication S-67, Mariners Guide to Accuracy of Electronic Navigational Charts (ENC)*, 0.5 ed.; IHO: Monaco, 2017.
15. International Hydrographic Organization. *S-57 Edition 3.1 Supplement No. 3, IHO Transfer Standard for Digital Hydrographic Data, Supplementary Information for the Encoding of S-57 Edition 3.1 ENC Data*; IHO: Monaco, 2014.
16. Mavraeidopoulos, B.A.K.; Pallikaris, A.; Oikonomou, E. Satellite Derived Bathymetry (SDB) and Safety of Navigation. *Int. Hydrogr. Rev.* **2018**, *17*, 7–20.
17. Calder, B.R. On Risk-Based Expression of Hydrographic Uncertainty. *Mar. Geod.* **2015**, *38*, 99–127. [CrossRef]
18. Specht, M.; Specht, C.; Mindykowski, J.; Dabrowski, P.; Masnicki, R.; Makar, A. Geospatial Modeling of the Tombolo Phenomenon in Sopot Using Integrated Geodetic and Hydrographic Measurement Methods. *Remote Sens.* **2020**, *12*, 737. [CrossRef]
19. Chénier, R.; Ahola, R.; Sagram, M.; Faucher, M.A.; Shelat, Y. Consideration of Level of Confidence within Multi-Approach Satellite-Derived Bathymetry. *ISPRS Int. J. Geo-Inf.* **2019**, *8*, 48. [CrossRef]
20. Knudby, A.; Ahmad, S.K.; Ilori, C. The Potential for Landsat-Based Bathymetry in Canada. *Can. J. Remote Sens.* **2016**, *42*, 367–378. [CrossRef]
21. Chénier, R.; Abado, L.; Sabourin, O.; Tardif, L. Northern Marine Transportation Corridors: Creation and Analysis of Northern Marine Traffic Routes in Canadian Waters. *Trans. GIS* **2017**, *21*, 1085–1097. [CrossRef]
22. Chénier, R.; Faucher, M.A.; Ahola, R.; Shelat, Y.; Sagram, M. Bathymetric Photogrammetry to Update CHS Charts: Comparing Conventional 3D Manual and Automatic Approaches. *ISPRS Int. J. Geo-Inf.* **2018**, *7*, 395. [CrossRef]
23. Greenaway, S.F.; Batts, A.; Riley, J. Are We Done Yet? An Empirical Estimator for Level of Effort for Seafloor Surveys—Including an Estimate for the Full Survey of U.S. Waters. *Mar. Geod.* **2020**, *43*, 87–104. [CrossRef]
24. Masetti, G.; Faulkes, T.; Kastrisios, C. Automated Identification of Discrepancies between Nautical Charts and Survey Soundings. *ISPRS Int. J. Geo-Inf.* **2018**, *7*, 392. [CrossRef]
25. Kasum, J. Updating Sea Charts and Navigational Publications. *J. Navig.* **2003**, *56*, 497–505. [CrossRef]
26. Bannari, A.; Kadhem, G. MBES-CARIS Data Validation for Bathymetric Mapping of Shallow Water in the Kingdom of Bahrain on the Arabian Gulf. *Remote Sens.* **2017**, *9*, 385. [CrossRef]
27. Stateczny, A.; Gronska-Sledz, D.; Motyl, W. Precise Bathymetry as a Step Towards Producing Bathymetric Electronic Navigational Charts for Comparative (Terrain Reference) Navigation. *J. Navig.* **2019**, *72*, 1623–1632. [CrossRef]
28. National Geospatial—Intelligence Agency. *Pub. No. 9 American Practical Navigator*, 2019 ed.; NGIA: Springfield, VA, USA, 2019.

29. Standard Safety. ECDIS—Understanding the Future of Navigation, Special Edition 2011. Available online: www.standard-club.com (accessed on 9 August 2019).
30. Rutkowski, G. ECDIS Limitations, Data Reliability, Alarm Management and Safety Settings Recommended for Passage Planning and Route Monitoring on VLCC Tankers. *TransNav Int. J. Mar. Navig. Saf. Sea Transp.* **2018**, *12*, 483–490. [CrossRef]
31. Pietrzykowski, Z.; Wielgosz, M. Navigation Safety Assessment in the Restricted Area with the Use of ECDIS. *TransNav Int. J. Mar. Navig. Saf. Sea Transp.* **2011**, *5*, 29–35.
32. Sabelis, H. Voyage Planning in ECDIS. *Int. Hydrogr. Rev.* **1999**, *76*, 41–48.
33. International Maritime Organization. *Resolution MSC.232 (82), Adoption of the Revised Performance Standards for Electronic Chart Display and Information Systems (ECDIS)*; IMO: London, UK, 2006.
34. International Maritime Organization. *Resolution A.817 (19), Performance standards for Electronic Chart Display and Information Systems (ECDIS)*; IMO: London, UK, 1995.
35. International Maritime Organization. *Resolution A.741 (18), International Management Code for the Safe Operation of Vessels and for Pollution Prevention (International Safety Management (ISM) Code)*; IMO: London, UK, 1993.
36. Department of Defense, United States of America. *Global Positioning System Standard Positioning Service Performance Standard*, 5th ed.; DoD: Washington, DC, USA, 2020.
37. Vujičić, S.; Mohović, R.; Tomaš, I.Đ. Methodology for Controlling the Ship's Path during the Turn in Confined Waterways. *Pomorstvo, Sci. J. Mar. Res.* **2018**, *32*, 28–35. [CrossRef]
38. Aylward, K.; Weber, R.; Man, Y.; Lundh, M.; MacKinnon, S. "Are You Planning to Follow Your Route?" The Effect of Route Exchange on Decision Making, Trust, and Safety. *J. Mar. Sci. Eng.* **2020**, *8*, 280. [CrossRef]

Journal of
Marine Science and Engineering

MDPI

Article

Human Factor in Navigation: Overview of Cognitive Load Measurement during Simulated Navigational Tasks

Dejan Žagar [1,*], Matija Svetina [2], Andrej Košir [3] and Franc Dimc [1]

1 Faculty of Maritime Studies and Transport, University of Ljubljana, 6320 Portorož, Slovenia; franc.dimc@fpp.uni-lj.si
2 Faculty of Arts, University of Ljubljana, 1000 Ljubljana, Slovenia; matija.svetina@ff.uni-lj.si
3 Faculty of Electrical Engineering, University of Ljubljana, 1000 Ljubljana, Slovenia; andrej.kosir@fe.uni-lj.si
* Correspondence: dejan.zagar@fpp.uni-lj.si

Received: 7 September 2020; Accepted: 30 September 2020; Published: 3 October 2020

Abstract: This paper is intended to give an overview of the experiments to evaluate the cognitive load of the officer on watch (OOW) during a collision avoidance maneuver in a full-mission simulator. The main goal is to investigate the possibilities of recording the biometric parameters of an OOW during a simulated collision avoidance maneuver. Potentially dangerous navigation errors known as human erroneous action (HEA) are induced by excessive cognitive load. Despite modern navigational aids on the ship's bridge, investigators of maritime incidents typically link the reason for incidents at sea with human factors, including high cognitive load. During the experimental tasks on the bridge, the biometric parameters of the OOW are recorded. Statistical tools are used to visualize the data and evaluate the cognitive load of the OOW. Biometric peaks of the OOW typically occur either during the collision avoidance maneuver or when the OOW has been exposed to disturbing factors that increase reaction time and cause potentially dangerous navigation. Assessing the cognitive load of OOWs in the simulator is challenging for several reasons: e.g., the environmental conditions of the simulator, the type of task to be simulated, and even the type of sensor used. After careful study of the available literature, an original experimental design using non-invasive biometric sensors is proposed.

Keywords: cognitive load; human factor; human erroneous action; marine simulator; disturbing factor; stress

1. Introduction

In recent years, the trend of the officer on watch's (OOW's) cognitive research has risen. As a result of a literature study in the maritime field, the experiments conducted during actual ships' operation are challenging due to uncontrolled environment parameters, like strong sunlight [1] or unexpected traffic situation [2], which is likely to affect the results. During the research, it has become apparent that the training and cognitive load research are rarely conducted experimentally by exposing participants to marine simulators' safe environment ([3], Tables 1–3). Despite the aim to contribute maritime safety, the simulator and technical equipment have limitations. Thus the generalization of the OOW's cognitive load recorded in the simulator, with the cognitive load during actual ships' maneuvering, is not always an easy task.

Nevertheless, the simulation is the safest way to expose the stakeholders to simulated danger without risking the collision, losing property, and marine pollution. The observed high cognitive load during experiments' is induced by simulated traffic or weather conditions or additional tasks of the OOW (e.g., *n*-back test) [4]. According to the analysis of maritime accidents in the merchant navy [5], 75% of the incidents are caused by HEA (human erroneous action), mainly due to the high cognitive load of the OOW. Despite the most modern electronic navigation equipment integrated with a modern ship's bridge, the OOW remains responsible for positioning and decision making during navigation and berthing [6,7]. In the available studies dealing with the cognitive load research on the ship's bridge, the OOWs' cognitive load is typically correlated with working (practical) experience. Studies, where the comparison between experienced officers vs. trainees [8] or students vs. experienced officers [9] reports, experienced participants perform specified tasks better. The findings are consistent with the theory of information processing in working memory [10]. As work has an impact on both human safety and health [11], the recommendation regarding workload is included in the International Maritime Organization (IMO) conventions [12,13], which proposed recommendations for the working environment and working hours to avoid high cognitive load generally, and thus HEA [14].

The goal of this paper is to investigate the impact of the distraction on the observed biometrical parameters. The cognitive load increases when certain situations occur; distractions such as fire alarms, dangerous overtaking of a ship, or modern distractions like checking social media websites, occupy the working memory, which cannot process all available navigational information at once, resulting in a longer response time. In a collision avoidance maneuver, even under normal circumstances, the OOW has but a limited amount of time available, so any increase in reaction time due to high cognitive load will result in a potentially dangerous HEA, resulting in minor failures, misunderstandings, major errors and even direct rule violations [14]. Cognitive load studies using the nautical simulator as a test environment are summarized in Section 2—Materials and Methods. The direct assessment reflects available maritime studies that have observed stress hormone levels or analyzed brain wave intensities to assess workload directly. Indirect approaches typically use eye tracker data, heart rate, or response time values to analyze an OOW's workload. For method comparison, studies from a road traffic area are also considered [7,15]. A preliminary testing proposal based on the literature study is described in Section 3, where OOW's biometric parameter measurement was conducted. By measuring pupil diameter, heart rate, blood pulse, electrodermal activity, wrist acceleration, the biometric equipment's connectivity is tested, together with the visualization of the complex biometric results to assess the influence of the disturbance factor. In Section 4, the conclusion is delineated, highlighting future research directions of the full experimental design for quantifying the cognitive load of OOWs during their navigation tasks.

2. Materials and Methods

2.1. Related Work

In this section, a decision on the most suitable methods of cognitive load measurement will be described as well as the typical experimental design. According to the available literature, two approaches for collecting an OOW's biometry are typically conducted: direct, and indirect (Table 1).

Table 1. Biometrical data collecting.

Direct Approach	Indirect Approach
Brain waves intensity (EEG)	Heart rate
Stress hormone rate (Cortisol)	Electro-Dermal Activity
-	Pupil diameter Electrocardiogram (ECG)
-	Body acceleration

The advantage of a direct approach is that it yields a more in-depth view of the cognitive process. Brain wave intensity measurements with an electroencephalograph (EEG) observe brain activity and extract cognitive load data in real-time. The challenge with using this method is the difficulty of interpretation of the recorded EEG brain wave signals due to the noise in the raw data. Nevertheless, this method provides insight into the mind process, from which the cognitive load [4] and psycho-physical state (emotions) [16] of the participants can be understood. Noise in the raw data is induced by invasive multi-wire sensors; and procedures in which sampling of the participant's stress hormone cortisol [8] was found disturbing and stressful, affecting the clarity of the final results. Comparison of the direct approach cognitive load studies involving a nautical simulator is challenging due to different research aims and different sensors used in the experiments (Table 2). Considering this fact and with a careful study of the available literature, a less invasive approach was proposed in our experiment. On the other hand, processing raw data with the machine learning (ML) support vector method [15] and the spectral decompositions method [2,4], was found useful in the post-processing phase, including data mining.

Table 2. Direct workload assessment.

PAPER	Main 2018	Liu 2017	Miklody 2017
Environment	Simulator	Simulator	Simulator
Task	Maneuvering	Overtaking	Maneuvering
Aim	Stress	Emotions	Workload
Sensor	Cortisol	EEG	EEG
Questionnaire	Nasa TLX	Profile	Profile
Statistical tool	ANOVA	-	-
Processing raw data	-	Support vector ML	Spectral decompositions ML
Reason for cognitive load	Weather	Traffic	Weather, *n*-back test

Sensors in the indirect approach are less invasive. Thus, noise in raw data due to wearable sensors is typically lower compared with expected with EEG or with the stress hormone level method. According to the available studies (Table 3), many different body markers support the indirect approach—e.g., reaction time [9], heart rate (HR), electro-dermal activity (EDA), electro-cardio graph (ECG), [17–19], pupil diameter, [1,20,21], and acceleration of the body's extremities. Even when measured indirectly, they can achieve a high degree of accuracy of up to 90%, indicating a level of cognitive load validated in several experiments tailored for key operators of different modes of transport [15]. As shown in Table 3, cognitive research in the maritime field in recent years has different aims, although a common denominator is based on cognitive research. In these studies, typically, eye-tracking glasses are used to reduce the disturbance factor caused by classical-type wired biometrical sensors. Eye-tracking manufacturers also provide high-tech software based on ML classifiers, which helps us gain additional information on cognitive load by data post-processing.

Table 3. Indirect approach of cognitive load assessment.

PAPER	Kim 2005	DiNocera 2015	Hareide 2019	Fredman 2018
Environment	Simulator	Simulator	Navy Ship	On-Road
Task	Awareness	Navigation	High speed	Car driving
Aim	Decision-time	Workload	Visual attention	Workload
Sensor	Timer	Eye-tracker	Eye-tracker	Camera
Questionnaire	Profile	Nasa-TLX	Profile	Profile
Statistical tool	t-test	ANOVA	Average	Classifier
Processing raw data	Custom SW	Toby-Studio	TobyPro2	ConvNet
Reason for cognitive load	Traffic	Traffic	High speed	n-back test
PAPER	**Saus 2012**	**Orlandi 2018**	**Kim 2007**	**Arenius 2010**
Environment	Simulator	Simulator	Simulator	Simulator
Task	Navigation	Berthing	Navigation	Navigation
Aim	Awareness	Workload	Performance	Human error
Sensor	ECG	EEG, ECG	ECG	Eye-tracker
Questionnaire	Personality test	NASA TLX	NASA-TLX	Interview
Statistical tool	Regression	ANOVA	-	-
Processing raw data	rMSSD *, Interplt. *	MatLab	t-test	-
Reason for Cog. load	Traffic	Familiarity	Alcohol	Emotions

* root mean of the squared successive differences, interpolation.

After the presented papers were carefully studied, it was decided to collect biometrical data with pupil diameter, heart rate, EDA, and wrist acceleration.

2.2. The Current Experiment

For the current experiment (Figure 1), five captains volunteered to participate. They were, on average, 48 years old and had 20.5 years of navigation experience. The primary source for data collecting are observables from biometrical sensors. The simulated ship's bridge log-files and data from interviews (questionnaire) are secondary sources of data. By comparing participants' experimental and control phases, the preliminary analyses and visualization of the biometrical data are performed (Figure 2). Like a computer processor, more tasks in the OOWs' working memory (during navigation) cause higher cognitive load and consequently prolong reaction times, which may eventually lead to a potentially dangerous situation [10].

Figure 1. Full mission simulator: Experienced participant during collision avoidance in the traffic separation zone wearing pupil-diameter sensor and wrist multi-sensor.

The experiment is carried out in the full mission marine simulator, Wärtsilä's TechSim5000, where the navigation bridge's design is comparable to that of the latest merchant ships. Four video projectors provide a 170° forward viewing angle during the voyage simulation, ensuring a high degree of realism. Side and aft views are afforded by LCD monitors on port and starboard bridge wings with a visual angle and zoom option. Environment conditions within the bridge simulator are recorded by a temperature sensor, humidity sensor, noise sensor, an illuminance sensor to measure light conditions and provide the same testing conditions for all the participants. There is only one OOW on the bridge during the simulation, to control the cognitive load-sharing effect, which typically appears in multi-participant's experiments [16]. Navigational tasks, devised by experienced captains, are designed for the northern Adriatic traffic separation zone [22,23]. The task is divided into two groups, easy and hard; each consists of 15 min of navigation. The easy task aims to collect reference biometrical observables (control phase). Hard navigation (experimental phase) consists of collision avoidance in the traffic separation zone, where several ships' routes cross when they head to the ports of Koper, Trieste, and Monfalcone. During the hard task, the OOW has experienced the disturbance factor—the fire alarm's sound—simulating a fire in the engine room. The aim is to determine if and how the alarm's sound affects the volunteering OOW's biometrical markers (and cognitive process) during the navigational task. More details on the participants are further explained in Section 3.

2.3. Raw Data Sampling

For biometrical data recordings, participants wear two sets of sensors: on the head and the wrist. Head-mounted "Pupil-Core" eye-tracking glasses measure average pupil-diameter, and the Empatica E4 wristband measures multiple parameters as presented on Table 4. Ship's parameters, e.g., ship's speed, the telegraph position, rudder position, ship's heading, and engine RPM, are recorded in the simulator's log file and at the end of the simulation extracted as a CSV-file. As per Table 4, the sampling rate of each sensor is different due to the Nyquist rate, wherein the sampling frequency is twice the maximum frequency of the signal being sampled, which prevents loss of the information.

Table 4. Sensor's sampling rate.

Type of Parameter	Manufacturer	Sampling Rate
PUPIL DIAMETER (mm)	PUPIL CORE	200 Hz
EDA (µS)	EMPATICA	4 Hz
BVP	EMPATICA	64 Hz
SHIPS' PARAMETERS	WäRTSILä	10 Hz
HEART RATE (BPM)	EMPATICA	1 Hz
ACCELERATION (m/s^2)	EMPATICA	32 Hz
TEMPERATURE (°C)	EMPATICA	4 Hz

3. Results

3.1. Experimental Procedure

The quantification of the human factor has been the subject of research in all transport sectors for many years (Tables 2 and 3). This particular study was launched last year when EMSA issued a report which found that 75% of marine accidents are caused by human factors [5]. Human factor deficiencies include lack of competence, inadequate supervision, inattention, and certain states of working memory of the OOW. The working memory in our brain consists of two parts from an engineer's point of view. The first part performs logical and cognitive operations. The second part is a limited temporary memory in which data from our body sensors is stored before the cognitive part processes it. Thus, cognitive load is a term that describes the mental effort involved in processing information. Mental effort causes typical biometric changes that can be measured [10].

The following study is an attempt to measure the rate of biometrical changes caused by mental effort during a simulated navigational task, the connectivity of the biometrical equipment, the impact of the chosen disturbance factor, and the visualization of the complex biometrical results. The aim is to isolate the weak points of the proposed experimental design before engaging in a full-scale experiment.

Each of the participants was given the same instruction before the experiment. The first simulation is collision avoidance under normal (typical) conditions, and the second simulation is affected by the disturbance factor (sound of fire alarm). In the particular experiment, the large amount of data was not obtained by a large number of participants, but by a wide scale of biometrical markers acquired on the small (N = 5) homogeneous sample of experienced OOW. For the determination of the change in the cognitive process in the working memory, we assume that a statistically significant change in biometric markers (e.g., heart rate and EDA) indicates the state of cognitive load of the OOW [10]. As seen in Figure 2, during the collision avoidance simulation, multiple data are processed in the brain (e.g., target distance, CPA, TCPA, speed, and course, etc.). After the disturbance factor is added, the cognitive process in working memory becomes high. Thus, the OOWs' reaction becomes significantly different than under the standard (control) conditions. The described negative effect reduces the time for collision avoidance, which is limited even in normal conditions. When the OOW is facing a lack of collision avoidance time, and with the appearance of the disturbance factor, typical measurable biometrical changes occur (Figure 3). Increased average values of the observables in the time when the disturbance factor is involved increase the pressures of the cognitive load. According to the literature study, invasive methods typically cause noise in the results. Consequently, we pay special attention to the selection of the least invasive wireless sensor devices. The environmental conditions (e.g., temperature, humidity, noise level, and light conditions) have a great influence on the biometric readings; therefore, they are controlled during the task. This means that the environmental parameters of the simulator are the same for all participants. The room temperature is set to 23 °C, the humidity to 38%, the noise level is 45 dB with a peak value of 75 dB during the alarm.

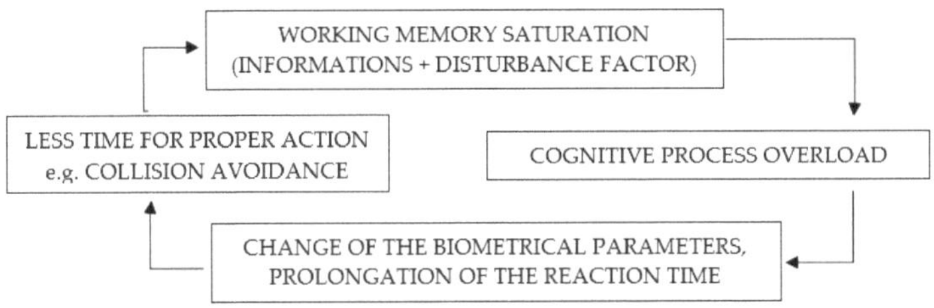

Figure 2. Information saturation with disturbance factor cause typical change of biomarkers.

3.2. Results

The typical data from the biometrical sensors (Figure 3) shows the moment where disturbance factor occur between T1 and T2. Again, the disturbance factor is the sound of the fire alarm with its red rotating light. Biometrical values in the demanding phase become higher compared with the control phase. We experienced issues if the right-handed participants wear the accelerometer on the left wrist. More accurate readings from the sensor are recorded when the accelerometer was on the working wrist. The average HR rate of participants in the test phase is significantly higher compared to navigation in the control phase. Blood volume pulse (BVP) shows higher values during the disturbance factor interval. The more the blood-vessels expand (vasodilation), the higher is the amplitude of the signal, which implies the brain's high cognitive process. The heart rate (HR) is denoted by the distance between the peaks (the Inter-Beat Interval (IBI)). However, the analysis of the individual biometric reaction shows that the participants vary in the degrees of reaction recorded by the biometric data. The differences seem to correspond to the level of experience of the individual participants in their profession. From the literature study, we assume that the biometric response is related to the personality of the participant (openness, conscientiousness, extraversion, comfort, and neuroticism), including a potentially post-traumatic stress response that occurred when the participant confronted with a situation remembers the dangerous experience from reality [10].

In the current experiment, we faced an unexpected challenge with pupil diameter measurement. Participants with correction glasses could not wear an eye-tracking sensor at the same time. Thus this part of the experiment was excluded from this research.

3.3. Post Processing and Data Visualization

In the post-processing phase, the log files from sensors are resampled, synchronized, and processed with Excel, E4-Connect firmware, and Python. This phase aims to visualize the complex biometrical results that determine the influence of the disturbance factors on the cognitive load. After the simulation starts, the navigational task begins, and the participant takes control over the ship's bridge (Figure 4). The moment of processing the available navigation information in the memory of the OOW from simulated bridge instruments causes the jump of biometric signals EDA, HR, and BVP, which indicate a high cognitive load. The acceleration of the wrist is high, which is indicated by the detection of the moving hand on the radar screen and on the electronic chart ECDIS. The biometrics during the disturbance factor (Figure 5) shows higher values during all disturbance factor sequences. An interesting observation is the gap of the BVP pulse shortly after the initiation of the disturbance factor. The gap indicates an expectation due to real-life experience that the alarm is not essential [24] and will be silenced shortly, or that the participant's working memory dismissing this sound as likely a false alarm [25]. That is not the same as ignoring an alarm but is yet another variable that likely, at times, affects the behavior of people in perilous situations. Responding to impending unexpected disasters, there is also a tendency toward disbelief that affects certain human personality types.

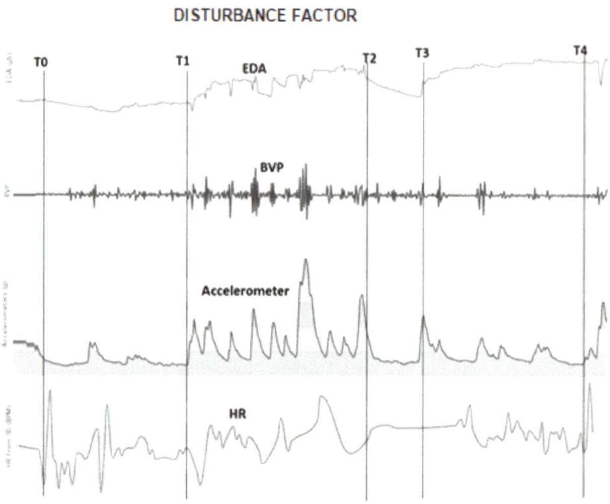

Figure 3. Biometric parameters during the typical simulation: EDA—Electro-Dermal activity, BVP—Blood-Volume pulse, HR—Heart-Rate. T0—Beginning of the collision avoidance scenario, T1—The moment when the disturbance factor occurs, T2—Disturbance factor stops, T3—Secondar0079 collision avoidance (easy task), T4—Simulation ends.

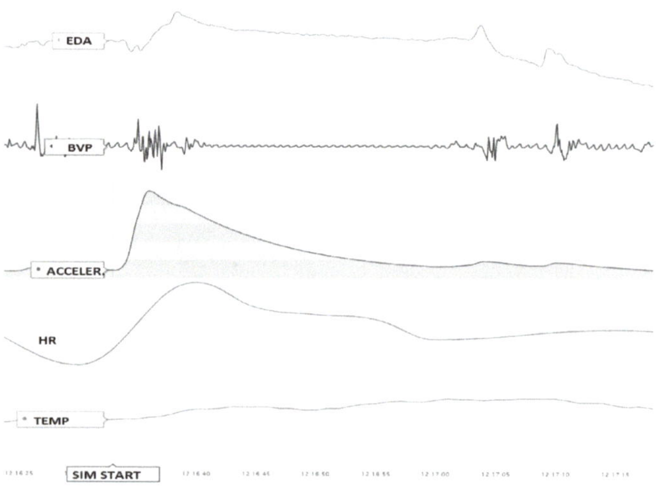

Figure 4. Biometric parameters when simulation starts. EDA—Electro-Dermal activity, BVP—Blood-Volume pulse, ACCELER—wrist acceleration, HR—Heart-Rate, TEMP—body temperature. In approximately 500 ms after the stimulants (start of the simulation), the biometrical parameters show significant response.

The EDA observation confirms the assumptions from the literature that participants have a measurable response to stress. In our case, collision avoidance and noise of the disturbance factor are reasons for stress and cause changes in skin conductance, so that the EDA varies (Figure 5). The data on the acceleration diagram shows the movement of the participant's wrist. The higher intensity and amplitude are observed at the moment of actual collision avoidance when the participant's wrist jumps quickly from one navigational instrument to another to avoid the dangerous obstacle safely.

4. Discussion

In typical cognitive load studies, including those conducted in a nautical simulator (Tables 2 and 3), the experimental design is based on the OOW's biometrical markers' readings. The authors face challenges regarding the experimental design and type of sensors used. The major shortcoming of the studies where the N is relatively small, in our case $N = 5$, the generalization of the results is not applicable. However, the particular experiment was not intended to include large numbers of participants. Instead, we assessed a small homogeneous sample of OOW with a wide scale of biometrical markers to indicate a broad spectrum of biological and behavioral responses to the task estimate tendencies that are likely to occur during the workload; studies with a larger number of selected markers are awaiting further research. The advantage of this proposed experimental method is applying modern wirelessly monitored sensors to achieve a relatively non-invasive approach, where pupil diameter BVP, EDA, and HR ratio as biometrical indicators for assessing the cognitive load are recorded. Non-invasive methods have the advantage of eliminating the biometric reaction caused by wearing a sensor, influencing the results. Unfortunately, at the beginning of the experiment, we found that most volunteers wore glasses. Participants with correction glasses could not, at the same time, wear an eye-tracking sensor. After we equipped the OOW with an eye-tracking sensor only and without the participant's correction glasses, the participants could not clearly see the navigation monitors. The conclusion was that even before starting the experiment, the OOW's biometry showed stress and indicate a high cognitive load. Consequently, the pupil-diameter measurement was excluded from the current experiment.

The current test shows a conspicuous difference in the average values of the biometrical markers between timelines in Figure 4, where disturbance factors occur. Readings of the parameters EDA and HR show higher average values when disturbance factors occur, which is promising for further research. The participants' self-evaluation reveals that the fire alarm's sound as a disturbance factor evoked a stress response, which affected the cognitive load during the collision avoidance maneuver and, consequently, the HEA (Figure 2). The assumption is that the mentioned fact is more evident in participants which in his career facing dangerous situations onboard which indicate to post-traumatic response, as explained to the participants during the self-evaluation interview.

Statistically, the distance between the upper and lower quartiles represents the interquartile range (IQR). The box plot visualization (Figure 6) shows a five-digit summary from bottom to top: the minimum line, the 1st (25%) quartile, the sample median, the 3rd (75%) quartile, and the maximum line. The dots below the minimum and above the maximum represent the samples that exceed the 1.5 of the IQR range either below the 1st quartile or above the 3rd quartile.

Observation of the HR rate (Figure 5) reveals a comparable participant's HR pattern during the collision avoidance (CA) task, with a median 79.05 BPM, which we assume is a typical condition. With disturbance factor simulation (DF), the HR median rises to 84.65 BPM compare to the CA task. What is interesting is the appearance of the outliers. There are two explanations for them. First, we consider whether there is a sensor issue or some similar hardware problem. If the pattern repeats, we consider the second explanation, which is the participants' characteristics. In the interview, participants confirmed that dangerous situations experienced on board in the past explained HR extremes (outliers) seen on the DF-P1 and DF-P5 (Figure 6). At zero visibility task (ZVT), the safety of navigation depends only on navigation instruments. We assume that the brain's cognitive process will increase due to the saturation of the working memory. The median, in this case, rises to 92 BPM, which indicates a higher cognitive load due to the lack of visibility.

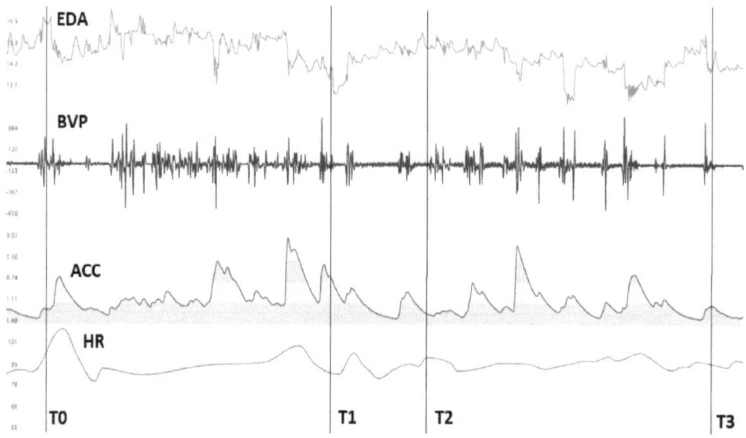

Figure 5. Biometric parameters during disturbance factor. EDA—Electro-Dermal activity, BVP—Blood-Volume pulse, ACC—wrist acceleration, HR—Heart-Rate. T0—beginning of the primary disturbance factor, T1—end of the primary disturbance factor, T2 beginning of the secondary disturbance factor, T3—end of the secondary disturbance factor.

The aim of the experiment was to determine the appearance of stress and high cognitive load with the biometrical markers. The composite of the biometrical data is a promising approach to the OOWs' cognitive load quantification, although to develop a mathematical-based model of the machine learning algorithm, which recognizes an OOWs' high workload, enough biometrical sample data (training data) must be provided. Thus, a full-scale experiment will be conducted.

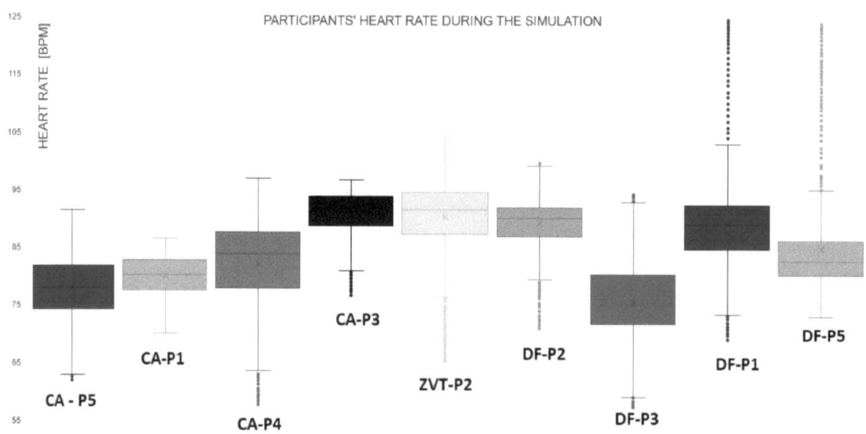

Figure 6. HR parameters during 20 min simulated navigational task. CA—collision avoidance task, ZVT—zero visibility task, DF—Disturbing factor task, P—participant.

5. Conclusions

This study attempts to represent a cross-section of the cognitive load measurement during a simulated navigational task with an experimental proposal. The simulated navigational task is approaching the pilot station trough the northern Adriatic traffic separation zone with moderate traffic density. The experiment assumes that the OOW's biometrical parameters change when the OOW faces a disturbance factor—the sound of a fire alarm—during collision avoidance. The disturbance factor in on real ships takes many forms; e.g., all sorts of navigational alarms and environmental conditions to the latest disturbance factor in the form of distraction from surfing on social media sites. For the present experiment, we chose an alarm indicating a fire in the engine room, which, we assumed, would induce stress and raise the cognitive load of experienced participants. The BVP, HR, and EDA are primary biometrical observables collected from a multi-sensor wristband on the OOW's wrist. The experiment reveals measurable biometrical responses, and even with a limited number of biometrical sensors indicates the state of stress and high cognitive load. The median of HR during collision avoidance at normal conditions is 79.05 BPM. At disturbance factor simulation, the HR rises for 5.6% to the 84.65 BPM. The highest rate, as expected, is at zero visibility task, where HR raised for 13.67% to 92 BPM. The results are encouraging for future research into cognitive load quantification.

We are aware that the present sample is small and generalization at this phase of the experiment is challenging. Additionally, the results (and temptation for generalization) can be attributed to the homogenous sample, where participants share a similar response due to navigational challenge. The methods were chosen on purpose. We aimed to test a wide range of biometric markers so that studies in the future can further elaborate on those that have previously proved useful. Thus, for a full-scale experiment, we prepare a combined quantitative/qualitative approach supported with a machine learning algorithm that will be developed to recognize the state of cognitive load of the participant during the simulation. The next step would be to transfer measurement into reality.

Author Contributions: Conceptualization, D.Ž., F.D., and M.S.; methodology, D.Ž.; software, D.Ž.; formal analysis, D.Ž. and M.S.; equipment resources, A.K.; writing—original draft preparation, D.Ž.; validation, M.S., A.K., and F.D. All authors have read and agreed to the published version of the manuscript.

Funding: The authors acknowledge the financial support of the Slovenian Research Agency (research core funding No. P2-0394, Modeling and Simulations in Traffic and Maritime Engineering).

Acknowledgments: The authors acknowledge vascular surgeon Črtomir Seušek MD, and researcher Blaž Luin PhD.

Conflicts of Interest: The authors declare no conflict of interest. The funders had no role in the design of the study; in the collection, analyses, or interpretation of data.

References

1. Hareide, O.S.; Ostnes, R. Maritime Usability Study by Analysing Eye Tracking Data. *J. Navig.* **2016**, *70*, 927–943. [CrossRef]
2. Miklody, D.; Moessmer, P.; Dettmann, T.; Klinkenberg, K.; Blankertz, B. Multi-timescale spectra as features for continuous workload estimation in realistic settings. In Proceedings of the 7th Graz Brain-Computer Interface Conference, Graz, Steiermark, Austria, 18–22 September 2017. [CrossRef]
3. Bertram, V.; Plowman, T. A Hitchhiker's Guide to the Galaxy of Maritime e-Learning. In Proceedings of the 2019 18th COMPIT Conference Proceedings, Rostov-on-Don, Russia, 20–21 September 2019; pp. 7–23, ISBN 978-3-89220-709-2.
4. Miklody, D.; Uitterhoeve, W.; Van Heel, D.; Klinkenberg, K.; Blankertz, B. *Maritime Cognitive Workload Assessment*; Symbiotic Interaction; Springer International Publishing: Berlin/Heidelberg, Germany, 2017; pp. 102–114.
5. EMSA Annual Overview of Marine Casualties and Incidents 2019. Available online: http://www.emsa.europa.eu/news-a-press-centre/external-news/item/3734-annual-overview-of-marine-casualties-and-incidents-2019.html (accessed on 22 December 2019).
6. Perkovič, M.; Gucma, L.; Bilewski, M.; Muczynski, B.; Luin, B.; Vidmar, P.; Lorenčič, V.; Batista, M. Laser-Based Aid Systems for Berthing and Docking. *J. Mar. Sci. Eng.* **2020**, *8*, 346. [CrossRef]
7. Sumpor, D. *Relevant Factors of Driver Reaction Time for Cognitive Motoric Tasks*; Fakultet Prometnih Znanosti: Zagreb, Croatia, 2019.
8. Main, L.C.; Wolkow, A.; Chambers, T.P. Quantifying the Physiological Stress Response to Simulated Maritime Pilotage Tasks: The Influence of Task Complexity and Pilot Experience. *J. Occup. Environ. Med.* **2017**, *59*, 1078. [CrossRef] [PubMed]

9. Kim, H.; Kim, H.J. Collision Scenario-based Cognitive Performance Assessment for Marine Officers, 2010. *Trans. Navig. Int. J.* **2010**, *4*, 73–77.
10. Haberlandt, K. *Cognitive Psychology*; Allyn & Bacon: Boston, MA, USA, 1997; ISBN -13 9780205139392.
11. Merchant Shipping. Safe Manning, Hours of Work and Watchkeeping. In *SOLAS Regulations*; SOLAS: Haven, CT, USA, 1997.
12. Maritime Labour Convention. *Consolidated Text Established by the International Labour Office, Including the Amendments of 2014 and 2016 to the Code of the Convention*; Maritime Labour Convention: Geneva, Switzerland, 2006.
13. *Work Hours and Rest Hours on Ships (Including STCW 2010 and Merchant Shipping (Safe Manning, Hours of Work and Watchkeeping); Regulations: 1997)*; SOLAS: Haven, CT, USA, 1997.
14. Barnett, M.L. Searching for the Root Causes of the Maritime Casualties (2005). *WMU J. Marit. Aff.* **2005**, *4*, 131–145. [CrossRef]
15. Fridman, L.; Reimer, B.; Mehler, B.; Freeman, W.T. Cognitive Load Estimation in the Wild (2018). In Proceedings of the 2018 CHI Conference on Human Factors in Computing Systems, Montreal, CQ, Canada, 21–26 April 2018; p. 652.
16. Liu, Y.; Lim, W.L.; Subramaniam, S.C.H.; Liew, S.H.P.; Krishnan GSourina, O.; Konovessis, D.; Ang, H.E.; Wang, L. EEG-based Mental Workload and Stress Recognition of Crew Members in Maritime Virtual Simulator: A Case Study. In Proceedings of the International Conference on Cyberworlds (CW), Chester, UK, 20–22 September 2017.
17. Saus, E.R.; Johnsen, B.H.; Eid, J.; Thayer, J.F. Who benefits from simulator training in relation to situation awareness during navigation training: Personality and heart rate variability. *Comput. Hum. Behav.* **2012**, *28*, 1262–1268. [CrossRef]
18. Orlandi, L.; Brooks, B. Measuring mental workload and physiological reactions in marine pilots: Building bridges towards redlines of performance. *Appl. Ergon.* **2018**, *69*, 74–92. [CrossRef] [PubMed]
19. Kim, H.; Yang, C.S.; Lee, B.W.; Young, Y.H.; Hong, S.K. Alcohol effects on navigational ability using ship handling simulator. *Int. J. Ind. Ergon.* **2007**, *37*, 733–743. [CrossRef]
20. Di Nocera, F.; Mastrangelo, S.; Baldauf, M.; Steinhage, A.; Kataria, A.; Proietta, S. Mental workload assessment using eye-tracking glasses in a simulated maritime scenario. In Proceedings of the HFESE, Groningen, The Netherlands, 14–16 October 2015; pp. 235–248.
21. Arenius, M.; Athanassiou, G.; Sträter, O. Systemic assessment of the effect of mental stress and strain on performance in a maritime ship-handling simulator. *IFAC* **2010**, *43*, 43–46. [CrossRef]
22. Vidmar, P.; Perkovič, M. Safety assessment of crude oil tankers. *Saf. Sci.* **2018**, *105*, 78–191. [CrossRef]
23. Vidmar, P.; Perkovič, M. Methodological approach for safety assessment of cruise ship in port. *Saf. Sci.* **2015**, *80*, 189–200. [CrossRef]
24. Maglić, L.; Zec, D. The Impact of Bridge Alerts on Navigating Officers. *J. Navig.* **2020**, *73*, 421–432. [CrossRef]
25. Maglić, L.; Valcic, S.; Gundić, A.; Maglic, L. Voice Communication Systems Impact on Navigating officers. MDPI, Journal of Marine Science and Engineering. *J. Mar. Sci. Eng.* **2020**, *8*, 197. [CrossRef]

Journal of
Marine Science and Engineering

MDPI

Article

Assessing Cyber Challenges of Maritime Navigation

Andrej Androjna [1,*], Tanja Brcko [1], Ivica Pavic [2] and Harm Greidanus [3]

1 Faculty of Maritime Studies and Transport, University of Ljubljana, 6320 Portorož, Slovenia; tanja.brcko@fpp.uni-lj.si
2 Faculty of Maritime Studies, University of Split, 21000 Split, Croatia; ipavic71@pfst.hr
3 European Commission, Joint Research Centre (JRC), 21027 Ispra, Italy; harm.greidanus@ec.europa.eu
* Correspondence: andrej.androjna@fpp.uni-lj.si

Received: 15 September 2020; Accepted: 1 October 2020; Published: 3 October 2020

Abstract: This paper provides a close investigation into the landscape of both cyber threats and actual incidents in the maritime sector, identifying the cyber trends and challenges as they relate to safe navigation and marine shipping. As an important subset of cyber threats that impact many maritime systems, the vulnerabilities of satellite navigation systems, in particular the Global Positioning System (GPS), receive special attention. For this article, a systematic literature review was conducted, complemented by the research and analysis of a specific spoofing event. Analyzing available resources, we might summarize that a shift in mind-set is essential to direct more attention and resources toward cybersecurity as well as the necessity for manufacturers to improve the cybersecurity of their products, as shipping systems currently remain vulnerable to cybercriminals. There is a need for multiple positioning, navigation, and timing (PNT) systems onboard maritime vessels to complement GPS-only navigation. The use of multiple satellite navigation constellations, public as well as private, in combination with the terrestrial components of an enhanced LOng-RAnge Navigation (eLoran) system and ports' laser-based aid system for berthing and docking should provide the shipping industry with the direly needed increased protection from cyber-attackers for the foreseeable future.

Keywords: maritime cyber; cybersecurity; the safety of navigation; shipboard systems; GPS jamming and spoofing

1. Introduction

Today's global maritime sector depends increasingly on digitalization, integration of operations, and automation. New opportunities arise—and cyber threats emerge. Cyber technologies have become essential, even critical, not just to the operation and management of numerous systems and processes onboard ships and in ports, but also for the safety, security, and protection of the ship, the crew, the cargo, and the marine environment. These technologies have integrated IT (Information Technology) and OT (Operational Technology) onboard ships through networking and connectivity to the internet [1–6]. In the World Economic Forum's Global Risks Report 2020, cyberattacks on critical infrastructure with a reference to shipping are rated the fifth top risk in 2020 [1]. According to Rizika [2]: "Cyber–attacks on the maritime industry's OT systems have increased by 900% over the last three years. There were 50 significant OT hacks reported in 2017, rising to 120 in 2018 and more than 300 in the previous year. This year will probably end with more than 500 major cybersecurity breaches, with substantially more going unreported". When deliberate disruptions are discovered, there are many incentives to keep that information quiet, mainly because of the maritime industry not being eager to reveal the weaknesses and vulnerabilities of their products or services.

Safety has always been a critical driver in regulations of maritime operations; and now that navigation systems have been increasingly reliant on cyber technologies to improve the effectiveness and safety of navigation, a need for safeguarding shipping from cyber threats has arisen [3–21].

Therefore, the International Maritime Organization (IMO) has already taken action and given ship owners and executives until 2021 to include cyber risk management into ship safety protocols. Owners run the risk of having ships detained if they have not included cybersecurity in the International Safety Management Code (ISM Code) on safety management onboard ships by 1 January 2021 [22,23].

The IT security system is known as cybersecurity. In contrast, the security system of the OT system is known as cyber safety, although both form part of the concept of cybersecurity [22]. Cybersecurity can be defined in brief as the "preservation of confidentiality, integrity and availability of information in the Cyberspace" [24]. However, the longer definition by the International Telecommunication Union (ITU) better reflects the wide scope: "Cybersecurity is the collection of tools, policies, security concepts, security safeguards, guidelines, risk management approaches, actions, training, best practices, assurance and technologies that can be used to protect the cyber environment and organization and user's assets. Organization and user's assets include connected computing devices, personnel, infrastructure, applications, services, telecommunications systems, and the totality of transmitted and/or stored information in the cyber environment" [25]. Indeed, "cybersecurity combines a multiplicity of disciplines from the technical to behavioural and cultural" [26].

Within this definition, 'cyber environment' comprises the interconnected networks of both IT and cyber-physical systems utilizing electronic, computer-based, and wireless systems, including the information, services, social, and business functions that exist only in cyberspace. On a ship, the computer-based systems will comprise a range of information technology components (for example, personal computers (PCs), laptops, tablet devices, servers, and networking components such as routers and switches) and operational technology (for example, control systems, sensors, actuators, radar) [27–29]. Current shipboard control systems contain significant levels of automation to perform complex functions such as navigation and propulsion control. The purpose of employing automated systems has been to reduce cost and improve performance. While automation offers excellent benefits, it also introduces a set of corresponding cybersecurity-related risks [22,30,31].

The global maritime industry systems depend on satellite navigation, especially GPS. Of particular concern is the relative ease [32] by which these systems can be jammed (through denial of reception by a competing signal) or spoofed (through deliberate introduction of a false signal). Satellite navigation is a vital part of a wide variety of the shipboard, port, and even oil rig systems [22,32–39], including the Automatic Identification System (AIS) that is vital for navigation safety [40].

This article is a close investigation of the landscape of both cyber threats and actual incidents in the maritime sector. Furthermore, it discusses the risks, the motives and likely entities behind the threats, and the impacts of an attack that can range far beyond the company being attacked. Finally, it recommends how cybersecurity could be improved in the maritime sector over time, and hopefully it might inspire further research work.

The paper is organized as follows: Section 2 describes the methodology, Section 3 presents an analysis of the cybersecurity landscape in the maritime sector and practical research into an AIS spoofing event. Section 4 discusses significant findings and provides some recommendation, and Section 5 contains a summary and conclusions.

2. Methodology

A systematic literature review was conducted, based on and structured according to documented guidelines [41–43] through which a comprehensive, explicit, reproducible, and implicit idiosyncratic method of data collection is followed. This method consists of ten steps that can be grouped into three main phases:

(1) Planning the review: The planning phase focused on defining a review question to guide the search: "What are the effects of cyberattacks and cybersecurity in the maritime domain?"
(2) Conducting a review: In the search phase, the relevant research databases, the keywords to be used during these searches, and the proper timeframe for the resulting documents to be included were identified. Data for the study were available in the databases such as Scopus, Web of Science,

Google Scholar, and open sources. The search keywords were determined from a knowledge domain analysis around the concept of "maritime cyber". The two main knowledge domains to be scanned were identified as "maritime cyber" and "cybersecurity". After the broad initial literature search, explicit inclusion and exclusion criteria—i.e., refined selection (e.g., document type, themes, research area) to identify relevant documents for this analysis—were applied. The documents were analyzed and synthesized according to contexts, methodological approaches, and outcomes. Our final list consisted of 171 documents (76 articles, 52 peer-reviewed journal papers, and 43 reports by specialized agencies) that covered the area of "maritime cyber", ranging from 2016 to 2020. While the results of this article are novel, a few earlier studies on this topic were also taken into account as references.

(3) Reporting and dissemination: In the next section, we report on our findings from the literature review.

The specific aspect of AIS/GPS spoofing is reinforced by an analysis in Chapter 3 by the Faculty of Maritime Studies and Transport, the University of Ljubljana, research and analysis [44] regarding a particular AIS spoofing event at Elba Island at the end of 2019.

3. Findings

This chapter demonstrates the unique challenges of maritime cybersecurity that include the issues with securing vessels at sea and the shore-based infrastructure supporting this industry. It presents findings of some of the possible cyberattack trajectories on maritime-related systems for navigation, propulsion, and cargo. Despite recent headlines in the media regarding the effects of cyberattacks in the maritime domain, there still seems to be a lack of understanding of cyber incidents on marine navigation systems [6]. To understand the current research being done, it is essential to apprehend its background, the working of the internet, its liabilities, and the methods which can be used to initiate attacks on the system [45]. Until 2010, the majority of cyberattacks were driven by an attempt to obtain personal or financial data. The nature of cyber is changing, and today, the maritime sector is experiencing highly sophisticated and complex attacks seeking to take the reins of its industrial control systems that are designed to be closed to the outer world [46].

3.1. *Regulatory Framework—Global*

The maritime cybersecurity legal issues are complex [40]. In 2017, the International Maritime Organization (IMO) adopted resolution MSC.428(98) on Maritime Cyber Risk Management in Safety Management System (SMS) [47]. The resolution stated that an approved SMS should take into account cyber risk management following the objectives and functional requirements of the International Safety Management Code (ISM Code) [48]. It further encourages administrations to ensure that cyber risks are appropriately addressed in safety management systems no later than the first annual verification of the company's Document of Compliance after 1 January 2021. IMO also developed guidelines on maritime cyber risk management that provide high-level recommendations on maritime cyber risk management to safeguard shipping from current and emerging cyber threats and vulnerabilities. These guidelines highlighted that effective cyber risk management should start at the senior management level [49]. Aligned with both IMO documents, the Guidelines on Cyber Security Onboard Ships were made to provide practical recommendations on maritime cyber risk management covering both cybersecurity and cyber safety [50]. In addition, IMO is preparing, in collaboration with the International Electrotechnical Commission (IEC), a new standard for maritime navigation and radio-communication equipment and systems: IEC 63,154 "Cybersecurity—General Requirements, Methods of Testing and Required Test Results" [3,51].

3.2. Regulatory Framework and Policy Priorities—EU

The 2013 EU cyber strategy aims to safeguard the EU's core values in cyberspace in light of the rapidly increasing growth of cyberspace and the threats to it. A safe cyberspace is essential to the digital single market that the EU sees as a vehicle for increased prosperity. Goals of the 2013 cyber strategy are increased resilience, decimated cybercrime, development of cyber defense policies and capabilities, increased EU autonomy in industrial resources for cybersecurity, and the fostering of a coherent international policy [52]. An EU Directive of 2013 seeks to legally criminalize cyberattacks and cybercrime [53].

The 2016 NIS directive [54,55] requires that EU Member States are adequately equipped to deal with cyber incidents—e.g., by Computer Security Incident Response Teams (CSIRT)—and that businesses in vital and ICT-dependent sectors that provide essential services take appropriate security measures. Information exchange and operational cooperation are required. The directive specifies operators of essential services for the maritime sector: passenger and freight water transport; ports and their facilities and operating entities; and vessel traffic services.

The EU has set up specialized entities such as the European Union Agency for Cybersecurity (ENISA), the European Cyber Crime Centre (EC3) at Europol, and the Computer Emergency Response Team (CERT-EU) [56], as well as launching initiatives to increase cybersecurity in various critical sectors. In particular, the Information Sharing and Analysis Centres (ISAC) are intended to be trusted entities to foster information sharing and good practices about physical and cyber threats and their mitigation [57]. The US also uses ISACs, as well as Information Sharing and Analysis Organizations (ISAO), based on US government regulations, to share cyber threat information between various stakeholders [58], while the maritime sector has three (MPS-ISAO, Maritime ISAC, maritime transportation system (MTS)-ISAC). However, in Europe the maritime sector lags behind in creating ISACs [57]. The European Cyber Security Organisation (ECSO) represents the counterpart to the European Commission for the Implementation of the Cyber Security Contractual Public–Private Partnership (cPPP). Their recent report on the transportation sector [59] aims to take a holistic approach of the implications of cybersecurity on the transport sector as a whole, but also treats the maritime sector separately. It notes that information sharing in the maritime sector falls dramatically short, and recommends the creation of a safe, confidential, and anonymous reporting center for the maritime sector.

The 2017 Joint Communication on Resilience, Deterrence and Defence [60] proposes, among other things, greater resilience, strategic autonomy, more skills, and a security by design approach.

The 2019 EU Cybersecurity Act [61] sets new objectives and tasks for ENISA as a prime tool to implement cybersecurity in the EU, and provides a framework for the establishment of European cybersecurity certification schemes for digital products, services, and processes.

Specifically for maritime, the 2014 European Union Maritime Security Strategy [62] recognizes cyber as one of the risks in the maritime domain. Its 2014 Action Plan [63] contains five actions on cyber, and its 2018 revised Action Plan [64] contains six. Progress reports [65–67] provide details on the implementation of these actions in the EU maritime sector, including work on risk assessments, response capacity, exercises, workshops, and working groups. ENISA has published studies dedicated to the maritime sector [68,69].

Most recently, the EU Security Union Strategy for 2020–2025 [70] notes that cyberattacks and cybercrime continue to rise. It calls for a whole-of-society approach to security, with sector-specific initiatives to tackle the specific risks faced by critical infrastructures such as in transport and maritime. The latest conclusions of the Council of the EU [71] underline that cybersecurity remains a shared responsibility of all players, and continue to call for improved cyber resilience, more effective responses to cyberattacks, further development of cybersecurity standards and ICT certification schemes, more cybersecurity research, and innovation capabilities to autonomously secure the EU economy and critical infrastructures.

The development of 5G is expected to boost connectivity, with significant impacts on the maritime sector. The cybersecurity of 5G will be essential, and the recent communication "Secure 5G deployment in the EU—Implementing the EU toolbox" [72] offers references and guidelines.

3.3. Cyber Trends and Challenges

This chapter presents findings of some of the possible cyberattack trajectories on maritime-related systems for navigation, propulsion, and cargo, as well as shore-based systems. It has identified some areas where a special targeted effort is required to ensure cyber and information security [73].

3.3.1. Types of Cyberattack

There may be nothing new about the need for ships to deliver cargo or patrol their country's coasts. However, the threats they are increasingly likely to encounter, invisible to any telescope, might place the maritime sector in uncharted waters [74]. According to Jones et al. [75], current threat implications of maritime-based cyberattacks include business disruption, financial loss, damage to reputation, damage to goods and environment, incident response cost, and fines or legal issues. For example, attackers could manipulate passenger lists, perform illegal activities (transports), breach sensitive cargo transports, cause engines failures, shut down vessels, or otherwise manipulate onboard control systems [74,76]. Bansal et al. [45] categorized the risks associated with an attack into three dependent factors: threats (who is attacking), vulnerabilities (the weaknesses they are attacking), and impacts (what the attack does).

There is a variety of methods that exist for those who seek to target the shipping industry [74,77]:

- Extortion/ransomware for allowing the vessel/port to restore operations;
- Digital piracy by shutting down the vessel/port;
- Espionage for gaining sensitive information that can be used by the competition;
- Defamation/litigation by causing ISPS Code noncompliance/delaying the vessel/causing disruption;
- Subversion of the supply chain;
- Terrorism;
- (H)Activism for conveying a message.

In the Danish Cyber and Information Security Strategy for the maritime sector, 2019–2022 [73], it is assessed that the general cyber threat is directed against maritime commercial businesses and does not currently pose a direct threat to maritime operations. The strategy on one side considers that the threat from cyber espionage and cybercriminals against the maritime sector is very high. In contrast, on the other side, it evaluates the threat from destructive cyberattacks, cyber activism, and cyber terrorism as low.

In the maritime domain, a rise in spear-phishing of vessels at sea has been noted. The BIMCO survey [78] presented several incidents where malicious software was introduced to ship systems unintentionally, often by third parties, to check or even update specific bridge equipment. Although the malware significantly degraded functionality of the onboard computer system, no essential vessel control system had been impacted [79]. Consequently, BIMCO's survey [78] shows that maritime companies are increasingly not only assessing their own systems and work practices in a bid to limit the likelihood of an attack but are also assessing the risk introduced across their supply chain. Respondents have also noted that a company's staff was its greatest cyber vulnerability; therefore, many cybersecurity measures remain firmly focused on reducing human error.

3.3.2. Ships Suffer Cyberattack

Ship onboard systems are susceptible to a cyberattack. There are reports [6,32,33,44,46,75,77,79–93] in which significant weaknesses of these systems have been identified. Modern technologies have integrated IT (Information Technology) and OT (Operational Technology) onboard ships through networking and connectivity to the internet [2,22,27,94–96]. However, there is no real segregation between the IT and OT networks. Any person can come in on the OT side and penetrate the IT side. We are seeing this now. Our analysis showed that there were many reports issued regarding anonymous hackers trying to disrupt ships' electronics/computers or to steal sensitive information. The impact of these kinds of attacks could be enormous. It has been noted that successful IT network hacks have their origins in the initial penetration of the OT system [2,22]. To gain remote access to the IT or OT systems, the satellite, 4G, or Wi-Fi connections of the vessel have to be breached [22,27,32].

Concerning security, time is a complicating factor. Technological vulnerability and its exposure change significantly during the lifetime of a vessel. The IT and OT will need dedicated technical improvements over time. The maintenance on IT and OT systems must be aligned with their dock-time, of course, or appropriate remote access management should be in place to ensure that only the vendor is able to perform updates [14].

According to the analysis in [73], there are three prevalent cyber and information security risks related to the maritime sector in general:

- Lack of timely response to technical vulnerabilities: A technology gap is identified between the IT and on ships and land-based systems. Land-based systems are usually better updated than the equivalent ship-based systems [97], which are, therefore, more susceptible to cyberattacks.
- No process in place for upgrades: There is a risk if the upgrading process of OT equipment does not match the standards associated with IT technologies.
- Securing critical systems: The potential consequences of a targeted attack to databases and registers based on older technology are lack of data integrity, loss of reputation, and a potential financial loss.

Two different experiments conducted by Svilicic et al. [98] and Hareide [6] demonstrated that cyberattacks against integrated navigational systems (INS), usually considered as an offline system, are relatively easily achievable. Disconnection from the internet prevents outside threats; however, the vulnerabilities arising from the unmaintained operating system could also be triggered by inside actors (the ship's crew), either unintentionally or maliciously. With the INS connection to the internet, the cyber threats would rise to a critical level, demanding instant action [3].

According to the BIMCO survey [93], the most vulnerable systems onboard ships are positioning systems (GPS, AIS, Radar), ECDIS, engine control, and monitoring. Like AIS, GPS for civilian use is not encrypted or authenticated, and each has been identified as potentially vulnerable to attack.

ECDIS can be compromised in order to modify files and insert malicious content [6,99]. An ECDIS compromise can take over the whole INS or display the vessel in a false position. A cyberattack can mislead a ship as, for example, in 2016 when two naval ships were misdirected in the Persian Gulf [100]. Another example happened in February 2017. Cybercriminals reportedly took control of the navigation systems of a German-owned 8250 TEU container vessel. The crew attempted to regain control and had to bring IT experts on board to solve the situation. The case serves as a "pre-warning" about hackers' abilities to gain control over the vessels to carry out, for instance, kidnap and ransom [101,102].

Another potential cyberspace vulnerability is the Voyage Data Recorder (VDR), from its connection to other ship systems that links to online services through satellite communications. However, the risks related to VDR weaknesses is, according to Kala [86], marginal, since VDRs do not directly control the movement of a vessel.

3.3.3. Offices Onshore

After the devastating cyber incident in June 2017 when the NotPetya malware attack, originating in Ukraine, infected the IT systems of the shipping giant Maersk and forced the company to shut down all devices and handle all operations manually, shipping offices onshore have realized that the shipping industry is not immune to cybercrime. Maersk was not explicitly targeted, and thus it was rather collateral damage. The attack resulted in significant interruptions to Maersk's operations and terminals worldwide, costing them up to USD 300 million [6,77,79,103].

Another wake-up call for the maritime industry were two major cyber incidents reported in 2018. The first happened in July 2018 when COSCO Shipping Lines fell victim to a cyberattack. The company's internet connection was disrupted within its offices in the Americas. After activation of COSCO's contingency plans, operations were back to normal after five days. Being aware of what happened to Maersk, they had taken proactive steps to minimize their risk of a cyberattack. The second incident occurred in October 2018 when the Australia-based ferry and defense shipbuilder Austal was hit by a cyberattack that penetrated their data management systems. The attackers managed to steal internal data and offered some of it for sale on the dark web in an apparent extortion attempt [103].

Carnival Corporation was the latest to fall victim to a ransomware attack on its IT systems in August 2020. The cybercriminals managed to download certain data files related to guests and employees' personal data, which could result in potential claims from guests, employees, shareholders, or regulatory agencies [104].

According to Hannemann [103], there are three key takeaways from these three cyberattacks. The first is related to IT hygiene, which is key to fighting cybercrime—a need to shift people's mind-set towards IT security. Second, every shipping manager needs to approach cybersecurity as an integral part of overall safety management. Response and recovery plans should be in place, updated, and tested frequently. Third, there is no zero cyber risk environment today, since new cyber threats and vulnerabilities are constantly emerging. Despite all precautions, vulnerabilities remain in the systems and networks, and attackers are constantly trying to find new tools to break through cyber defenses.

3.3.4. Ports, Terminals, and Supply Chains

Ports are an integral node of maritime transportation and the land transport chain. They rely on information from both shipping lines and land-side logistics companies. The lack of clear standards and requirements addressing critical maritime infrastructure demonstrates a compelling need for standardized policies for assessing, containing, and mitigating cyber risks. Legacy IT systems and an expanding Internet of Things (IoT) contribute to making ports vulnerable [105]. Roughly half, only, of the world's shipping ports understand or are aware of their problems and vulnerabilities concerning cybersecurity [59]. As valves in global economic arteries, the port infrastructures' protection against cyber risks is an absolute imperative [97].

As for this infrastructure, critical systems (ship's cargo handling, container tracking) may be penetrated, as occurred in the port of Antwerp where hackers gained access to the port's terminal operating system and trafficked drugs [77]. Hackers working with drug-smuggling criminals infiltrated the computerized cargo tracking system to identify the shipping containers in which consignments of drugs had been hidden. The gang then drove the containers from the port, retrieved the drugs, and covered their tracks. The criminal activity continued for two years from June 2011, until it was stopped by investigative authorities [46,75].

Another example is the hacking into the Port of San Francisco when their Electronic Information System "moved" the port in cyberspace twenty miles north, which became problematic in the foggy weather [106].

The analysis showed that successful cyber penetration into the port infrastructure information system allows attackers movement or the theft of illicit cargo. This system is vulnerable and it is easy to see how it can be exploited by cybercriminals that will continue to do the unexpected [7,107]. We can be certain that the nature of attacks of this sort will evolve [46,73].

Aids to Navigation (AtoN) are an integral and critical part of marine infrastructure. Traditionally AtoNs have been physical objects such as lighthouses, buoys, and beacons. The introduction of virtual AtoNs that appear on AIS INS displays via the AIS communication system is an achievement of the modern digital era. Though still under development, virtual AtoNs play a vital role in enhancing navigational safety. The AIS system is not as secure as it should be and can, therefore, be disrupted by malignant action.

3.3.5. Jamming and Spoofing

This chapter presents the vulnerability of the Global Navigation Satellite System (GNSS) to jamming and spoofing activities of state and non-state actors that can cause significant damage to major economies and everyday consumers alike. GNSS refers to a space-based system such as the US Global Positioning System (GPS/NAVSTAR—a military capability which civilian use), Russia's Global Navigation Satellite System (GLONASS), the European Union's Galileo System (a civil project intended for civil government and commercial use), and China's Beidou Navigation Satellite System. In addition, there are regional systems such as India's Navigation Indian Constellation (NavIC) and Japan's Quazi-Zenith Satellite System (QZSS).

GNSS Jamming is the deliberate transmission of signals on frequencies used by GNSS in an effort to prevent receivers from locking-on to authentic GNSS Signals. GNSS jamming requires relatively little technical knowledge and can be conducted by merely drowning out genuine signals with random or disruptive noise. *GNSS Spoofing* refers to the transmission of simulated false GNSS satellite ephemeris and timing information which coerces the victim receiver into calculating incorrect positioning and, in some cases, timing information [108–110]. GNSS spoofing is quite different from GNSS jamming. While navigation systems sound alarms when they recognize jammers, spoofing systems create false signals that confuse even state-of-the-art GNSS systems, leading to more severe consequences. To mitigate most of the vulnerabilities of one navigational system, it is recommended to use more of them [3].

The threat of GNSS spoofing has been known to the maritime industry for years. In 2017, the incident at Gelendzhik Airport received attention in international media. At least 20 vessels in the vicinity of the Black Sea Novorossiysk Commercial Sea Port reported that their AIS traces erroneously showed their position as Gelendzhik Airport, around 32 km inland. There were a large number of vessels involved, and all of the ships' tracking systems placed them in the same nonsensical location. That led to informed speculation that the incident could be attributed to one of the space superpower states' testing of satellite navigation spoofing technology. Whether it was as part of its electronic warfare arsenal or they were simply using it as an anti-drone measure for very VIP protection, it is somewhat suggestive that it was GNSS spoofing of defense development inflicted on a civilian scenario.

In his research, Bergman [87] has found ships in various parts of the world reporting locations thousands of miles away and circling at precisely 20 knots. It is unclear if these errors were the result of the ships' AIS system or some fault or influence on GPS receivers, a real mystery of some form of GPS interference.

In another event, in July 2019, a British oil tanker, the Stena Impero, was seized by Iranian forces after being spoofed to cause the vessel to shift its course into Iranian waters. As a consequence, the vessel, cargo, and the crew had become more than pawns in a geopolitical war [111]. Many of the shipping companies operating ships in the region have also instructed their vessels to transit Hormuz only at high speed and during the daylight hours. Nevertheless, we should not forget that one-third of the world's seaborne oil—some 17 million barrels per day—passes through the strait, making it one of the most important oil trading routes in the world.

AIS is a critical safety system designed to provide a ship's position and course to neighboring ships to prevent collision [3]. The manipulation of the AIS system that was observed not long ago, on 3 December 2019, as illustrated in Figure 1, might have been spoofing.

An Italian AIS base station experienced a ship-spoofing situation near Elba Island that was visible in the European Maritime Safety Agency (EMSA) Maritime Application, SSN Ecosystem GUI [89]. The first investigation provided by the Italian Coast Guard indicated 870 different vessels were created at two different moments (13:13 and 13:28) with a duration of 3 min in the first transmission and 2 min in the second. All the tracks appeared in an area of 28 $\times$ 21 nautical miles between Elba Island and

Corsica with different routes and speeds, rendering the monitoring of the maritime traffic in the area impossible and impacting real vessel transmissions.

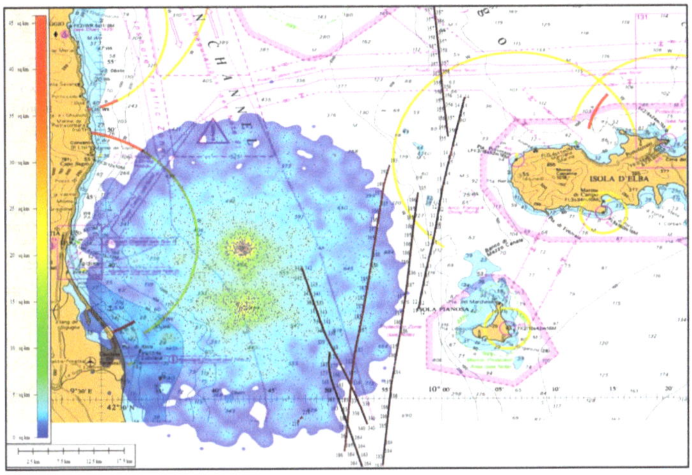

Figure 1. Automatic Identification System (AIS) spoofing analysis near Elba Island.

At the Faculty of Maritime Studies and Transport, University of Ljubljana, we have thoroughly investigated the situation to support EMSA's analysis and found 3742 fake ships (861 false tracks with MMSI 24480XXXX) that generated together 5133 reports, as illustrated in Figure 1. By using the European Marine Observation and Data Network (EMODnet) method [112], a traffic density map (TDM) is created by using ship positioning data collected from Terrestrial and Satellite AIS data sources, the maritime infrastructure, and the SafeSeaNet Ecosystem Graphical Interface (SEG) application. It showed, according to Perkovič [44], shipping density of up to 45 ships/km^2. The analyses pointed out that the AIS spoofing generator was located in the Elba Island area and that spoofing was conducted by a transmitter tuned onto AIS channel A, occupying around 54% of the available slots. Data were further processed to see whether it affected traffic safety. Only one vessel can be seen to have deviated slightly, although it cannot be proven that the change in course was due to the spoofing.

As evident through such examples, deliberate disruptions have affected shipping in international waters and engaged in innocent passage through territorial waters. Disorders to GNSS-enabled positioning and navigation have therefore become a global phenomenon, and to tackle a global problem, the GNSS community requires a global solution [35].

Another concern of the satnav community is related to the risks of being covertly tracked and of being exposed to malware through a satellite communication channel. The two-way communication that some GNSS systems now offer opens up that possibility. These risks have in particular been discussed in connection with the now officially operational BeiDou system [90,91].

3.3.6. Autonomous Ship

Remotely piloted and even more so autonomous marine vessels are going to revolutionize maritime operations and will be undoubtedly more vulnerable to cyberattacks [66,67]. Navigation issues and cyber risks should be taken into serious consideration by IMO when creating rules for autonomous shipping. It is expected that the maritime attack surface will continue to expand and autonomous ships will form a prominent piece of the future maritime landscape, underscoring the growing reliance on interconnected information systems [74,113].

3.3.7. Cyber and Social Media

Analysis has shown that there is a need to consider the impact of social media on the crew's life at sea. The usage of social networking services can unintentionally cause many problems, such as the leakage of information on confidential shipping matters, diffusion of marine accident information, and the exposure of personal private information or photos [114]. Communication and socialization with the outside world, particularly with friends and family has become commonplace for the modern seafarer. From the perspective of the human element, cyber wellness is a critical component of their overall health and personal well-being. However, with that privilege also comes the responsibility to protect oneself and ensure that the ship's safety and security are not compromised. These shared objectives are achieved by employing good personal cybersecurity practices [115].

3.3.8. Hybrid Threats

In recent years, the West and the world's democratic societies have become increasingly exposed to hybrid threats. Hybrid threats are multi-faceted subversive or coercive activities that aim to undermine a state, frustrate its decision making processes, erode the trust in its institutions, stay below a threshold of detectability, and remain difficult to attribute. A threat is hybrid if it is part of a concerted set of mutually reinforcing actions. Hybrid threat campaigns are long-term and of varying intensity over time [70,116]. As a critical domain for economy, resources, trade, transport, security, and defense, the maritime domain, including ports, finds itself a target of hybrid threats [105]. The fact that much of the (maritime) critical infrastructure is privately owned complicates the construction of a strong and coordinated response to hybrid threats [117,118].

Cyberattacks are one of the main vectors of hybrid threats [117,119], and GNSS spoofing would also seem to be an attractive attack vector [118]. Foreign direct investments to gain ownership of critical infrastructure or technology (e.g., ports, hardware and software companies that supply the maritime industry) are part of the arsenal [119,120].

Since 2016, the EU and NATO have been cooperating on hybrid threats and cybersecurity [121], with these two topics prominently figuring in a common set of proposals [122,123]. In relation to that cooperation, the European Centre of Excellence for Countering Hybrid Threats (Hybrid CoE) was set up in April 2017 in Helsinki to encourage strategic dialogue and carry out research and analysis on hybrid threats [124]. To this end, common exercises and workshops are carried out [125]. Following a closer look at hybrid threats versus sea lines of communication, the Hybrid CoE recently published a Handbook on Maritime Hybrid Threats [126].

Due to their low-level and dispersed nature, hybrid threats can only be fully recognized by bringing together the information about occurrences of disruptions that individually may seem minor but are part of a larger whole, thereby creating situational awareness [70]. The sharing of information on disruptions is therefore crucial, just like it is the case for cyberattacks outside the hybrid context. For adequate business continuity, operators should develop a resilience against hybrid attacks. The EU plans to identify sectoral hybrid resilience baselines [70]—one sector would be the maritime. Sectorial approaches, including maritime, are proposed in several other EU policies [127,128]. Risk analysis and enhancing of the overall resilience of EU critical maritime infrastructure with regard to cyber and hybrid threats, among others, are slated for action in the updated EU Maritime Security Strategy Action Plan [64]. A proper defense against hybrid threats requires all-of-society awareness and resilience [118,129,130].

3.3.9. Environmental Pollution

According to research, the side effects of disasters caused by a hacked port system or deluded onboard ship system could include environmental threats. The key to the security of the waterways is agility and constant paradigm-shifting to out-manoeuvre those who want to damage or disrupt the maritime transportation system (MTS) [69,131]. One might expect that cybercriminals, terrorists, and rogue states will, at some point, begin holding the environment to ransom—cyber-induced environmental pollution [2]. It is a possible scenario that hackers would over-ride ship systems in ports, open valves and initiate leaks, or dump hazardous materials, ballast water, fuel, or oil.

4. Discussion

Analyzing available resources, we may summarize that there are severe consequences posed by cybersecurity risks, ranging from ship accidents caused by hacking e-Navigation to massive operational and economic disruption, to a port or shipping companies' activities and business. Cybercrime is a growing threat to the shipping industry that may have severe repercussions, as the Maersk, COSCO, Austal, Carnival, and other incidents have demonstrated. The present security situation in regard to hybrid threats further exposes the risk. As per BIMCO [78], it is not a question of whether an organization will be affected by a cybersecurity incident, but of when. Perhaps the organization has already been affected without knowing it. It is the connectedness of multiple and diverse systems that exacerbates the cyber risks, and connectivity will only continue to increase—e.g., with the introduction of the Internet of Things (IoT) and 5G in the maritime domain [59,67,132]. Therefore, maintaining the operational safety of these systems is of paramount importance. There continue to be numerous advancements in the field of hardware and software network security with which cyber administrators need to keep up. Though with the increased demand of the internet, it is only going to become more difficult; therefore, IT managers of maritime specialty may become necessities.

The maritime sector has always focused on safety and security issues, such as aspects of seafarer safety, transport, and cargo security. Now, the essence of navigational safety and maritime security include building a significant security culture which needs to be developed even further to include the cyber and information security domain, *maritime cybersecurity*. Maritime cybersecurity includes people, technologies, and processes capable of preventing cyber and information security breaches [73]. Hareide et al. [6] and Fitton et al. [133] note that maritime cybersecurity can be understood as a part of maritime security concerned with the protection from cyber threats of all aspects of maritime cyber systems that should include *technology, information,* and the *human* factor to understand and mitigate cyberattacks. Hasratyan et al. [59] mention the new concept of 'cyber seaworthiness' recently introduced in the shipping insurance industry.

Following the BIMCO 2020 survey [78], we found that there is room for improvement in regard to the *human* factor in maritime cybersecurity training that is seen by many as the first line of defense against the most common cyber incidents. The survey is quite encouraging, as 88% of respondents indicated that their company offers some sort of cybersecurity training; however, improvement is necessary, as only 22% of respondents received high-quality training. A 2020 ECSO report estimates that the maritime industry lacks 50,000 to 100,000 trained people in the cyber field [59]. It is a fact that cybersecurity training and awareness are paramount to the maritime industry. People need to be aware of the threats they encounter, not only on work IT systems but to their private devices as well. Training courses, refreshers, and adequate software protection should be offered and modified to fit the crews' needs. The crew on the ships might rotate quite often, meaning that seafarers potentially often use systems they are unfamiliar with, which might increase the potential for human error.

It was noted, concerning *technology* and *information*, that a shift in mind-set is essential to direct more attention and resources toward cybersecurity [134–136]. In the event of the worst scenario, appropriate contingency plans should be in place. Such plans should also include the use of alternative modes to ensure safe and reliable operations in the cyber non-secure environment [75,137]. Manufacturers should also improve the cybersecurity of their products, not leaving critical shipping systems vulnerable to cybercriminals [60].

National defense forces are already in the hunt for alternate position, navigating, and timing (PNT) systems (to GPS) within two years. Even though military commanders would prefer equipment that does not need to rely on GPS because their current systems are increasingly being jammed or spoofed, the costly GPS III program is going to be developed in the USA. It will be designed to include improved anti-jam features, including the higher-fidelity M-Code signal for military users only [81]. In the meantime, the US Air Force is looking for solutions that could provide alternative signal sources that might even be from a non-US GNSS system. Lockheed U-2 high altitude reconnaissance jet pilots wear GPS-enabled smartwatches (Garmin D2 Charlie) for navigation backup in case their GNSS signal becomes unavailable [80]. Watches are capable of receiving multiple navigation signals including GPS, GLONASS, and BeiDou. Indeed, the combination of multiple GNSS systems in a smart receiver makes spoofing much more difficult, and now that Galileo is available worldwide, it is already used in over a billion smartphones [138]. The questions we might be asking here are

- Is it not a bit unusual that a superpower state does not have its alternative navigation system for their own defense forces in order not to rely on the signal sources of their rivals?
- Would it not be wise to develop and put into practice GPS-enabled smartwatches for mariners in case their GPS signal onboard becomes unavailable?
- Is there any viable non-GNSS alternative or backup system for vulnerable GPS navigation?

At this stage, we might find an answer only to the third question. IMO recognizes the need for multiple PNT systems onboard maritime vessels. However, there has been no widely available substitute system for GPS navigation, although, now Galileo is becoming a viable alternative. IMO has developed the e-Navigation concept to increase maritime safety and security via means of electronic navigation, based on (at least) two different independent sources of the PNT system to make it robust and fail-safe. The most viable terrestrial system providing PNT services that meets IMO's requirements an enhancement of LOng-RAnge Navigation (LORAN)—eLoran. Not precisely as accurate as GPS, it can provide sub-5 m (in the Netherlands) to sub-10 m (in the United Kingdom) horizontal positioning accuracy. It meets the availability, integrity, and continuity performance requirements for maritime harbor entrance and approach manoeuvres, aviation non-precision instrument approaches, land-mobile vehicle navigation, and location-based services [83–85].

Galileo offers the Public Regulated Service (PRS) which is encrypted to protect against spoofing [139]. However, that is only available for use by authorities. However, in the near future, it will offer (free) Open Service Navigation Message Authentication (OS NMA) that will provide a low level of protection against spoofing, and a (paid) Commercial Authentication Service (CAS) that is encrypted like PRS—but for commercial use—while also providing higher accuracy [140].

Perkovic et al. [37–39] presented an impressive laser-based aid system for berthing and docking that is providing accuracy within 2 cm even when the GNSS satellite signal is obstructed by Ship to Shore (STS) cranes, or redirected. This system could be one of the core subsystems required for collision avoidance [141,142] of autonomous surface vessels, which still require solutions for a variety of technical problems [113].

The answer to the third question could eventually come about by a public–private partnership. Humphreys [143] predicts that companies such as Elon Musk's SpaceX and Amazon's Project Kuiper, maintaining networks with hundreds of low-Earth-orbit satellites, will eventually become a vital component of the GNSS ecosystem. The new GNSS will pick up the slack in the event of malfunctions or attacks.

5. Conclusions

To summarize, the importance of cybersecurity has been recognized, and there will be a series of new cybersecurity openings in the future through which hackers can attack if systems are not adequately protected. It is expected that cybersecurity challenges will expand as autonomous ships will form a prominent piece of the future maritime landscape, underscoring the growing reliance on interconnected information systems [74]. New, unexpected circumstances may lead to increased cyber risks, as the year 2020 is now seeing with COVID-19, with reports of 400 percent increase in attempted hacks since February 2020 [144]. Moreover, the global security landscape that gives rise to hybrid threats is not expected to improve soon.

The maritime industry has shown that it is neither immune to cyberattacks nor completely prepared to combat the risks involved in using some obsolete or modern digital systems. Maintaining seaworthiness given the impact of digital technologies requires robust cybersecurity policy/strategies, cyber-secure maritime technology, a shift in mind-set, and new insurance offers that specifically cover maritime cyber risks. A modest beginning is to have the right people on board and have established trust between maritime stakeholders. A company under attack must have the confidence to share the information with others in the sector, allowing them to bolster their defenses.

GNSS signals are indispensable to safe and efficient navigation and are an integral component of maritime operations. To interfere with them jeopardizes maritime safety and security at sea. This paper demonstrates the GNSS vulnerabilities that impact many maritime systems. The use of multiple GNSS constellations (public and private), intelligent processing on the receiver side, and encrypted signals will in the near future provide increased robust defense against jamming and spoofing. The paper also recommends that the maritime community implement an eLoran system as a terrestrial augmentation to space-based GNSS capabilities. Terrestrial signals could be coded and authenticated, further increasing security and, subsequently, safeguarding GNSS satellites by making them less attractive targets. Locally, the positioning systems can be complemented by the harbor's laser-based aid system for berthing and docking, so that the regional shipping industry can be better protected from cyberattacks for the foreseeable future. As presented, building resilience against cybercriminals is a never-ending battle.

Author Contributions: Conceptualization, A.A. and I.P.; methodology, A.A. and T.B.; data collection, A.A., T.B., and I.P.; EU aspects, H.G.; validation, A.A., T.B., and I.P.; formal analysis, A.A. and I.P.; data curation, A.A., T.B., and I.P.; writing—original draft preparation, A.A.; internal review, H.G. All authors have read and agreed to the published version of the manuscript.

Funding: The authors acknowledge the financial support of the Slovenian Research Agency (research core funding No. P2-0394, Modelling and Simulations in Traffic and Maritime Engineering).

Conflicts of Interest: The authors declare no conflict of interest.

List of Acronyms

AtoN	Aids to Navigation
AIS	Automatic Identification System
BeiDou	China's BeiDou Navigation Satellite System
CERT-EU	Computer Emergency Response Team
cPPP	contractual Public-Private Partnership
CSIRT	Computer Security Incident Response Teams
ECDIS	Electronic Chart Display and Information System
EC3	European Cyber Crime Centre at Europol
eLORAN	Enhanced LOng-RAnge Navigation
EMODnet	European Marine Observation and Data Network
EMSA	European Maritime Safety Agency
ENISA	European Union Agency for Cybersecurity
ECSO	European Cyber Security Organisation
Galileo	European Union's GNSS
GLONASS	Russia's Global Navigation Satellite System
GNC	Guidance, Navigation and Control
GNSS	Global Navigation Satellite System
GPS, GPS/NAVSTAR	Global Positioning System (USA's GNSS)
Hybrid CoE	European Centre of Excellence for Countering Hybrid Threats
IEC	International Electrotechnical Commission
IMO	International Maritime Organization
INS	Integrated Navigational System
IoT	Internet of Things
ISAC	Information Sharing and Analysis Centres

ISAO	Information Sharing and Analysis Organizations
ISM Code	International Safety Management Code
IT	Information Technology
ITU	International Telecommunication Union
LORAN	LOng-RAnge Navigation
MSC	Maritime Safety Committee
MTS	Maritime Transportation System
NavIC	India's Navigation Indian Constellation
OT	Operational Technology
PC	Personal Computer
PNT	Positioning, Navigation and Timing
QZSS	Japan's Quazi-Zenith Satellite System
SCADA	Supervisory Control And Data Acquisition
SEG	SafeSeaNet Ecosystem Graphical Interface
SMS	Safety Management System
SOLAS	Safety Of Life At Sea
SSN Ecosystem GUI	The SafeSeaNet Ecosystem Graphical User Interface (GUI)
STS	Ship to Shore
TDM	Traffic Density Mapping
TEU	Twenty-foot equivalent unit
VDR	Voyage Data Recorder
VIP	Very Important Persons

References

1. World Economic Forum. Wild Wide Web—Consequences of Digital Fragmentation. *The Global Risks Report 2020*, 15th Ed. ed. January 2020. Available online: https://www.weforum.org/reports/the-global-risks-report-2020 (accessed on 20 June 2020).
2. Maritime Cyber-Attacks Increase by 900% in Three Years. Available online: https://www.marineinsight.com/shipping-news/maritime-cyber-attacks-increase-by-900-in-three-years/# (accessed on 20 July 2020).
3. Middleton, A. Hide and Seek: Managing Automatic Identification System Vulnerabilities: Proceedings of the Marine Safety and Security Council, Coast Guard. *J. Saf. Secur. Sea* **2014**, *71*, 48–50.
4. Chybowski, L.; Gawdzinska, K.; Laskowski, R. Assessing the Unreliability of Systems during the Early Operation Period of a Ship—A Case Study. *J. Mar. Sci. Eng.* **2019**, *7*, 213. [CrossRef]
5. Dobryakova, L.A.; Lemieszewski, L.S.; Ochin, E.F. GNSS spoofing detection using static or rotating single-antenna of a static or moving victim. *IEEE Access* **2018**, *6*, 79074–79081. [CrossRef]
6. Hareide, O.S.; Jøsok, Ø.; Lund, M.S.; Ostnes, R.; Helkala, K. Enhancing Navigator Competence by Demonstrating Maritime Cyber Security. *J. Navig.* **2018**, *71*, 1025–1039. [CrossRef]
7. Kaleem Awan, M.S.; Al Ghamdi, M.A. Understanding the Vulnerabilities in Digital Components of an Integrated Bridge System (IBS). *J. Mar. Sci. Eng.* **2019**, *7*, 350. [CrossRef]
8. Lee, E.; Mokashi, A.J.; Moon, S.Y.; Kim, G. The Maturity of Automatic Identification Systems (AIS) and Its Implications for Innovation. *J. Mar. Sci. Eng.* **2019**, *7*, 287. [CrossRef]
9. Polatidis, N.; Pavlidis, M.; Mouratidis, H. Cyber-attack path discovery in a dynamic supply chain maritime risk management system. *Comput. Stand. Interfaces* **2018**, *56*, 74–82. [CrossRef]
10. Kalogeraki, E.; Apostolou, D.; Polemi, N.; Papastergiou, S. Knowledge management methodology for identifying threats in maritime/ logistics supply chains. *Knowl. Manag. Res. Pract.* **2018**, *16*, 508–524. [CrossRef]
11. Kessler, G.C.; Craiger, J.P.; Haass, J.C. A Taxonomy Framework for Maritime Cybersecurity: A Demonstration Using the Automatic Identification System. *Trans. Nav. Int. J. Mar. Navig. Safety Sea Trans.* **2018**, *12*, 429–437. [CrossRef]
12. Lee, Y.C.; Park, S.K.; Lee, W.K.; Kang, J. Improving cybersecurity awareness in maritime transport: A way forward. *J. Korean Soc. Mar. Eng.* **2017**, *41*, 738–745.
13. Lewis, S.; Maynard, L.; Chow, C.E.; Akos, D. Secure GPS Data for Critical Infrastructure and Key Resources: Cross-Layered Integrity Processing and Alerting Service. *Navig. J. Inst. Navig.* **2018**, *65*, 389–403. [CrossRef]

14. Shapiro, L.R.; Maras, M.-H.; Velotti, L.; Pickman, S.; Wei, H.-L.; Till, R. Trojan horse risks in the maritime transportation systems sector. *J. Trans. Secur.* **2018**, *8*, 1–19. [CrossRef]
15. Svilicic, B.; Brčić, D.; Žuškin, S.; Kalebić, D. Raising Awareness on Cyber Security of ECDIS. *TransNav Int. J. Mar. Navig. Saf. Sea Trans.* **2019**, *13*, 231–236. [CrossRef]
16. Svilicic, B.; Kamahara, J.; Celic, J.; Bolmsten, J. Assessing Ship Cyber Risks: A Framework and Case Study of ECDIS Security. *WMU J. Mar. Aff.* **2019**, *18*, 509–520. [CrossRef]
17. Svilicic, B.; Kamahara, J.; Rooks, M.; Yano, Y. Maritime Cyber Risk Management: An Experimental Ship Assessment. *J. Navig.* **2019**, *72*, 1108–1120. [CrossRef]
18. Svilicic, B.; Rudan, I.; Frančić, V.; Mohović, Đ. Towards a Cyber Secure Shipboard Radar. *J. Navig.* **2019**, *73*. [CrossRef]
19. Tam, K.; Jones, K. MaCRA: A model-based framework for maritime cyber-risk assessment. *WMU J. Mar. Aff.* **2019**, *18*, 129–163. [CrossRef]
20. Tsimplis, M.; Papadas, S. Information Technology in Navigation: Problems in Legal Implementation and Liability. *J. Navig.* **2019**, *72*, 833–849. [CrossRef]
21. Zăgan, R.; Raicu, G.; Pazara, R.H.; Enache, S. Realities in Maritime Domain Regarding Cyber Security Concept. *Adv. Eng. Forum* **2018**, *27*, 221–228. [CrossRef]
22. Assessing the Cyber Risks of Maritime Navigation. Available online: https://www.kennedyslaw.com/media/3288/kennedys_assessingthecyberrisksofmaritimenavigation.pdf (accessed on 20 May 2020).
23. Lessons Learned from Maritime License to Operate at Risk? Available online: https://www.kongsberg.com/digital/resources/stories/2019/10/maritime-license-to-operate-at-risk/ (accessed on 4 July 2020).
24. ISO. Information Technology—Security Techniques—Guidelines for Cybersecurity. ISO/IEC 27032:2012, 07/2012. Available online: https://www.iso.org/obp/ui/#iso:std:iso-iec:27032:ed-1:v1:en (accessed on 24 August 2020).
25. International Telecommunication Union. Overview cybersecurity. In *ITU-T X.1205 Recommendation;* International Telecommunication Union: Geneva, Switzerland, 2008; p. 2.
26. European Commission. *Cybersecur. Eur. Digit. Single Mark.* **2017**, *2*, 13. [CrossRef]
27. Maritime-License-to-Operate-at-Risk-KPMG-and-KONGSBERG.Pdf. Available online: https://assets.kpmg/content/dam/kpmg/no/pdf/2019/09/Maritime-license-to-operate-at-risk-KPMG-and-KONGSBERG.pdf (accessed on 2 August 2020).
28. Jensen, L. Challenges in Maritime Cyber-Resilience. *Technol. Innov. Manag. Rev.* **2015**, *5*, 35–39. [CrossRef]
29. Code of Practice - Cyber Security for Ships. Available online: https://assets.publishing.service.gov.uk/government/uploads/system/uploads/attachment_data/file/642598/cyber-security-code-of-practice-for-ships.pdf (accessed on 22 June 2020).
30. Babineau, G.; Jones, R.; Horowitz, B. A System-Aware Cyber Security Method for Shipboard Control Systems with a Method Described to Evaluate Cyber Security Solutions. In Proceedings of the 2012 IEEE Conference on Technologies for Homeland Security (HST), Waltham, MA, USA, 13–15 November 2012; pp. 99–104. [CrossRef]
31. Masala, C.; Tsetsos, K.A. Cyber risks and threats: Demanding challenge for the maritime industry. In *Look Out 2016 Maritime Domain Cyber: Risks, Threats & Future Perspectives;* Lampe & Schwartze KG: Bremen, Germany, 2015; pp. 11–26.
32. Glomsvoll, O.; Bonenberg, L. GNSS Jamming Resilience for Close to Shore Navigation in the Northern Sea. *J. Navig.* **2017**, *70*, 33–48. [CrossRef]
33. Direnzo, J.; Goward, D.A.; Roberts, F.S. The Little-Known Challenge of Maritime Cyber Security. In Proceedings of the 2015 6th International Conference on Information, Intelligence, Systems and Applications (IISA), Corfu, Greece, 6–8 July 2015; pp. 1–5. [CrossRef]
34. Ziebold, R.; Romanovas, M.; Gewies, S. Experimental Evaluation of the Impact of Jamming on Maritime Navigation. In Proceedings of the 29th International Technical Meeting of the Satellite Division of the Institute of Navigation (ION GNSS+ 2016), Portland, OR, USA, 12–16 September 2016; pp. 3461–3480. [CrossRef]
35. Thombre, S.; Bhuiyan, M.Z.H.; Eliardsson, P.; Gabrielsson, B.; Pattinson, M.; Dumville, M.; Fryganiotis, D.; Hill, S.; Manikundalam, V.; Pölöskey, M.; et al. GNSS threat monitoring and reporting: Past, present, and a proposed future. *J. Navig.* **2018**, *71*, 513–529. [CrossRef]
36. Elsobeiey, M.E. Accuracy Assessment of Satellite-Based Correction Service and Virtual GNSS Reference Station for Hydrographic Surveying. *J. Mar. Sci. Eng.* **2020**, *8*, 542. [CrossRef]

37. Perkovic, M.; Gucma, M.; Luin, B.; Gucma, L.; Brcko, T. Accommodating larger container vessels using an integrated laser system for approach and berthing. *Microprocess. Microsyst.* **2017**, *52*, 106–116. [CrossRef]
38. Perkovič, M.; Gucma, L.; Bilewski, M.; Muczynski, B.; Dimc, F.; Luin, B.; Vidmar, P.; Lorenčič, V.; Batista, M. Laser-Based Aid Systems for Berthing and Docking. *J. Mar. Sci. Eng.* **2020**, *8*, 346. [CrossRef]
39. Gucma, L.; Bak, A.; Jankowski, S.; Zalewski, P.; Perkovic, M. Laser docking system integrated with Pilot Navigation Support System, a background to high precision, fast and reliable vessel docking. In Proceedings of the 17th Saint Petersburg International Conference on Integrated Navigation Systems, St. Petersburg, Russia, 31 May–2 June 2010.
40. Mileski, J.; Clott, C.; Galvao, C.B. Cyberattacks on Ships: A Wicked Problem Approach. *Marit. Bus. Rev.* **2018**, *3*, 414–430. [CrossRef]
41. Tranfield, D.; Denyer, D.; Smart, P. Towards a Methodology for Developing Evidence-Informed Management Knowledge by Means of Systematic Review* Introduction: The Need for an Evidence-Informed Approach. *Br. J. Manag.* **2003**, *14*, 207–222. [CrossRef]
42. Grant, M.J.; Booth, A. A Typology of Reviews: An Analysis of 14 Review Types and Associated Methodologies. *Health Inf. Libr. J.* **2009**, *26*, 91–108. [CrossRef]
43. Milner, K.A. Systematic Reviews. *Oncol. Nurs. Forum* **2015**, *42*, 89–93. [CrossRef]
44. Perkovič, M. *AIS Spoofing Near Elba Island Analysis and Research Data*; University of Ljubljana, Faculty of Maritime Studies and Transport: Ljubljana, Slovenia, 2020.
45. Bansal, M.; Kaur, J.; Kaur, A.; Raina, C.K. Cyber Security: Impact and Preventions. *IJSRCSEIT* **2017**, *2*, 1096–1100.
46. The Risk of Cyber-Attack to the Maritime Sector. Glob. Mar. Pract. Available online: https://www.marsh.com/uk/insights/research/the-risk-of-cyber-attack-to-the-maritime-sector.html (accessed on 26 June 2020).
47. IMO Resolution MSC.428 (98). 2017. Available online: http://www.imo.org/en/OurWork/Security/Guide_to_Maritime_Security/Documents/Resolution%20MSC.428(98).pdf (accessed on 20 May 2020).
48. ISM. *International Safety Management Code*; IMO Publishing: London, UK, 2018.
49. Guidelines on Cyber Risk Management. Maritime Safety Committee: 2017, MSC-FAL (1/Circ.3). pp. 1–6. Available online: http://www.imo.org/en/OurWork/Security/Guide_to_Maritime_Security/Documents/MSC-FAL.1-Circ.3%20-%20Guidelines%20On%20Maritime%20Cyber%20Risk%20Management%20(Secretariat).pdf (accessed on 16 July 2020).
50. The Guidelines of Cyber Security Onboard Ships (2018). Available online: https://www.bimco.org/about-us-and-our-members/publications/the-guidelines-on-cyber-security-onboard-ships (accessed on 15 May 2020).
51. International Electrotechnical Commission. *Maritime Navigation and Radiocommunication Equipment and Systems-Cybersecurity-General Requirements, Methods of Testing and Required Test. Results*; IEC 63154 ED1; IEC: Geneva, Switzerland, 2019.
52. European Commission; High Representative. Cybersecurity Strategy of the European Union: An Open, Safe and Secure Cyberspace. JOIN(2013) 1 final, Brussels. 7 February 2013. Available online: https://eur-lex.europa.eu/legal-content/EN/TXT/?uri=celex:52013JC0001 (accessed on 15 July 2020).
53. Directive 2013/40/EU of the European Parliament and of the Council of 12 August 2013 on Attacks Against Information Systems and Replacing Council Framework Decision 2005/222/JHA. Available online: https://eur-lex.europa.eu/legl-content/EN/ALL/?uri=CELEX:32013L0040 (accessed on 15 July 2020).
54. Directive 2016/1148 of the European Parliament and of the Council of 6 July 2016 Concerning Measures for a High Common Level of Security of Network and Information Systems Across the Union (NIS Directive). Available online: https://eur-lex.europa.eu/eli/dir/2016/1148/oj (accessed on 15 July 2020).
55. European Commission. Communication Making the Most of NIS—towards the Effective Implementation of Directive (EU) 2016/1148 Concerning Measures for a High Common Level of Security of Network and Information Systems across the Union. COM(2017) 476 final/2, Brussels. 4 October 2017. Available online: https://eur-lex.europa.eu/legal-content/EN/TXT/?uri=COM:2017:476:FIN (accessed on 15 July 2020).
56. European Commission. Strengthening Europe's Cyber Resilience System and Fostering a Competitive and Innovative Cybersecurity Industry. COM(2016) 410 final, Brussels. 5 July 2016. Available online: https://eur-lex.europa.eu/legal-content/EN/ALL/?uri=COM:2016:0410:FIN (accessed on 15 July 2020).
57. ENISA. Information Sharing and Analysis Centres (ISACs)—Cooperative Models. Available online: https://doi.org/10.2824/549292 (accessed on 17 July 2020).

58. What is an ISAC or ISAO? How These Cyber Threat Information Sharing Organizations Improve Security. Available online: https://www.csoonline.com/article/3406505/what-is-an-isac-or-isao-how-these-cyber-threat-information-sharing-organizations-improve-security.html (accessed on 8 September 2020).
59. Hasratyan, N.; Olesen, N. Transportation Sector Report—Cyber Security for Road, Rail, Air, and Sea. European Cyber Security Organisation. Available online: https://www.ecs-org.eu/documents/publications/5e78cb9869953.pdf (accessed on 17 July 2020).
60. European Commission; High Representative. Joint Communication on Resilience, Deterrence and Defence: Building strong cybersecurity for the EU. JOIN(2017) 450 final, Brussels. 13 September 2017. Available online: https://eur-lex.europa.eu/legal-content/en/TXT/?uri=CELEX:52017JC0450 (accessed on 17 July 2020).
61. Regulation (EU) 2019/881 of the European Parliament and of the Council of 17 April 2019 on ENISA (the European Union Agency for Cybersecurity) and on Information and Communications Technology Cybersecurity Certification and Repealing Regulation (EU) No 526/2013 (Cybersecurity Act). Available online: https://eur-lex.europa.eu/eli/reg/2019/881/oj (accessed on 17 July 2020).
62. Council of the EU. European Union Maritime Security Strategy. 11205/14, Brussels. 24 June 2014. Available online: http://register.consilium.europa.eu/doc/srv?l=EN&f=ST%2011205%202014%20INIT (accessed on 17 July 2020).
63. Council of the EU. European Union Maritime Security Strategy Action Plan. 17002/14, Brussels. 16 December 2014. Available online: https://ec.europa.eu/maritimeaffairs/sites/maritimeaffairs/files/docs/body/20141216-action-plan_en.pdf (accessed on 17 July 2020).
64. Council of the EU. Revised European Union Maritime Security Strategy (EUMSS) Action Plan. Annex to 10494/18, Brussels. 26 June 2018. Available online: https://data.consilium.europa.eu/doc/document/ST-10494-2018-INIT/en/pdf (accessed on 17 July 2020).
65. European Commission; High Representative. On the implementation of the EU Maritime Security Strategy Action Plan. Joint Staff Working Document SWD(2016)217 Final. Available online: https://ec.europa.eu/maritimeaffairs/sites/maritimeaffairs/files/docs/body/swd-2016-217_en.pdf (accessed on 17 July 2020).
66. European Commission; High Representative. Second report on the implementation of the EU Maritime Security Strategy Action Plan. Joint Staff Working Document SWD(2017)238 Final. Available online: https://ec.europa.eu/maritimeaffairs/sites/maritimeaffairs/files/swd-2017-238_en.pdf (accessed on 17 July 2020).
67. European Commission; High Representative. Report on the implementation of the revised EU maritime security strategy action plan. In *Joint Staff Working Document*; European Commission: Brusselsl, Belgium, 2020; in draft.
68. European Union Agency for Cybersecurity. *Port Cybersecurity—Good Practices for Cybersecurity in the Maritime Sector*; ENISA: Athens, Greece, 2019. [CrossRef]
69. Cimpean, D.; Meire, J.; Bouckaert, V.; Stijn, V.C.; Pelle, A.; Hellebooge, L. *Analysis of Cyber Security Aspects in the Maritime Sector*; ENISA: Athens, Greece, 2011.
70. European Commission. Communication on the EU Security Union Strategy. COM(2020) 605 Final. Available online: https://eur-lex.europa.eu/legal-content/EN/TXT/?uri=CELEX:52020DC0605 (accessed on 17 July 2020).
71. Council of the EU. Shaping Europe's Digital Future—Council Conclusions. 8711/20, Brussels. 9 June 2020. Available online: https://data.consilium.europa.eu/doc/document/ST-8711-2020-INIT/en/pdf (accessed on 18 July 2020).
72. European Commission. Secure 5G Deployment in the EU—Implementing the EU Toolbox. COM(2020) 50 Final. Available online: https://eur-lex.europa.eu/legal-content/EN/ALL/?uri=COM:2020:0050:FIN (accessed on 18 July 2020).
73. Cyber and Information Security Strategy for the Maritime Sector 2019–2022. Available online: https://www.dma.dk/Documents/Publikationer/Cyber%20and%20Information%20Security%20Strategy%20for%20the%20Maritime%20Sector.pdf (accessed on 27 July 2020).
74. Modernized Maritime Industry Transports Cyber threats to Sea. Available online: https://www.csoonline.com/article/3410236/modernized-maritime-industry-transports-cyberthreats-to-sea.html (accessed on 27 July 2020).
75. Jones, K.; Tam, K.; Papadaki, M. Threats and Impacts in Maritime Cyber Security. *Eng. Technol. Ref.* **2016**, *1*. [CrossRef]

76. Caponi, S.; Belmont, K. Maritime Cybersecurity: A Growing Threat Goes Unanswered. *Intellect. Prop. Technol. Law J.* **2015**, *27*, 16.
77. Lagouvardou, S. Maritime Cyber Security: Concepts, Problems and Models. Master' Thesis, Technical University of Denmark, Copenhagen, Denmark, 2018. Available online: https://pdfs.semanticscholar.org/3158/103669fe46911b52e55dc7afe82237994036.pdf (accessed on 3 August 2020).
78. Safety at Sea and BIMCO cybersecurity white paper—IHS Markit 2020 Cyber Security Survey. Available online: https://ihsmarkit.com/info/0819/cyber-security-survey.html?utm_medium=website&utmsource=sas-news-article-1&utm_campaign=cyber-security-whitepaper (accessed on 20 May 2020).
79. Maritime Meets Cyber Security. Available online: https://www.maritime-executive.com/blog/maritime-meets-cyber-security (accessed on 3 August 2020).
80. Why Are U-2 Jet Pilots Wearing Garmin Satellite Navigation Smartwatches? Available online: https://arstechnica.com/gadgets/2020/03/why-are-u-2-jet-pilots-wearing-garmin-satellite-navigation-smartwatches/ (accessed on 30 April 2020).
81. SASC Wants Alternative GPS by 2023. Available online: https://breakingdefense.com/2020/06/sasc-wants-alternative-gps-by-2023/ (accessed on 29 July 2020).
82. Silgado, D.M. Cyber-Attacks: A Digital Threat Reality Affecting the Maritime Industry. *World Marit. Univ. Diss.* **2018**, 9–26.
83. Bartlett, S.; Offermans, G.; Shue, C. Enhanced Loran. A Wide-Area Multi-Application PNT Resiliency Solution. *GPS World* **2015**, *26*, 58–64.
84. Johnson, G.; Swaszek, P.; Hartnett, R.; Shalaev, R.; Wiggins, M. An Evaluation of eLoran as a Backup to GPS. In Proceedings of the 2007 IEEE Conference on Technologies for Homeland Security, Woburn, MA, USA, 16–17 May 2007; pp. 95–100. [CrossRef]
85. E-Loran: The PNT Technology Which Is More Accurate and Less Vulnerable—Sea News Global Maritime News. Available online: https://seanews.co.uk/features/e-loran-the-pnt-technology-which-is-more-accurate-and-less-vulnerable/ (accessed on 2 August 2020).
86. Kala, N.; Balakrishnan, M. Cyber Preparedness in Maritime Industry. *Int. J. Sci. Technol. Adv.* **2019**, *5*, 19–28.
87. New GPS' circle Spoofing' Moves Ship Locations Thousands of Miles—GPS World/SkyTruth/RNTF. Available online: https://www.gpsworld.com/new-gps-circle-spoofing-moves-ship-locations-thousands-of-miles/ (accessed on 3 August 2020).
88. Drougkas, A.; Sarri, A.; Kyranoudi, P.; Zisi, A. European Union Agency for Cybersecurity. *Port. Cybersecur. Good Pract. Cybersecur. Marit. Sect.* **2019**, 12–46.
89. EMSA. AIS spoofing incident. In Proceedings of the 6th HLSG for Governance of the Digital Maritime System and Services, Brussels, Belgium, 20 January 2020. Available online: https://ec.europa.eu/transparency/regexpert/index.cfm?do=groupDetail.groupMeeting&meetingId=18913 (accessed on 14 May 2020).
90. Now Operational, BeiDou Could Conceal Cybersecurity Threat—Inside GNSS. Available online: https://insidegnss.com/now-operational-beidou-could-conceal-cybersecurity-threat/ (accessed on 3 August 2020).
91. Wilson, J. *China's Alternative to GPS and Its Implications for the United States*; U.S. China Economic and Security Review Commission: Washington, DC, USA, 2017.
92. Ship Automation/Control System—KONGSBERG. Available online: http://www.shippipedia.com/ship-automation-control-system/ (accessed on 4 August 2020).
93. IHS-BIMCO-Survey-Findings—Story in Numbers. Available online: https://cybersail.org/wp-content/uploads/2017/02/IHS-BIMCO-Survey-Findings.pdf (accessed on 4 August 2020).
94. The importance of Cyber Security Risk Management in Shipping. Available online: https://www.shippingandfreightresource.com/cyber-security-risk-management-in-shipping/# (accessed on 4 July 2020).
95. The Future of Maritime Cybersecurity. Available online: https://www.maritimecyberadvisors.com/l/the-future-of-maritime-cybersecurity2/ (accessed on 3 July 2020).
96. Detect and Address Cyber Risk in the Maritime Industry. Available online: https://home.kpmg/no/nb/home/campaigns/2019/10/detect-and-address-cyber-risks-in-the-maritime-industry.html (accessed on 10 November 2019).
97. Trimble, D.; Monken, J.; Sand, A. A Framework for Cybersecurity Assessments of Critical Port Infrastructure. In Proceedings of the 2017 International Conference on Cyber Conflict (CyCon US), Washington, DC, USA, 7–8 November 2017; pp. 1–7.

98. Svilicic, B.; Rudan, I.; Jugovi, A. A Study on Cyber Security Threats in a Shipboard Integrated Navigational System. *J. Mar. Sci. Eng.* **2019**, *7*, 364. [CrossRef]
99. Preparing for Cyber Battleships–Electronic Chart Display and Information System Security. Available online: https://www.nccgroup.com/uk/our-research/preparing-for-cyber-battleships-electronic-chart-display-and-information-system-security/ (accessed on 27 May 2020).
100. The Story You Aren't Being Told About Iran Capturing Two American Vessels. Available online: https://www.mintpressnews.com/the-story-you-arent-being-told-about-iran-capturing-two-american-vessels/212937/ (accessed on 21 May 2020).
101. Hackers Took 'Full Control' of a Container Ship's Navigation Systems for 10 Hours. Available online: https://rntfnd.org/2017/11/25/hackers-took-full-control-of-container-ships-navigation-systems-for-10-hours-ihs-fairplay/ (accessed on 21 January 2020).
102. Shipping Must Confront Onboard Systems' Cyber Vulnerabilities. Available online: https://safetyatsea.net/news/2017/shipping-must-confront-onboard-systems-cyber-vulnerabilities/ (accessed on 21 January 2020).
103. Key Takeaways from 3 Recent Cyber Attacks in Shipping. Available online: https://www.dualog.com/blog/key-takeaways-from-3-recent-cyber-attacks-in-shipping (accessed on 24 July 2020).
104. Carnival Hit by Cyber Attack: Hackers Steal Personal Information of Cruise Giant's Passengers and Staff. Available online: https://www.thisismoney.co.uk/money/markets/article-8640269/Carnival-hit-ransomware-attack-Hackers-steal-passenger-information.html (accessed on 17 August 2020).
105. Filitz, J. Maritime port systems cyber security vulnerability. *NMIO Tech. Bull.* **2019**, *13*, 22–27.
106. Kramek, J. The critical infrastructure gap: US port facilities and cyber vulnerabilities. In *Federal Executive Fellows Policy Papers 16*; Brookings: Washington, DC, USA, 2013; pp. 414–430. [CrossRef]
107. Newberry, M.E. Maritime Critical Infrastructure Cyber Risk: Threats, Vulnerabilities, and Consequences. Proceedings of the Marine Safety and Security Council, Coast. Guard. *J. Saf. Secur. Sea* **2014**, *71*, 42–45.
108. Gunther, C. Design of maritime cybersecurity systems. In *Look Out 2016 Maritime Domain Cyber: Risks, Threats & Future Perspectives*; Lampe & Schwartze KG: Bremen, Germany, 2015; pp. 27–46.
109. Above Us Only Stars—Exposing GPS Spoofing in Russia and Syria. Available online: https://www.c4reports.org/aboveusonlystars (accessed on 27 July 2020).
110. Jie, H.; Presti, L.; Motella, B.; Pini, M. GNSS Spoofing Detection: Theoretical Analysis and Performance of the Ratio Test Metric in Open Sky. *ICT Express* **2016**, *2*, 37–40. [CrossRef]
111. Seized UK Tanker Likely 'Spoofed' by Iran. Available online: https://lloydslist.maritimeintelligence.informa.com/LL1128820/Seized-UK-tanker-likely-spoofed-by-Iran (accessed on 20 July 2019).
112. Traffic Density Mapping Service—Methodology. EMS—Ref. Ares (2019)4005069—24/06/2019. Available online: http://www.emsa.europa.eu/related-projects/tdms.html (accessed on 26 June 2020).
113. Autonomous Shipping Concepts. Available online: https://www.norclub.no/blog/autonomous-shipping-concepts/ (accessed on 7 January 2020).
114. Cyber risk and Cybersecurity Countermeasures Supplement. P & I Loss Prevention Bulletin-Vol.48_Full.Pdf. Available online: https://www.piclub.or.jp/wp-content/uploads/2020/05/Loss-Prevention-Bulletin-Vol.48_Full.pdf (accessed on 24 July 2020).
115. Cyber Awareness. Available online: https://www.american-club.com/files/cyber_awareness_comic.pdf (accessed on 4 July 2020).
116. Giannopoulos, G.; Smith, H.; Theocharidou, M. *The Landscape of Hybrid Threats—A conceptual Model*; Joint Research Centre, Centre of Excellence for Countering Hybrid Threats: Helsinki, Finland, 2020.
117. Savolainen, J. Hybrid Threats and Vulnerabilities of Modern Critical Infrastructure—Weapons of Mass Disturbance (WMDi)? Hybrid CoE Working Paper 4. Available online: https://www.hybridcoe.fi/publications/hybrid-threats-and-vulnerabilities-of-modern-critical-infrastructure-weapons-of-mass-disturbance-wmdi/ (accessed on 10 August 2020).
118. Kremidas-Courtney, C. Countering Hybrid Threats in the Maritime Environment. Center for International Maritime Security. Available online: http://cimsec.org/countering-hybrid-threats-in-the-maritime-environment/36553 (accessed on 12 June 2020).
119. European Commission; High Representative. Increasing resilience and bolstering capabilities to address hybrid threats. JOIN(2018) 16 Final. Available online: https://eur-lex.europa.eu/legal-content/EN/ALL/?uri=JOIN:2018:16:FIN (accessed on 15 June 2020).

120. Regulation (EU) 2019/452 of the European Parliament and of the Council of 19 March 2019 Establishing a Framework for the Screening of Foreign Direct Investments Into the Union. Available online: https://eur-lex.europa.eu/eli/reg/2019/452/oj (accessed on 15 June 2020).
121. Joint Declaration by the President of the European Council, the President of the European Commission, and the Secretary General of the North Atlantic Treaty Organization. Warsaw. 8 July 2016. Available online: https://www.consilium.europa.eu/en/press/press-releases/2016/07/08/eu-nato-joint-declaration/ (accessed on 2 August 2020).
122. EU; NATO. Common Set of Proposals for the Implementation of the Joint Declaration by the President of the European Council, the President of the European Commission and the Secretary General of the North Atlantic Treaty Organization. Available online: https://data.consilium.europa.eu/doc/document/ST-15283-2016-INIT/en/pdf (accessed on 2 August 2020).
123. EU; NATO. Common Set of New Proposals on the Implementation of the Joint Declaration signed by the President of the European Council, the President of the European Commission and the Secretary General of the North Atlantic Treaty Organization. Available online: https://www.consilium.europa.eu/media/31947/st14802en17.pdf (accessed on 2 August 2020).
124. What is Hybrid CoE? The European Centre of Excellence for Countering Hybrid Threats. Available online: https://www.hybridcoe.fi/what-is-hybridcoe/ (accessed on 9 September 2020).
125. Hybrid Threats Against Harbours: Workshop at EDA. Available online: https://www.eda.europa.eu/info-hub/press-centre/latest-news/2018/05/30/hybrid-threats-against-harbours-workshop-at-eda (accessed on 9 September 2020).
126. Lohelia, T.; Schatz, V. Handbook On Maritime Hybrid Threats—10 Scenarios and Legal Scans. Hybrid CoE Working Paper. Available online: https://www.hybridcoe.fi/publications/handbook-on-maritime-hybrid-threats-10-scenarios-and-legal-scans/ (accessed on 9 September 2020).
127. European Commission; High Representative. On the Implementation of the Joint Framework on Countering Hybrid Threats—A European Union Response. JOIN(2017) 30 Final. Available online: https://eur-lex.europa.eu/legal-content/en/ALL/?uri=CELEX:52017JC0030 (accessed on 9 September 2020).
128. European Commission; High Representative. Joint Framework on Countering Hybrid Threats—A European Union Response. JOIN(2016) 18 Final. Available online: https://eur-lex.europa.eu/legal-content/EN/TXT/?uri=CELEX:52016JC0018 (accessed on 9 September 2020).
129. Braw, E. From Schools to Total Defence Exercises: Best Practices in Greyzone Deterrence. Available online: https://rusi.org/sites/default/files/20191115_newsbrief_vol39_no10_braw_web.pdf (accessed on 6 September 2020).
130. Council of the EU. Complementary Efforts to Enhance Resilience and Counter Hybrid Threats. Council Conclusions, 14972/19. Available online: https://data.consilium.europa.eu/doc/document/ST-14972-2019-INIT/en/pdf (accessed on 4 August 2020).
131. Radgowski, J.; Tiongson, K. Cyberspace—The Imminent Operational Domain. Proceedings of the Marine Safety and Security Council. *Coast. Guard J. Saf. Secur. Sea* **2014**, *71*, 18–22.
132. Jaskolka, J.; Villasenor, J. Securing cyber-dependent maritime systems and operations. *NMIO Tech. Bull.* **2017**, *12*, 4–6.
133. Fitton, O.; Prince, D.; Germond, B.; Lacy, M. *The Future of Maritime Cyber Security*; Lancaster University: Lancashire, UK, 2015.
134. Secretary of the Navy: Cybersecurity Readiness Review. Available online: https://www.navy.mil/strategic/CyberSecurityReview.pdf (accessed on 27 July 2020).
135. Tam, K.; Jones, K. Factors Affecting Cyber Risk in Maritime. In 2019 International Conference on Cyber Situational Awareness, Data Analytics and Assessment (Cyber SA). *IEEE* **2019**, *3*, 1–8.
136. Navy Culture Must Be Adapted to Fit the Information Age. Available online: http://cimsec.org/navy-culture-must-be-adapted-to-fit-the-information-age/40594 (accessed on 27 July 2020).
137. Naval Dome—Maritime Cyber Defense Solution. Available online: https://navaldome.com/aapa-video-2020-07.html (accessed on 24 July 2020).
138. GSA Celebrates 1 Billion Galileo Smartphone Users. Available online: https://www.gsa.europa.eu/newsroom/news/gsa-celebrates-1-billion-galileo-smartphone-users (accessed on 8 September 2020).
139. PRS. Available online: https://www.gsa.europa.eu/security/prs (accessed on 8 September 2020).

140. How do We Ensure GNSS Security Against Spoofing? Available online: https://www.gpsworld.com/how-do-we-ensure-gnss-security-against-spoofing (accessed on 8 September 2020).
141. Van Cappelle, L.E.; Chen, L.; Negenborn, R.R. Survey on Short-Term Technology Developments and Readiness Levels for Autonomous Shipping. In *Computational Logistics*; Cerulli, R., Raiconi, A., Voß, S., Eds.; Springer: Berlin/Heidelberg, Germany, 2018.
142. Spange, J. Autonomous Docking for Marine Vessels Using a Lidar and Proximity Sensors. Ph.D. Thesis, Norwegian University of Science and Technology, Trondheim, Norway, 2016.
143. How Vulnerable is, G.P.S.? Available online: https://www.newyorker.com/tech/annals-of-technology/how-vulnerable-is-gps (accessed on 6 August 2020).
144. Maritime Cyberattacks Up by 400 Percent. Available online: https://maritime-executive.com/article/report-maritime-cyberattacks-up-by-400-percent (accessed on 20 September 2020).

Journal of
Marine Science and Engineering

MDPI

Article

The Navigator's Aspect of PNC before and after ECDIS Implementation: Facts and Potential Implications towards Navigation Safety Improvement

Maro Car [1], David Brčić [2,*], Srđan Žuškin [2] and Boris Svilicic [2]

1 Nautical Department, University of Dubrovnik, Ćira Carića 4, 20000 Dubrovnik, Croatia; mcar1@unidu.hr
2 Faculty of Maritime Studies, University of Rijeka, Studentska ulica 2, 51000 Rijeka, Croatia; szuskin@pfri.hr (S.Ž.); svilicic@pfri.hr (B.S.)
* Correspondence: brcic@pfri.hr; Tel.: +385-98-633-538

Received: 1 October 2020; Accepted: 22 October 2020; Published: 26 October 2020

Abstract: The global maritime digitalization reflects on navigation and paperless vessels with Paper Navigational Charts (PNC) nowadays superseded by Electronic Chart Display and Information System (ECDIS). Considering the system implementation and its acceptance as a sole navigational means, opinions of navigators differ. Although the ECDIS mandatory implementation ended in 2018, some navigators have been still favouring PNCs, pointing out their advantages over ECDIS navigation. These standpoints may have an impact on the safety of navigation in terms of acceptance, interpretation, and understanding of the system as well as on conflict of standpoints of decisive navigational ranks, the latter reason being found as one of the real problems. The presented study has focused on a specific period, soon after the transitional period completion, aiming to determine the views of traditional navigation advocates, their arguments in the present maritime navigation paperless era and to identify potential problems emerging from the conflict of two navigational means. The research has induced two independent, internationally distributed questionnaires, dedicated to navigational ranks. The first survey has referred to the period from 2012 to 2018, marking the transition to ECDIS navigation. The second survey was conducted after the implementation period completion date. The answers were analysed and discussed from the navigational ranks' perspective, considering their competitiveness and the level of ECDIS education. The research results have indicated and confirmed that PNCs could not entirely be ruled out, at least at this stage. Besides definitive questionnaire answers, the findings have been supported with categorised comments as interpreted from the first survey questionnaire results. The paper aims to present the future of the PNCs, including possibilities of fusion with modern means. The proposed suggestions have been directed towards the benefits of maritime navigation safety, referring especially to disagreement between navigational ranks in terms of particular means acceptance.

Keywords: maritime navigation; electronic chart display and information system; paper navigational charts; electronic navigational charts; ECDIS EHO

---

1. Introduction

> Acceptance of new technology is simply a generation problem, and it is only fairly related to navigational safety. —Unknown reviewer

The tangible development of the electronic chart systems started almost four decades ago. In the 1980s, International Hydrographic Organization (IHO) created a digital data exchange committee that laid down the foundation of the future electronic chart systems, thus enabling the beginning of a long process of PNC digitisation. One of the essential dates was 1st of July 2012, marking the beginning of the transitional period and the time when the ECDIS system was officially recognized as a system that meets chart carriage requirements. The transitional period lasted for six consecutive years, after which it became a mandatory navigational aid for most International Convention for the Safety of Life at Sea (SOLAS) vessels [1].

In the wake of global maritime digitalisation, traditional navigational means and the role of PNCs still appear to occupy an inevitable role. Besides recognised system- and data-centred ECDIS issues, the reasons for retaining PNCs include the feedback of navigators as central system stakeholders. The standpoints and views differ depending on experience, rank, and navigator's engagement with the system. To determine viewpoints and the way modern navigators perceive PNCs in the paperless era, the authors conducted two independent segments of research. The first survey was conducted in the period from the year 2014 to 1st of July 2018, the date of the completion of the transitional period for most SOLAS vessels. A second research was conducted after this date in the same year. Both researches were internationally distributed among the eligible maritime navigational ranks. The answers were analysed, summarized, and subsequently presented. Both expected and debatable findings emerged from the research. The latter, together with the provided observations, represented the motivation for further activities.

This paper is structured as follows. After the brief introduction on the topic, general features on paper and Electronic Navigational Charts (ENCs) have been presented, with an emphasis on the possible future outcomes of PNCs. Previously related research achievements are summarized, both on the research topic and the system in general. In the methodology chapter, the utilized system of research methods has been described. Research results have been presented subsequently and discussed in the following section. Based on previous findings and current research results, a rudimentary safety-related model has been defined, providing insight into potential system-related threats and possible risks which can be still considered to work in favour of traditional means. The paper concludes with summarized findings and desired future research.

2. Background and Previous Research

The SOLAS Convention requires installation of one ECDIS system on board vessels engaged in international voyages, other than non-tanker cargo ships with a gross tonnage of less than 10,000 built before 1st of July 2013. [1]. As a back-up arrangement imposed by carriage requirements, the following means have been recognized: (i) an Appropriate folio of PNCs (APC); (ii) autonomous system equipped with an independent emergency power supply; and *iii)* chart Radar [2–4].

The advantages of ECDIS navigation over traditional means can be summarized as workload reduction; task automation; and, as stated by the International Maritime Organization [5], the system usage contributing to the safety of navigation, particularly referring to predefinition and settings of safety parameters. Contrary to benefits, the system has been considered as a nonautonomous, still dependent on external factors such as sensors and hydrographic data, and prone to failures, malfunctions, and inability of proper performance nevertheless the root cause. The proper education and the lack of knowledge have been further recognised as drawbacks. This current-state pros and cons outcome, together with the desired path towards navigational safety have been the motivation for the research and for the results obtained and presented in continuation.

For ECDIS to be accepted as a system which meets charts carriage requirements, the following conditions have to be met: (i) the system has to be type-approved. (ii) it is necessary to employ up-to-date ENCs; (iii) the system software has to be maintained and compatible with the latest IHO standards; and, (iv) the system needs to have adequate and independent back-up arrangements [3,5–9].

During the ECDIS transitional period, several issues regarding the system have been recognized [10], although the current scientific collection of findings has been relatively scarce. The analyses of ECDIS-related accidents have emphasized the need for a more efficient system operation [11,12].

The previously conducted surveys and research results on justification of the system have indicated potential problems on different levels, generally related to the system, installation, maintenance, positioning, handling, navigation, insufficient operator's knowledge, ENCs production, displayed chart symbols without their features, etc. [10,13–15].

The summarized problems and difficulties (Figure 1) [13] have served as a further step towards the research. An overview of the existing ECDIS-related training has suggested some potential improvements in the training for deck officers, as well as in the increase of operational awareness [16–18].

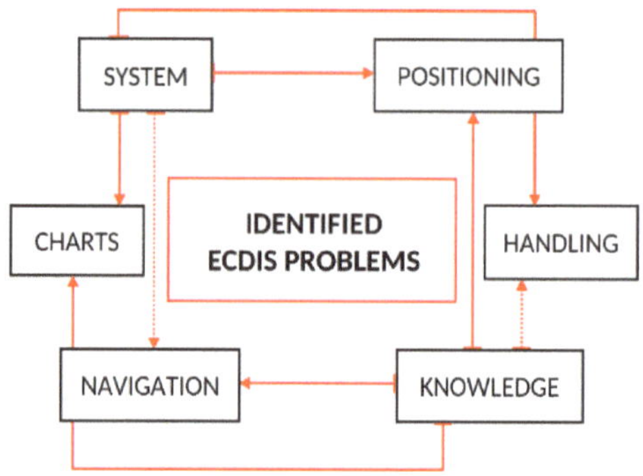

Figure 1. Summarized Electronic Chart Display and Information System (ECDIS)-related issues adapted and modified from [13].

There is no evidence that the system will be better accepted if system operators are more experienced seafarers [19]. Considering the particular navigational rank engagement and interaction with the system, a continuous need to adopt ECDIS Education and Training (EET) for an individual rank has appeared. The risks of overreliance, lack of situational awareness, and other subtle problems should not be disregarded [13,15,20].

The navigator has to be aware of the system limitations and should always cross-check displayed information on the ECDIS with available and suitable sources [21] due to the risk that the safety of navigation is compromised by infrequently utilizing basic safety settings [22,23]. Furthermore, lack of standardization of the system settings, display, functions, and terminology among system manufacturers has showed a negative impact on the safety of navigation [24]. In Figure 2, a share of answers [20] referring to opinions on advantages of both navigational means has been presented, with the justification of opinions related to the main features.

To prepare for the future of PNCs, some critical issues on the subject have been introduced to the IHO and hydrographic offices [25]. The National Oceanographic and Atmospheric Association (NOAA) has already announced the end of the production of traditional PNCs and of the implementation of an online application, which will provide the users with an option to create their own, custom made charts [26].

Hydrographic offices have been considering whether to reduce or altogether discontinue the printing of the PNCs. As a consequence, even for ECDIS back-up arrangements, the world will not be covered entirely with PNCs. Concerning the ENCs world coverage and their features [27], the quality of the ENC data, in particular the depth hydrographic information, may vary, given that the source of surveys is mostly the same as on PNCs. There are approximately 75% of the navigable waters covered with ENCs where seafarers need to exercise caution or a high degree of caution due to the possible existence of uncharted dangers for navigation [25,28,29]. Recent efforts to provide a valuable voyage-planning tool for the vessels operating in the Arctic waters [29] has highlighted the importance of a proper ENC coverage, especially in the remote areas.

Inevitable reduction in the PNCs' production and the mandatory transfer to electronic means have been further problem-addressing points of the research.

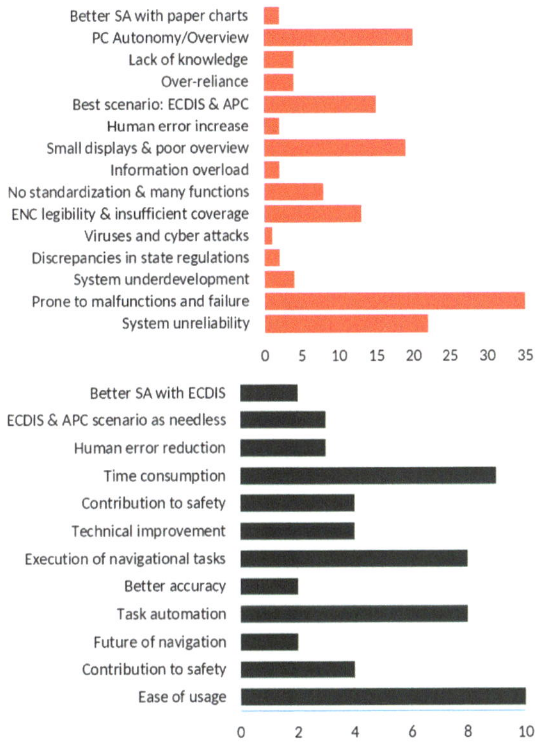

Figure 2. Counts (x-axis) of feature comments (y-axis) on advantages (lower image: system supporters) and drawbacks (upper image: traditional supporters) of ECDIS navigation compared with traditional means adopted and modified from [20].

3. Research Methodology

The research methodology has used two interrelated, separated parts: (i) the segment of *ECDIS* Experience, Handling, and Opinion (*EHO*) research and the (ii) Human Machine Interface (*HMI*) survey.

The *ECDIS EHO* research has started in the first years of the implementation period to improve educational processes and to develop an appropriate curriculum to increase the level of knowledge of the Officers of the Watch (OOW), being the true system end-users. One of the research tools has been an international questionnaire consisting of introductory and topic-related questions, providing feedback received from navigational ranks and, to a lesser extent, from apprentice officers and other system stakeholders.

The previous research results and findings have yielded several scientific, educational and practical contributions, reflecting, among other things, in educational process improvements [13,16,18–20,22–24]. Recognized key observations have served as an additional factor for the proposed study. While the distribution and analysis of the questionnaire ended on 1st of July 2018, the research continued further in different segments. On November 2018, the *HMI survey* was distributed to the international maritime professionals, containing 19 introductory and topic-related questions. Representing a continuation of the research, a dedicated online survey was distributed to the navigational ranks who have obtained the *Generic ECDIS certificate* at least. The concept of the ECDIS EHO survey methodology has been presented in Figure 3, with the HMI segment added as well as incorporated after the implementation period.

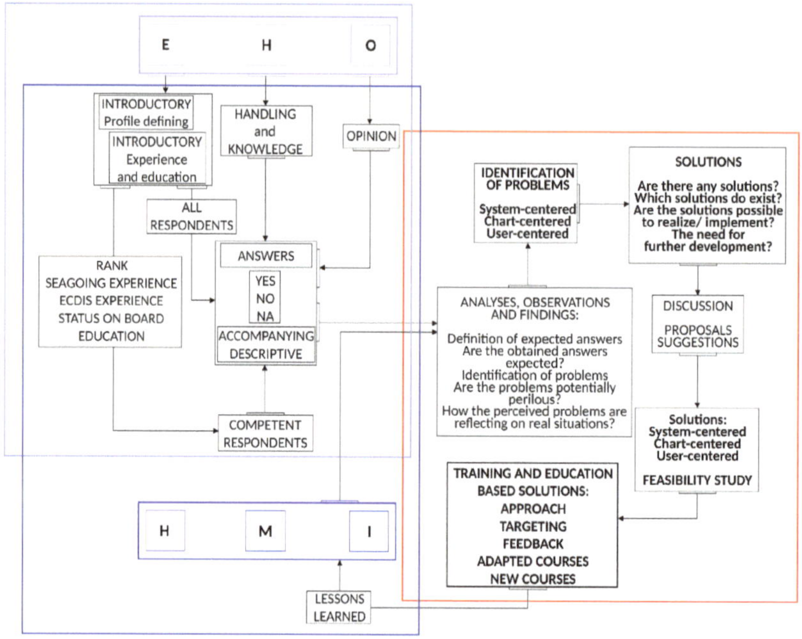

Figure 3. The concept of the ECDIS Experience, Handling, and Opinion (EHO) methodology with the Human Machine Interface (HMI) segment adapted, modified, and complemented on the basis of [5].

The subject of this research has been the opinion and the mindset of current navigational ranks, which have had the opportunity to work with both types of the navigational means. The research has focused on eligible navigational ranks: masters, staff captains, chief officers, safety officers, 1st officers navigation, second officers, and the third officers. The following introductory questions have been used in both the surveys to categorise the respondents' profiles:

- rank,
- working experience,
- holding the ECDIS Generic Training Certificate, and
- holding the type specific/ECDIS manufacturer-approved equipment-specific training.

The following topic-related questions from *EHO* and *HMI* segments/questionnaires have been analysed, with the *EHO* part providing a possibility of accompanying comments to possible *YES/NO* answers:

- *Do you agree with the withdrawal of paper charts from service if certain conditions are met regarding the ECDIS system, i.e., there is no further obligation to possess the same? You can explain the YES/NO answer if you want to.* (abbreviated further as *Q1 EHO*),
- *Do you think that it is still necessary to keep navigational paper charts after implementation of the mandatory ECDIS system?* (abbreviated further as *Q2 HMI*).

At this point, the collected answers have been regrouped in the main research categories', containing introductory and topic-related questions, i.e., their responses.

Respondents have been categorized according to navigational ranks, years of sea experience, and ECDIS-related certificates' holders. Summarized results have been given for the *Q1 EHO* and *Q2 HMI* questions. The EHO segment comments have been categorised and discussed in the respective section. The obtained results, analysis, and discussion have been presented as follows.

4. Results

A total of 402 maritime active seafarers have taken part in the research, with the working experience ranging from one to more than 20 years.

The ECDIS EHO survey (Figure 4) has included answers from 269 respondents: 100 masters, 8 staff captains, 77 chief officers, 4 safety officers, 67 second officers, and 13 third officers. As for their active working experience, 5 categories have been defined: (i) the 6% respondents have had less than 5 years of working experience, (ii) 14% have had 5 to 10 years of experience, (iii) 40% have had experience from 10 to 20 years, (iv) 40% of respondents have had from 10 to 20 years of working experience, and (v) 37% of respondents have had more than 20 years of working experience. The remaining 3% have not specified this answer.

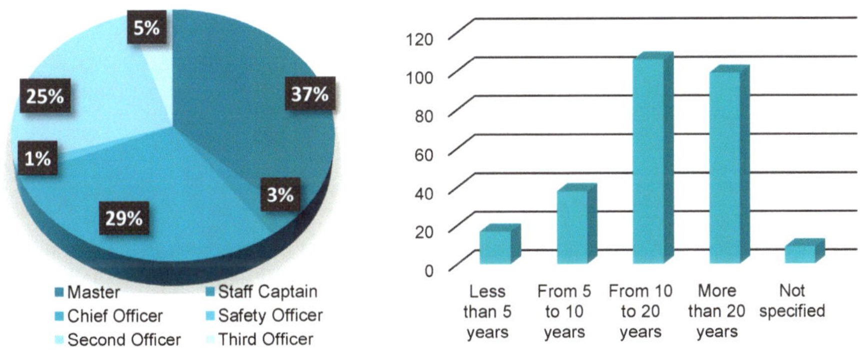

Figure 4. Navigational ranks of *ECDIS EHO* participants (left) and their sea experience (in years) (right). Source: Authors.

The *HMI survey* (Figure 5) has included answers from 133 respondents: 44 masters, 10 staff captains, 35 chief officers, 10 1st officers navigation, 23 second officers, 8 third officers, and 3 undefined respondents. Active working experience of the participants has been divided into five categories: respondents with less than five years of working experience (12%), respondents from 5 to 10 years of working experience (27%), respondents from 10 to 15 years of working experience (27%), respondents from 15 to 20 years of working experience (15%), and respondents with more than 20 years of working experience (19%).

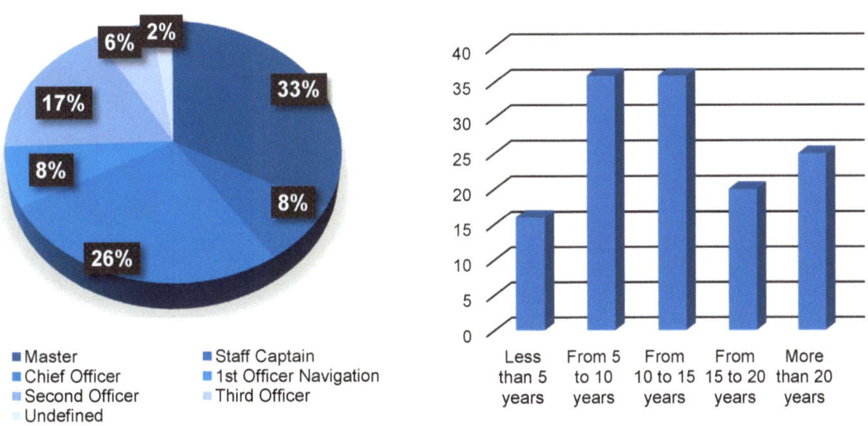

Figure 5. Navigational ranks of *HMI* participants (left) and their sea experience (in years) (right). Source: Authors.

During the transitional period, it has been noted that the *generic ECDIS certification* from the participants in the *ECDIS EHO* survey has increased from 61% to 100% in the *HMI survey*. The *type-specific (familiarization) ECDIS training* certification has risen from 47% in the *ECDIS EHO* survey to 90% in the *HMI survey*, as shown in Figure 6.

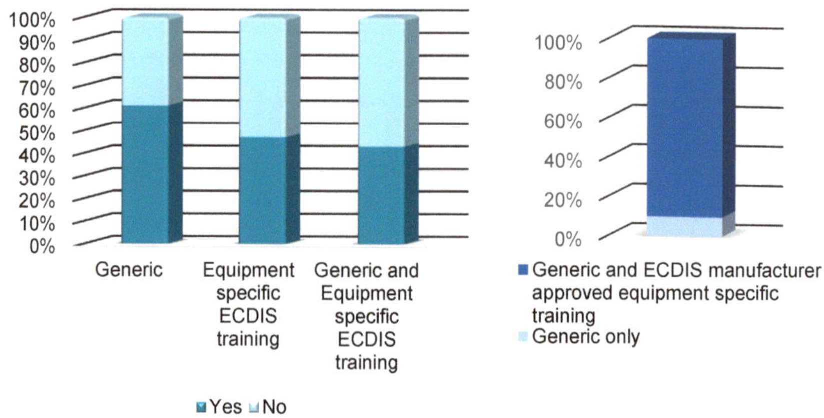

Figure 6. Presentation of answers referring to the possession of the generic training certificate, type-specific training certificate, and both certificates according to *EHO* (left) and *HMI* survey (right) results. Source: Authors.

The *ECDIS EHO* participants have had an option to specify the system model on which they had attended the certification course. *Transas* has been the most commonly utilized system for the generic training certification, while for the type-specific training certification, *SAM Electronics* was the prevailing one.

The total share of answers on (*Q1 EHO*) and (*Q2 HMI*) is shown in Figure 7, together with the trend of the respondents' opinions. In the *EHO* research, there have been a few cases where participants have not given an answer, thus making a 4% share of all responses. In general, more than half (52%) of the respondents have agreed with the paper chart withdrawal from service if certain conditions have been met regarding the ECDIS system (*ECDIS EHO*). In comparison, almost 73% of the *HMI survey* participants have agreed to the PNC removal. There has been a positive, increasing trend for more than 20% in favour of the PNC removal after the mandatory ECDIS implementation.

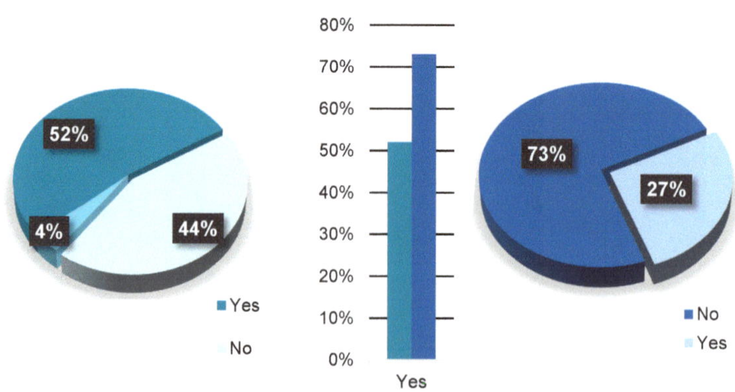

Figure 7. *ECDIS EHO* participants' opinions regarding Paper Navigational Chart (PNC) removal (left), the trend of affirmative PNC removal answers (centre), and *HMI* participants' opinions regarding the necessity of the PNCs (right). Source: Authors.

Despite all efforts, system modernization, professional certifications, training, and hands-on experience of the participants, there has still been a considerable share of the respondents (27%) who have favoured PNCs and traditional means of navigation and their opinions cannot be ruled out.

5. Further Analyses and Discussion: Towards the Unwanted Chain of Errors Avoidance

As a technical achievement, the ECDIS has been conventionally accepted as a primary navigational means, having brought many benefits to seafarers. Generally speaking, this fact has been expected. Nevertheless, it would be reasonable to elaborate on the opinions of the seafarers from both the researches who have favoured PNCs and traditional navigational means. There is a division of seafarers [30], as to whether they have developed their navigational skills before or after the electronic means have taken place globally, requiring their roles and working experience to be considered. The overall results have been subsequently analysed by years and participants' rank.

A first analysis has been made on the basis of the questionnaire submission year. The results in terms of three answering options (yes/no/no answer) have been plotted using a 100% stacked column chart (Figure 8). This allowed comparison of the percentage of contribution for each category, showing that the most significant share on the PNC removal was in 2018, equal to more than 65%.

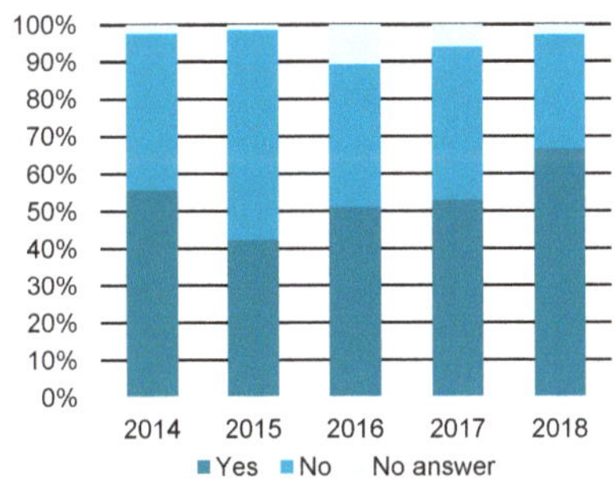

Year	2014	2015	2016	2017	2018
Yes	49	33	24	9	26
No	37	44	18	7	12
Na *	2	1	5	1	1

* *No answer*

Figure 8. *ECDIS EHO* survey participants' opinions regarding PNC removal divided by years. Source: Authors.

The second analysis regarded both the navigational rank and the questionnaire submission year (Table 1 and Figure 9, respectively).

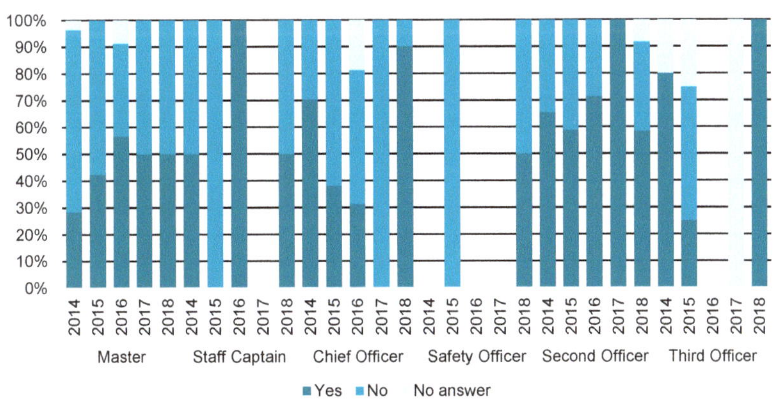

Figure 9. ECDIS EHO participants' opinions regarding PNC removal divided by years and rank (Q1: Do you agree with the fact of the withdrawal of the paper charts from the service if certain conditions are met regarding the ECDIS system, i.e., there is no further obligation to possess the same?) Source: Authors.

Table 1. *ECDIS EHO* participants' opinions regarding PNC removal divided by years and rank.

Year	2014	2015	2016	2017	2018	2014	2015	2016	2017	2018
Yes	8	14	13	4	4	1	0	1	0	2
No	19	19	8	4	4	1	1	0	0	2
No answer	1	0	2	0	0	0	0	0	0	0
Rank			**Master**					**Staff Captain**		
Year	**2014**	**2015**	**2016**	**2017**	**2018**	**2014**	**2015**	**2016**	**2017**	**2018**
Yes	19	8	5	0	9	0	0	0	0	1
No	8	13	8	3	1	0	2	0	0	1
No answer	0	0	3	0	0	0	0	0	0	0
Rank			**Chief Officer**					**Safety Officer**		
Year	**2014**	**2015**	**2016**	**2017**	**2018**	**2014**	**2015**	**2016**	**2017**	**2018**
Yes	17	10	5	5	7	4	1	0	0	3
No	9	7	2	0	4	0	2	0	0	0
No answer	0	0	0	0	1	1	1	0	1	0
Rank			**Second Officer**					**Third Officer**		

Source: Authors.

When comparing years 2014 and 2018, an even opinion or slight increase among all navigational ranks, with an exception of second officers, has been noted regarding the favouring of electronic navigation means, as opposed to traditional navigation. When analysing the first six months of the year 2018 (*EHO*) and the time frame before mandatory system implementation, there were still 50% of the masters, who have not agreed with the PNC removal. Staff captains and safety officers were equally divided (50%). A 90% of the chief officers and a significant share of the second officers (42%) did not agree with the PNCs withdrawal from the service. Although represented in a relatively small share, the results indicated that 100% of the third officers would like to withdraw the PNCs.

The general percentage of answers on *Q2 HMI* regarding the navigational rank is shown in Figure 10. The relative share of two possible answers (yes/no) has been provided by each participant's rank. Approximately 60% of masters, staff captains, and third officers would like to withdraw PNCs from the service after the mandatory system implementation. The majority of the second officers (78%) and chief officers (86%), as well as all of the 1st officers navigation share the same opinion. The exception have been unspecified participants (2%).

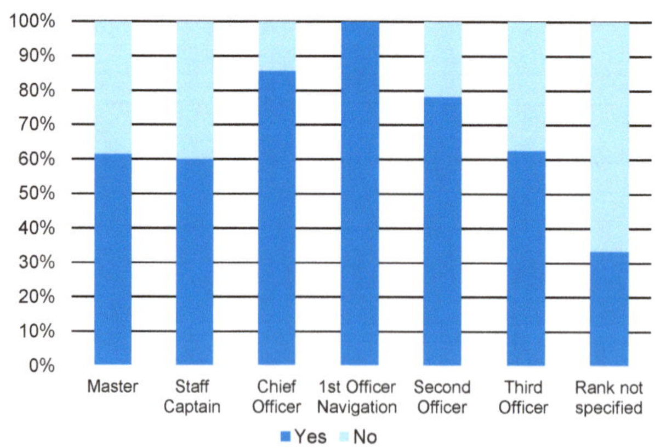

Rank	Yes	No
Master	27	17
Staff Captain	6	4
Chief Officer	30	5
1st Officer Navigation	10	0
Second Officer	18	5
Third Officer	5	3
Rank not specified	1	2

Figure 10. HMI survey participants' opinions regarding PNC removal (Q2: Do you think that it is still necessary to keep navigational paper charts after implementation of the mandatory ECDIS system?). Source: Authors.

Masters and staff captains have favoured the PNC removal for an additional 10%, while there has been approximately 20% of the increase of the second officers who share the same opinions (in 2018, as derived from both surveys). There has also been a 4% decrease in opinions of the chief officers (from 90% to 86%), and 38% of the third officers who have favoured the PNC removal after the mandatory system implementation. The total share of the answers from participants who have supported the PNC removal decreased from 100% to 62%. Since the individual results have not been totally uniform, more specific research should be conducted, providing a larger, and therefore, more representative samples of participants' categories. So far and it is, generally speaking, logical, the senior ranks have been mostly more reluctant to discontinuation of the PNCs usage, as opposed to junior officers. The latter have been in total consensus about the PNCs' complete removal from the service. Another consequential and perhaps underrated question has emerged or, rather to say, it has been confirmed: To what extent will the traditional supporters ignore the usage and the features of the system as a result of their standpoints regarding the non-acceptance, or even resistance to the means? Although seemingly negligible, this issue refers to the desired synergy between traditional and system supporters, i.e., their ability for joint and successful achievement of safe navigation.

Several potential issues and uncertainties can be identified on the basis of these results. The categorized issues have been shown in Figure 11. In their nature, they can be divided in two major groups: (i) system and data problems (grey coloured rectangles and orange arrows) and (ii) operational (navigator's) issues (blue coloured rectangles and orange arrows). Both can individually lead to potential precarious and difficult situations (red coloured rectangle), e.g., overreliance as a navigator's problem, or the ENC information quality as a data problem leading to difficulties or risks in system handling and the execution of navigation tasks.

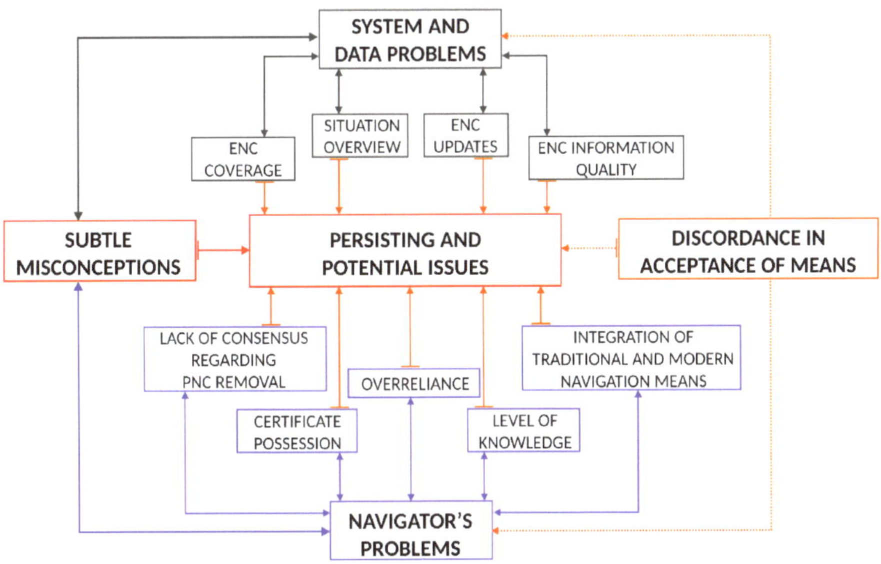

Figure 11. The model of ENC cause-and-effect chain of adverse events. Source: Authors.

A combination of (i) (grey coloured arrows) and (ii) (blue coloured arrows) has been pointed out (red arrowed colour) as a most unwanted scenario, i.e., an operator with a low level of knowledge experiencing a system problem. As for discrepancies in the system acceptance, they could lead to intentional non-usage of the system and its features, nevertheless its role. This, unwanted chain of errors leading to potential hazards has been found, especially in this instant of time, to pose a serious threat to the navigation venture.

Subsequent findings can be elaborated as follows:

- the trend of digital means' acceptance has increased, as expected,
- even though the system has been generally accepted among the end-users, there have still been specific issues which the system users have been pointing out, with similar key points confirmed by the respective organisations,
- like any other technical instrument, the system is prone to failures, what has been expectedly recognised by the end-users,
- the system users have identified the handling-independent problem of the ENC coverage,
- the ENC updates are dependent on satellite communication systems and internet availability,
- the quality of the displayed ENC information differs, depending on their depth accuracy level,
- in certain instances, the PNCs can offer a better overview of some navigable waters (coastal navigation, port approaches) due to their size,
- besides handling the system and familiarizing with, it is necessary to obtain, gain, and maintain a certain level of knowledge and skills to operate with the system properly,
- a problem of overreliance can emerge in terms of exaggerated confidence in the system,
- there are still OOWs without system-approved equipment specific training.
- opinions regarding PNC removal from the service widely differs between research participants,
- a recognized and justified need to incorporate traditional navigational means with modern navigational means to contribute to the safety of navigation has appeared.

Modern technology has speeded up the process of navigation planning, and facilitated easier monitoring of the vessel's movement and navigation. Even though most of the respondents agreed about the PNC removal, it would be prudent not to disregard suggestions from traditional navigation advocates. According to the presented analyses, a minimum set of PNCs should be kept on board as a back-up arrangement in a case of emergency as a worst-case scenario, especially when taking into consideration the *ECDIS EHO* participants' legitimate comments and concerns about issues on the system. Those have been addressed not only as system crash downs, screen freeze, and issues related to system and ENC updates but also as the announcements of certain hydrographic offices regarding the PNCs removal from the service [26]. There has been more than one-fourth of the respondents—*after* the transitional period completion—who have still been favouring the PNCs. As for navigational charts, the potential problem is standing in front of the considerable number of modern seafarers who have still been stranded in the traditional navigation era.

6. Conclusions

Digital navigation is replacing traditional means. Paperless navigation relies on present and future maritime navigational trends, with integrated navigational systems, e-navigation, and coordinating cloud data exchanges between vessels and the shore as most common examples.

The research carried out in this paper has focused on the opinions and arguments of eligible officers as the central stakeholders in the process of the maritime navigation, elaborating their level of acceptance of new navigational means. This research aimed at addressing potential problems arising from the transition to electronic charts to improve the safety of navigation. The participants' answers have been collected and analysed during and after the transitional period, considering their response through the survey periods and their current engagement with the ECDIS system.

The results have pointed out measurable differences in opinions on a particular navigational rank and trends of opinions through five consecutive years. As seen from the navigational ranks' standpoints, even though there has been a positive, increasing, and finally expected trend regarding the PNC removal from the service, except for the chief officers, there have still been more than one-fourth of the research participants who have, in one-way, been favouring the PNCs and traditional navigation means. Senior navigational ranks and overall commanding officers have still been reasonably reserved towards new technologies. The reasons can be found in existing system- and chart-related issues and insufficient trust in electronic equipment. Here, traditional equipment has been considered as an irreplaceable support. Apart from technical issues, the updating level of knowledge is essential, adapted to the particular rank and its actual engagement with the system, rather than relying on formal, prevalent courses. The research has indirectly confirmed the unwanted possible consequences of overreliance which could lead to the lack of situational awareness.

The research results brought up the complex matter referring to the particular approach of navigational ranks of different standpoints and, consequently, their mutual relationship aimed to provide a safe navigation. The discordance in accepting of navigational means could have a negative impact on conducting tasks in the ECDIS navigation and the synergy of the bridge team. Further, the integration of traditional and modern means appears to contribute to the safety of navigation.

The research findings have indicated a need for further analysis of the answers given by particular respondents' groups, especially the ranks which have been less represented in the research, e.g., the third officers. A further continuation of the research has been planned in order to engage as many navigational ranks as possible, in order to maintain and improve this feedback mechanism.

In pace of time and of the new generations of seafarers, it is to expect that they will categorically accept the modern navigation means. However, it has been evident that the PNCs cannot be entirely discarded. Even though the introduction of the ECDIS has not necessarily implied complete PNCs withdrawal, it can be assumed that it will happen. With the rise of new chart-creation technologies, it can be expected that the electronic charts and their appropriate accuracy will further improve. Nevertheless, it seems that the time has not yet come.

Author Contributions: Conceptualization, D.B. and M.C.; methodology, D.B., M.C., and S.Ž.; validation, D.B., S.Ž., B.S., and M.C.; formal analysis, D.B. and S.Ž.; investigation, M.C. and D.B.; resources, S.Z., D.B., and B.S.; data curation, M.C.; writing—original draft preparation, M.C.; writing—review and editing, M.C., D.B., S.Ž., and B.S.; visualization, M.C. and D.B.; supervision, D.B.; project administration, D.B. and B.S.; funding acquisition, B.S. All authors have read and agreed to the published version of the manuscript.

Funding: This research was financially supported, and the APC was funded by the University of Rijeka research project *Cyber Security of Maritime ICT-Based Systems* (grant number: uniri-tehnic-18-68).

Acknowledgments: This study has been supported by the University of Rijeka under the Faculty of Maritime Studies project *ECDIS EHO* and by the University of Rijeka research project *Cyber Security of Maritime ICT-Based Systems* (uniri-tehnic-18-68). The authors are grateful to all the navigational ranks, officers of the navigational watch, and other ECDIS stakeholders for their time and willingness to fulfil the surveys and discussions. The authors believe that their answers and opinions have an immense significance for the appropriateness of the research deliverables.

Conflicts of Interest: The authors declare no conflict of interest.

References

1. International Maritime Organization, IMO. *International Convention for the Safety of Life at Sea, 1974 as Amended*; IMO: London, UK, 2014.
2. International Hydrographic Organization (IHO). Latest IHO Standards that Apply to ECDIS Equipment and Data. 2020. Available online: https://iho.int/en/standards-in-force (accessed on 24 April 2020).
3. International Maritime Organization, IMO. *Performance Standards for Electronic Chart Display and Information Systems (ECDIS)*; Resolution A.817 (19); IMO: London, UK, 1995.
4. International Maritime Organization, IMO. *Adoption of New and Amended Performance Standards*; Resolution MSC.64(67); IMO: London, UK, 1996.
5. International Maritime Organization, IMO. *Adoption of the Revised Performance Standards for Electronic Chart Display and Information Systems (ECDIS)*; Resolution MSC.232(82); IMO: London, UK, 2006.
6. International Hydrographic Organization, IHO. *IHO Transfer Standard for Digital Hydrographic Data*; As Supplemented, Version 3.1; IHO: Monaco, 2014.
7. International Hydrographic Organization, IHO. *S-52: Specifications Chart Content and Display Aspects of ECDIS*; Version 6.1; IHO: Monaco, 2015.
8. International Hydrographic Organization, IHO. *IHO Data Protection Scheme*; Version 1.2; IHO: Monaco, 2015.

9. International Hydrographic Organization, IHO. *Information on IHO standards related to ENC and ECDIS*; Version 2.1; IHO: Monaco, 2020.
10. Weintrit, A. International recent issues about ECDIS, E-navigation and safety at sea: Introduction. In *International Recent Issues about ECDIS, E-Navigation and Safety at Sea*; Weintrit, A., Ed.; Taylor & Francis Group Ltd.: Boca Raton, FL, USA, 2011; pp. 9–12.
11. Turna, İ.; Öztürk, B.O. A causative analysis on ECDIS-related grounding accidents. *Ships Offshore Struc.* **2019**, *15*, 792–803. [CrossRef]
12. Marine Accident Investigation Branch, MAIB. Available online: https://www.gov.uk/government/organisations/marine-accident-investigation-branch (accessed on 15 September 2020).
13. Brčić, D.; Kos, S.; Žuškin, S. Partial structural analysis of the ECDIS EHO research: The handling part. In Proceedings of the 24th International Symposium on Electronics in Transport, Ljubljana, Slovenia, 29–30 March 2016; pp. 80–87.
14. International Hydrographic Organization, IHO. *IHO Report on the Results of the ECDIS Survey Conducted by BIMCO and Denmark*; IHO: Monaco, 2014.
15. Weintrit, A.; Stawicki, K. Operational requirements for Electronic Chart Display and Information Systems (ECDIS). Risk of overreliance on ECDIS. *Transp. Probl.* **2008**, 3, 67–74.
16. Brčić, D.; Žuškin, S.; Barić, M. Observations on ECDIS education and training. In Proceedings of the 12th International Conference on Marine Navigation and Safety of Sea Transportation, Gdynia, Poland, 21–23 June 2017; pp. 29–36.
17. Cole, J.S.; Asyali, E. Role of ECDIS training on improving situational awareness. In Proceedings of the 19th IAMU AGA Conference (IAMU 19), Barcelona, Spain, 17–19 August 2018; Grifoll, M., Martínez de Osés, F.X., Castells, M., Martin, A., Eds.; Universitat Politecnica de Catalunia/International Center for Numerical Methods in Engineering: Barcelona, Spain, 2018; pp. 165–172.
18. Žuškin, S.; Brčić, D.; Šabalja, Đ. A contribution to improving the standards of ECDIS training. *Pomorstvo* **2013**, *27*, 131–148.
19. Brčić, D.; Žuškin, S. Towards paperless vessels: A master's perspective. *J. Marit. Transp. Sci.* **2019**, *55*, 183–199. [CrossRef]
20. Brčić, D.; Žuškin, S.; Valčić, S.; Rudan, I. ECDIS transitional period completion: Analyses, observations and findings. *WMU J. Marit. Aff.* **2019**, *18*, 359–377. [CrossRef]
21. Rutkowski, G. ECDIS Limitations, data reliability, alarm management and safety settings recommended for passage planning and route monitoring on VLCC tankers. *TransNav Int. J. Mar. Navig. Saf. Sea Transp.* **2018**, *12*, 483–490. [CrossRef]
22. Brčić, D.; Kos, S.; Žuškin, S. Navigation with ECDIS: Choosing the proper secondary positioning source. *TransNav Int. J. Mar. Navig. Saf. Sea Transp.* **2015**, *9*, 317–326. [CrossRef]
23. Žuškin, S.; Brčić, D.; Kos, S. Partial structural analysis of the ECDIS EHO research: The safety contour. In Proceedings of the 7th International Conference on Maritime Transport, Barcelona, Spain, 27–29 June 2016; Martinez de Oses, F.X., Castells Sanabra, M., Eds.; Universitat Politecnica De Catalunya—UPC: Barcelona, Spain, 2016; pp. 246–262.
24. Car, M.; Vujičić, S.; Žuškin, S.; Brčić, D. Human machine interface: Interaction of OOWs with the ECDIS system. In Proceedings of the 1st International Conference of Maritime Science & Technology (NAŠE MORE 2019), Dubrovnik, Croatia, 17–18 August 2019; Koboević, Ž., Ed.; pp. 74–86.
25. International Hydrographic Organization, IHO. *The Future of the Paper Nautical Charts*; Preliminary Report; IHO: Monaco, 2019.
26. National Oceanic and Atmospheric Administration (NOAA). *Sunsetting Traditional NOAA Paper Charts. End of Paper and Raster Nautical Chart Production Introduction of NOAA Custom Charts*; National Oceanic and Atmospheric Administration: Washington, DC, USA. Available online: https://www.noaa.gov/ (accessed on 29 January 2020).
27. Kristić, M.; Žuškin, S.; Brčić, D.; Valčić, S. Zone of confidence impact on cross track limit determination in ECDIS passage planning. *J. Mar. Sci. Eng.* **2020**, *8*, 566. [CrossRef]
28. International Hydrographic Organization. *Mariners Guide to Accuracy of Electronic Navigational Charts (ENC)*; Version 1.0.0; IHO: Monaco, 2020.

29. Li, Z.; Jonas, W.; Ringsberg, J.W.; Rita, F. A voyage planning tool for ships sailing between Europe and Asia via the Arctic. *Ships Offshore Struc.* **2020**. [CrossRef]
30. International Maritime Organization, IMO. *Model Course 1.27, Operational Use of Electronic Chart Display and Information System*; IMO: London, UK, 2010.

Publisher's Note: MDPI stays neutral with regard to jurisdictional claims in published maps and institutional affiliations.

Journal of
Marine Science and Engineering

MDPI

Article

Vessel Multi-Parametric Collision Avoidance Decision Model: Fuzzy Approach

Tanja Brcko [1,*], Andrej Androjna [1], Jure Srše [1] and Renata Boć [2]

1 Faculty of Maritime Studies and Transport, University of Ljubljana, 6320 Portoroz, Slovenia; andrej.androjna@fpp.uni-lj.si (A.A.); jure.srse@fpp.uni-lj.si (J.S.)
2 Faculty of Navigation, Maritime University of Szczecin, 70-500 Szczecin, Poland; r.boc@am.szczecin.pl
* Correspondence: tanja.brcko@fpp.uni-lj.si

Abstract: The application of fuzzy logic is an effective approach to a variety of circumstances, including solutions to maritime anti-collision problems. The article presents an upgrade of the radar navigation system, in particular, its collision avoidance planning tool, using a decision model that combines dynamic parameters into one decision—the collision avoidance course. In this paper, a multi-parametric decision model based on fuzzy logic is proposed. The model calculates course alteration in a collision avoidance situation. First, the model collects input data of the target vessel and assesses the collision risk. Using time delay, four parameters are calculated for further processing as input variables for a fuzzy inference system. Then, the fuzzy logic method is used to calculate the course alteration, which considers the vessel's safety domain and International Regulations for Preventing Collisions at Sea (COLREGs). The special feature of the decision model is its tuning with the results of the database of correct solutions obtained with the manual radar plotting method. The validation was carried out with six selected cases simulating encounters with the target vessel in the open sea from different angles and at any visibility. The results of the case studies have shown that the decision model computes well in situations where the own vessel is in a give-way position. In addition, the model provides good results in situations when the target vessel violates COLREG rules. The collision avoidance planning tool can be automated and serve as a basis for further implementation of a model that considers the manoeuvrability of the vessels, weather conditions, and multi-vessel encounter situations.

Keywords: fuzzy logic; decision model; COLREG; collision avoidance

Citation: Brcko, T.; Androjna, A.; Srše, J.; Boć, R. Vessel Multi-Parametric Collision Avoidance Decision Model: Fuzzy Approach. *J. Mar. Sci. Eng.* **2021**, *9*, 49. https://doi.org/10.3390/jmse9010049

Received: 10 December 2020
Accepted: 2 January 2021
Published: 5 January 2021

Publisher's Note: MDPI stays neutral with regard to jurisdictional claims in published maps and institutional affiliations.

1. Introduction

Decision-making and responsiveness are the navigator's primary activities in avoiding collisions at sea. Due to the reduced number of the crew on the bridge, the amount of information required per person has increased, which adds a burden to the decision-making process. The decision is also influenced by the traffic situation, weather conditions, and, finally, the navigator's experience. Vessel avoidance has additional peculiarity as the navigator has extensive knowledge of his vessel and limited information of the vessels in the vicinity, which means that he/she makes decisions in an uncertain environment [1]. The automation of navigation devices has brought a new approach to maritime safety in maritime affairs and, at the same time, changed the nature of human error [2]. An essential cognitive aspect of the problem of automation is: How does the human brain process certain information? How much data is a person able to receive at one time? How should the information be displayed so that a person can receive it in the correct form and use it for further decision-making?

The improvement of maritime safety in the 1970s was mainly due to the upgrade of navigation radar with the Automatic Radar Plotting Aid (ARPA), a support decision-making tool for collision avoidance at sea. Above all, it shortened the time of collision risk assessment and increased the navigator's situation awareness. Although ARPA contains a lot of information about vessels in the area and a Trial Manoeuvre tool to simulate collision avoidance using a time delay command, there are standard limitations and errors that radars have, especially regarding the processing time of the received signal [3]. According to the International Maritime Organization (IMO) requirements, the ARPA radar must display, in one minute and with 95% accuracy, the relative motion (relative course and speed) and DCPA (Distance to Closest Point of Approach) of the target vessel; within three minutes, record the overall trend of the target vessel—relative and true course, relative and true speed, DCPA, and TCPA (Time to Closest Point of Approach). Therefore, it is advisable to use these data from the Automatic Identification System (AIS) device since they are updated, on average, every 30 s (depending on the vessels' speed and course change). Both ARPA radar and AIS are currently integrated into the Electronic Chart Display and Information System (ECDIS) and the navigator has a lot of information available on a single screen. However, the Trial Manoeuvre tool for planning collision avoidance at sea still requires manual adjustments which, in turn, means extending decision-making time in a situation that requires a dynamic decision-making process. There are also no integrated COLREG rules in the system itself, and their application is left to the knowledge of the person steering the vessel.

To relieve the navigator of the glut of information of today's technology used onboard vessels, in 2009, IMO issued a Strategy for the Development and Implementation of E-Navigation (MSC85-Report, Annexes 20 and 21). The goal of E-Navigation development is to improve maritime safety by integrating existing and new navigation devices in a structured manner, simplifying processes to prevent information overload and increase safety by aggregating information into those which are genuinely relevant for the navigator [4]. The task of the navigation decision system, besides its information function, is to supply solutions—determination of safe vessel trajectories in the process of collision avoidance. Decision systems consist of several components, including components for collision risk assessment and avoidance manoeuvre calculation. Quantitative methods of calculating the collision risk could primarily include the calculation of the CPA point. As it represents the distance to the closest point of approach, it is the first indicator of the possibility of a collision or entry into the area of the safe vessel's domain. A vessel's safety domain is the sea area around the vessel, which must remain free from other vessels and fixed installations. In some ARPA radars, the PAD method (Predicted Area of Danger) can also be found to show the collision estimate, which, unlike the CPA method, also takes into account the dimensions of both vessels, course, and speed [5,6]. Most of the early vessel's safety domain developments were created by statistical and analytical methods, and oval and elliptical shapes predominated. In recent times, however, we find models that change dynamically according to different navigation situations: vessel size, traffic density, relative speed, type of navigation situation, weather conditions, visibility, etc. [7,8]. According to Cockcroft [9], the size of a vessel's domain cannot be quantified, but it is suggested that in degraded visibility, this is limited to 2 M. However, it may be lower at low speeds in heavy traffic, in an overtaking situation, or when the observed vessel is expected to sail aft. In practice, this area is determined subjectively by the navigator, or it is determined by the shipowner or another person/institution responsible for it. If there is a risk that another vessel will enter the vessel's domain, an appropriate collision avoidance manoeuvre is required in accordance with the COLREG rules.

Many researchers are engaged in the development of the collision avoidance systems. Models of research may be divided into three main categories: mathematical models and algorithms, soft computing (the evolutionary algorithms, neural networks, fuzzy logic, and expert systems), and a combination of all—a hybrid navigation system [10]. The decision model, which determines the appropriate collision avoidance manoeuvre based on fuzzy logic, was introduced by several authors, among them Perera [6], who primarily shed light on situations occurring on the high seas when a vessel is in a critical position in relation to another vessel and must perform the collision avoidance manoeuvre. The simulation was tested using MATLAB's Fuzzy Logic Toolbox using the Mamdani fuzzy inference system. In setting rules (144 rules in total), the author specified five input parameters: the region where the target vessel is located; the relative course of the target; the level of encounter risk; the distance to the target, and the relative speed of approach. Based on these parameters, the model decided on the need to change course or speed based on COLREG rules 13, 14, and 15. In the following article [11], they tried to solve the problem of avoiding multiple vessels by combining fuzzy logic and the graphical probability model—Bayesian networks. Later, fewer input parameters were used to find the appropriate avoidance manoeuvre for the fuzzy inference system: distance to the target vessel, azimuth, relative course, and speed of approach [12]. Selection of the navigation strategy in traffic separation scheme, using a decision model based on a fuzzy logic algorithm was proposed by Wu [13], who analysed the dynamic characteristics of the navigation process. With a similar fuzzy logic approach the risk of collision with static and moving objects was calculated [14,15]. Zhuo et al. [16] calculated, in their model, the start time of the manoeuvre by altering the course by 30° in relation to the target vessel at 0.8 nautical miles (M). A similar approach to collision avoidance was taken by Su et al. [17] who calculated the position of own vessel to initiate avoidance for different rudder deviations. An attempt to find the optimal avoidance path was presented by Pietrzykowski [18], where an optimal control method was used for vessel motion: a multi-stage fuzzy control in combination with the Dijkstra algorithm for determining the shortest path. Using a combination of fuzzy logic and neural networks, Liu [19] calculated the direction and magnitude of own vessel course change and the time at which the alteration began. In doing so, they used input parameters with different navigation situations (COLREG rules 13, 14, and 15) and the speed ratio between the own and the target vessel. The vessel's trajectory in collision avoidance situations was also the basis for research by Szłapczyński [20], who used the technique of evolutionary algorithms that allow the navigator to predict the trajectory of the target vessel and thus plan its manoeuvre. In a congested traffic area, this method would also allow the VTS (Vessel Traffic System) operator to coordinate the movements of all vessels. More complex hybrid systems for autonomous navigation were presented by Lee et al. [21] and Hu [22]. They, in addition to using fuzzy logic, also introduced a Virtual Force Field (VFF) known in the field of mobile robotics.

In this paper, a collision avoidance decision model is proposed, based on fuzzy logic which calculates course alteration using four parameters. The model is considered as an upgrade of the current Trial Manoeuvre tool in ARPA radar, which is used to plan collision avoidance manoeuvres. The model structure and the rule-based system are built using a database of correct solutions, obtained by manual radar plotting method. This presents a novel approach in determining fuzzy parameters and rules. Additionally, two important segments of collision avoidance are considered in the model:

- the direction of avoidance (in compliance with COLREG rules 8 and 19), and
- the minimum course alteration (in compliance with COLREG rule 8).

The quality of the execution of the avoidance manoeuvre is, thus, not only in achieving a safe distance for the passage of vessels but also in the process of manoeuvring.

The paper is structured as follows. Section 2 provides a brief introduction to the methodology of fuzzy logic; Section 3 describes the design of the decision model in a two-step sequence, which also represents the originality of this study. Each step is described in more detail in sub-sections defining input parameters, the fuzzy inference system with its fuzzy rules and the output decision of the decision model. Section 3 presents six selected case studies that simulate different navigation situation encounters. In Section 4, the presented case studies are discussed, followed by the conclusion.

2. Fuzzy Logic Methodology

The literature review presented several methods for calculating the collision avoidance manoeuvre. The proposed decision model uses fuzzy logic, which belongs to the techniques of artificial intelligence. The advantage of fuzzy logic is that decisions can be made based on inaccurate data that cannot be described in mathematical notation because they are expressed in words. Zadeh [23] justified the numerical methods of fuzzy logic by claiming that humans perceive events in their environment inaccurately and without a precise distribution law. Fuzzy logic thus imitates the human way of thinking, which can solve complex tasks, although they may also contain a great deal of uncertainty. Other advantages to using fuzzy logic for a decision model are its ease of use and transparency, which were two essential features when choosing this technique. Fuzzy computing has also proven to be a very widespread technique in other areas of transport; some such examples were cited by Teodorović [24]: solving the problem of vehicle routing, route selection or timetable optimisation in various transport industries, regulation of traffic lights, etc.

Fuzzy inference system (FIS) is the process of formulating the mapping from a given input to output using fuzzy logic. It is one of the main elements of the fuzzy logic system. The FIS type in this paper is "Mamdani". The use of the IF-THEN rules is organised with the "AND (min)" and "OR (max)" operators. Therefore, the basic tasks of the fuzzy inference system (FIS) are fuzzification, fuzzy reasoning and defuzzification. Fuzzification is a procedure in which the input data is placed in an appropriate set, which is at the same time a linguistic variable, and the grade of membership is determined. The mathematical logic theory assumes only correct and incorrect statements; in fuzzy logic, however, an element belongs to a set with a certain grade of membership (μ_x). Fuzzy sets have different shapes; in the literature, triangular and trapezoidal are commonly used, represented by mathematical notation:

$$\mu(x) = \begin{cases} 0, & x < x_0; \\ \dfrac{x - x_0}{x_1 - x_0}, & x_0 \le x < x_1; \\ 1, & x_1 \le x \le x_2; \\ \dfrac{x_3 - x}{x_3 - x_2}, & x_2 < x \le x_3; \\ 0, & x \ge x_3. \end{cases} \tag{1}$$

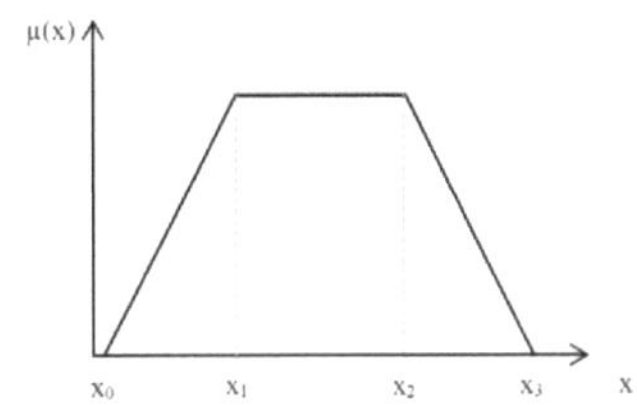

$$\mu(x) = \begin{cases} 1, & x < x_0; \\ \dfrac{x_1 - x}{x_1 - x_0}, & x_0 \le x \le x_1; \\ 0, & x > x_1. \end{cases} \tag{2}$$

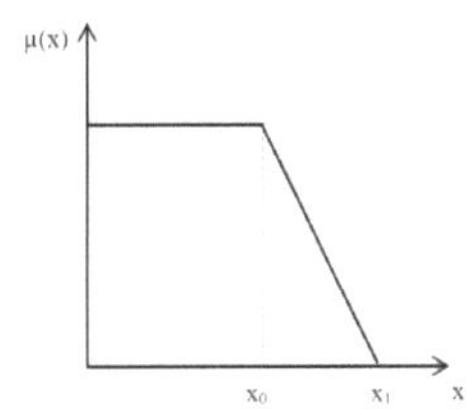

$$\mu(x) = \begin{cases} 0, & x < x_0; \\ \dfrac{x - x_0}{x_1 - x_0}, & x_0 \le x \le x_1; \\ 1, & x > x_1. \end{cases} \tag{3}$$

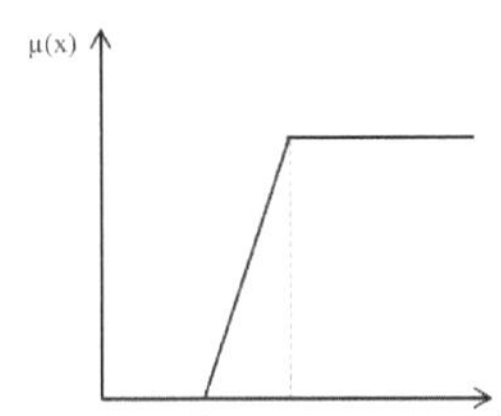

$$\mu_M(x) = \begin{cases} \dfrac{x - x_0}{x_1 - x_0}, & x_0 \le x \le x_1; \\ \dfrac{x_2 - x}{x_2 - x_1}, & x_1 < x \le x_2 \\ 0, & x < x_0 \text{ ali } x > x_2 \end{cases} \tag{4}$$

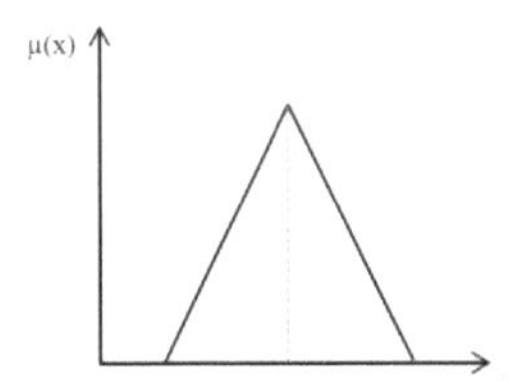

Fuzzy reasoning or the process of mapping input data to output decision is performed with a base of rules formed by IF–THEN conditional statements. Defuzzification is the last step in FIS and is a conversion of fuzzy output quantities into a crisp output quantity. For each unit of input and output data, a degree of membership in the corresponding fuzzy set is assigned. A common mapping is multiple inputs to one output, but there can also be multiple outputs. The greater their number, the greater the number of fuzzy rules in the system which, unfortunately, also affects its transparency. The decision model uses the Mamdani fuzzy inference type. A characteristic of this type is that both inputs and outputs are interpreted with linguistic variables connected by the fuzzy operators "min" and "max". The value of a linguistic variable is also called fuzzy value, which describes belonging to a certain fuzzy set [25]. Fuzzy logical operators are defined as follows:

- AND ("min")—fuzzy cross section or conjunction,
- OR ("max")—fuzzy union or disjunction,
- NOT—fuzzy complement.

The membership function of the output data D has a form (Figure 1):

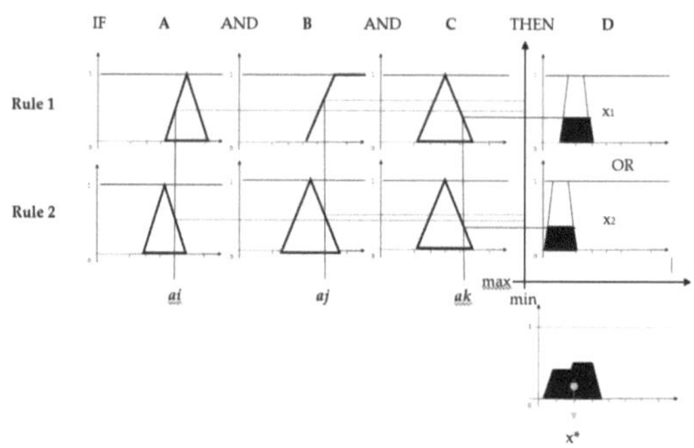

Figure 1. Graphical interpretation of Mamdani fuzzy inference system [26].

$$\mu_D(x) = \max_n \{\min [\mu_{An}(ai), \mu_{Bn}(aj), \mu_{Cn}(ak)]\}, n = 1, 2, 3, \ldots, N \tag{5}$$

which means that the input data are combined with the "min" function:

$$\mu_{A \cap B \cap C}(x) = \min [\mu_A(ai), \mu_B(aj), \mu_C(ak)] \tag{6}$$

and the output data is combined with the "max" function:

$$\mu_D(x) = \max [\mu_D(x_1), \mu_D(x_2), \ldots, \mu_D(x_n)] \tag{7}$$

The latter calculates the sum of all outputs, thus obtaining a shape representing a fuzzy set of all output data. A commonly used data sharpening method is the "centre of gravity" method, which is calculated by

$$x^* = \int_0^\infty \frac{x \cdot \mu_D(x)\,dx}{\mu_D(x)\,dx}. \tag{8}$$

The input parameters and the output decision, the composition of the fuzzy membership functions for each input and output, and the fuzzy rules that perform the process of fuzzy reasoning are presented in the following sections.

3. Design of the Proposed Collision Avoidance Model

Avoidance manoeuvre is a process that requires planning and observation of dynamic navigation situations in an appropriate timeframe. Vessel movement prediction is an important part of planning, with extrapolation of the vessel's trajectory within a time delay.

As on the radar, the navigation situation in the model is presented from the own vessel perspective. The observed vessel is considered as target vessel. The proposed decision model calculates the decision using two steps. In step one, the model collects input data of own and target vessels to assess the collision risk. To predict the target vessel position at the time of collision avoidance manoeuvre, time delay calculation is used to obtain the new relative position of a target vessel. At this point four parameters are calculated for further processing as input variables in a fuzzy inference system. The second step is calculation of the decision (course alteration), using fuzzy logic methodology.

3.1. FIS Input Parameters Calculation

In this paper the calculation of the appropriate course alteration is influenced by four dynamic parameters as input variables of the FIS:

1. DCPA—Distance to Closest Point of Approach,
2. AP—Action Point distance to the target vessel,
3. RB—Relative Bearing of a target vessel,
4. V_o—Own vessel Velocity.

To calculate input parameters used in a fuzzy inference system, the predicted position of a target vessel is calculated using a time delay [27]. A time delay of the navigation situation is a set of functions that calculate the relative position of the target vessel for the desired time delay if the course and speed of own and target vessel do not change in time. This method is called "Static Calculation of a Trial Manoeuvre". The new relative position of the target vessel is calculated as follows:

$$d_{tN} = \left| \frac{X_{tN}}{\sin \omega_{tN}} \right| [M], \tag{9}$$

$$RB = \omega_t - C_o \, [^\circ], \tag{10}$$

where d_{tN} is the new distance to target vessel, X_{tN} is the new relative position coordinate of the target vessel, ω_{tN} is a new azimuth of a target vessel, and RB is a new relative bearing of the target vessel.

Collision risk assessment parameters DCPA and TCPA are calculated with the following [28]:

$$DCPA = \left| \frac{(X_t \cdot V_{ry}) - (Y_t \cdot V_{rx})}{V_r} \right| [M], \tag{11}$$

$$TCPA = -\frac{(Y_t \cdot V_{ry}) + (X_t \cdot V_{rx})}{V_r^2} \cdot 60 \, [min], \tag{12}$$

where X_t and Y_t are relative position coordinates of the target vessel, V_{rx} and V_{ry} are relative velocity vector components, and V_r is the relative velocity of the approach (Figure 2).

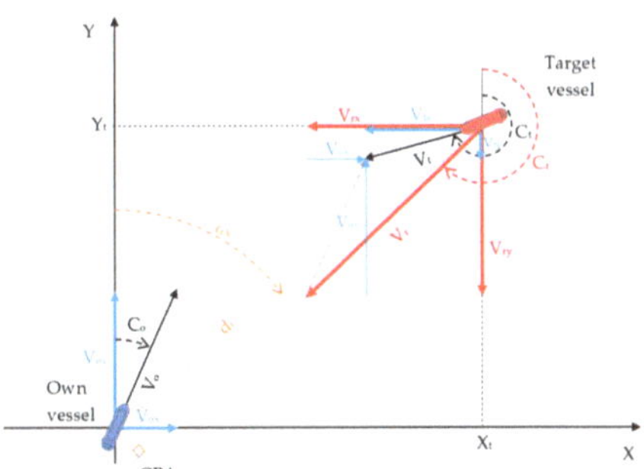

Figure 2. Graphical representation of the collision risk calculation in a relative coordinate system.

The AP and RB parameters are time variables, DCPA and V are considered as static data (assuming both vessels keep speed and course). Since fuzzy reasoning is conditioned by uncertainty, the velocity of the target vessel, relative velocity of the approach, and TCPA of the target vessel were not used as input parameters when modelling the fuzzy inference system since the consideration of the AP parameter nullifies the influence of these variables.

According to fuzzy logic theory, a parameter must be assigned to an appropriate fuzzy set. In the following sub-sections, parameters are presented in detail, and for each value, a degree of membership in the corresponding fuzzy set is assigned.

3.1.1. DCPA Parameter

DCPA is a dynamic parameter. If its value is lesser than a predetermined safe vessel domain, there is a risk of collision with the target vessel or object. Of particular importance is the position of the CPA (Closest Point of Approach), whether the own vessel will encounter the target vessel from the starboard side, port side, or through the centre [26].

The position of the CPA affects the size of the course alteration. Different directions of approach require a different course alteration; a larger alteration is already noticeable with the difference of the DCPA value +/− 0.3 M. Three trapezoidal membership functions were created for the DCPA parameter: negative, centre, and positive. Since DCPA data obtained from navigational devices do not provide information about the position of the CPA relative to own vessel (Figure 3), the positive or negative value of the DCPA is determined according to the following rule: if azimuth of the target vessel increases with time (time of observation), the position of the CPA is "positive" and the DCPA parameter has a positive value; if the azimuth of the target ship decreases with time, the CPA position is "negative" and the DCPA parameter has a negative value.

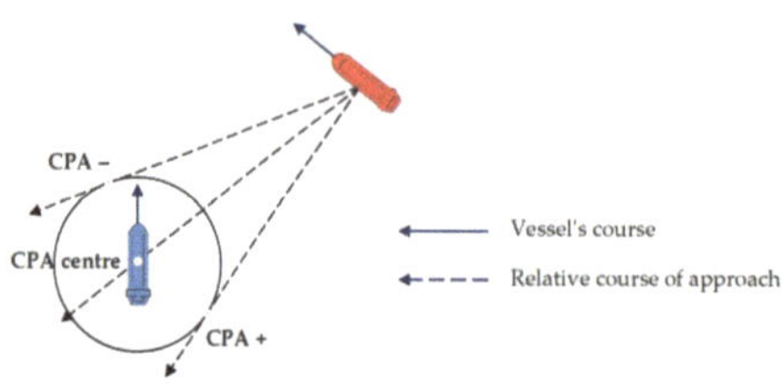

Figure 3. Position of Closest Point of Approach (CPA) for a different relative course of approach.

The following equations represent the DCPA membership functions:

$$\mu_{Negative}(d) = \begin{cases} 0, & d < -1 \\ \frac{d}{0.2} + 5, & -1 \le d \le -0.8 \\ 1, & -0.8 \le d \le -0.3 \\ -0.5 - \frac{d}{0.2}, & -0.3 \le d \le -0.1 \\ 0, & d > -0.1 \end{cases} \quad (13)$$

$$\mu_{Centre}(d) = \begin{cases} 0, & d < -0.5 \\ \frac{d}{0.2} + 2.5 & -0.5 \le d \le -0.3 \\ 1 & -0.3 \le d \le 0.3 \\ 2.5 - \frac{d}{0.2} & 0.3 \le d \le 0.5 \\ 0, & d > 0.5 \end{cases} \quad (14)$$

$$\mu_{Positive}(d) = \begin{cases} 0, & d < 0.1 \\ \frac{d}{0.2} - 0.5, & 0.1 \le d \le 0.3 \\ 1 & 0.3 \le d \le 0.8 \\ 5 - \frac{d}{0.2}, & 0.8 \le d \le 1 \\ 0, & d > 1 \end{cases} \quad (15)$$

These values were chosen since direct collisions are represented by values in the interval of −0.3 and +0.3 M, with a degree of membership equal to 1; for other values up to −0.5 or +0.5, the degree of membership decreases or increases linearly. The trapezoidal shape of a membership function was chosen because several elements in the individual fuzzy set represent the core of the membership function, i.e., grade of membership in the fuzzy set equals 1. This also applies to fuzzy sets for other parameters.

3.1.2. Action Point Parameter

This parameter is a calculated new distance to the target vessel using time delay simulation in step 2 (d_{tN}). According to Cockcroft [9], there are four stages of a close-quarters situation: from the moment when there is no risk of collision (stage one) to the extreme situation when the stand-on vessel is required to take action to avoid collision (stage four). Intermediate stages are collision avoidance by give-way vessel (stage two) on a safe passing distance to the stand-on vessel, which is 5 to 8 M on the open sea. This numbers are also related to visibility of ship lights on open sea according to COLREGs [29]; and collision avoidance by stand-on vessel, at the maximum distance 2 to 3 M from the give-way vessel [9]. Some shipowners are very specific in determining the AP. The CMA CGM safety management system sets the distance 6 M for the navigational situations covered by COLREG regulations 14 and 15 [30]. In the case of overtaking, the minimum distance to the observed vessel must be 2 M.

Based on the proposed distances in the literature, three trapezoidal membership functions were formed describing the distance to the target vessel the moment the own vessel starts the avoidance manoeuvre:

- Near—covers the area from >0 to 5 M distance, with varying degrees of membership,
- Middle—this area covers a distance between 2.5 and 9 M, with varying degrees of membership,
- Far—is a fuzzy set for all distances above 6 M taking values between 6 and 7.5 M with varying degrees of membership.

$$\mu_{Near}(d) = \begin{cases} 1 & d < 4 \\ 5 - d, & 4 \le d \le 5 \\ 0, & d > 5 \end{cases} \quad (16)$$

$$\mu_{Middle}(d) = \begin{cases} 0, & d < 2.5 \\ \frac{d}{1.5} - 1.6, & 2.5 \le d \le 4 \\ 1 & 4 \le d \le 7.5 \\ 6 - \frac{d}{1.5}, & 7.5 \le d \le 9 \\ 0, & d > 9 \end{cases} \quad (17)$$

$$\mu_{Far}(d) = \begin{cases} 0, & d < 6 \\ \frac{d}{1.5} - 4, & 6 \le d \le 7.5 \\ 1, & d > 7.5 \end{cases} \quad (18)$$

3.1.3. Relative Bearing Parameter

In the decision model, RB is measured in a clockwise direction, 0–360°. In practice, an angle of the approaching vessel is rarely expressed in degrees and, rather, in words according to general directions onboard (bow, stern, portside, starboard side). Therefore, the RB as an input parameter of the fuzzy inference system consisting of 8 fuzzy sets with corresponding trapezoidal membership functions, which are described by the words: Starboard Bow, Starboard Bow/Beam, Starboard Beam, Starboard Quarter, Stern, Port Quarter, Port Beam and Port Bow. The following values were adjusted during simulations until they took the following forms:

$$\mu_{S\_Bow}(deg) = (0,\ 0,\ 60,\ 70) \quad (19)$$

$$\mu_{S_B/B}(deg) = (40,\ 50,\ 80,\ 90) \quad (20)$$

$$\mu_{S\_Beam}(deg) = (70,\ 80,\ 140,\ 150) \quad (21)$$

$$\mu_{SQ}(deg) = (120,\ 130,\ 180,\ 190) \quad (22)$$

$$\mu_{Stern}(deg) = (160,\ 170,\ 220,\ 230) \quad (23)$$

$$\mu_{PQ}(deg) = (200,\ 210,\ 260,\ 270) \quad (24)$$

$$\mu_{P\_Beam}(deg) = (240,\ 250,\ 300,\ 310) \quad (25)$$

$$\mu_{P\_Bow}(deg) = (280,\ 290,\ 360,\ 360) \quad (26)$$

3.1.4. Velocity Parameter

An important parameter that influences the degree of the course alteration is own vessel's speed (velocity). The lower it is, the greater the alteration of the course is required for vessels to meet the safe distance. The velocity parameter consists of three fuzzy sets with corresponding trapezoidal membership functions:

- Low,
- Normal,
- High.

The fuzzy sets were defined according to a database of correct solutions, where three different own vessel's velocities were used: 10, 15, and 20 kn. In the fuzzy set "Low", vessels with speeds up to 12 kn were included, such as fishing boats, vessels approaching harbours, pilot stations, etc. The fuzzy set "Normal" includes vessels with speeds between 8 and 16 kn, which is normally the half ahead or full ahead speed of an average bulk carrier. "High" includes vessels with speed higher than 16 kn: high speed boats, container ships, etc. The following values were adjusted during simulations until they took the following forms:

$$\mu_{Low}(v) = \begin{cases} 1, & v < 8 \\ 3 - \frac{v}{4}, & 8 \le v \le 12 \\ 0, & v > 12 \end{cases} \quad (27)$$

$$\mu_{Normal}(v) = \begin{cases} 0, & v < 5 \\ \frac{v}{3} - 1.6 & 5 \le v \le 8 \\ 1 & 8 \le v \le 16 \\ 5 - \frac{v}{4}, & 16 \le v \le 20 \\ 0, & v > 20 \end{cases} \quad (28)$$

$$\mu_{High}(v) = \begin{cases} 0, & v < 12 \\ \frac{v}{4} - 3, & 12 \le v \le 16 \\ 1, & v > 16 \end{cases} \quad (29)$$

3.2. Decision Calculation

The most common way of collision avoidance is the alteration of the course. Such a manoeuvre is also the most appropriate, as COLREG rule 8 dictates:

"(b) Any alteration of course and/or speed to avoid collision shall, if the circumstances of the case admit, be large enough to be readily apparent to another vessel observing visually or by radar; a succession of small alterations of course and/or speed should be avoided". [29]

From the interpretation of rule 8, it can be concluded that the manoeuvre must be noticeable, which is not the case if the speed is reduced by a few knots. In addition, the uncertainty of a navigator who observes such a manoeuvre from another vessel can trigger wrong decisions. Cockcroft [9] recommends that the course alteration should be at least 30°, but it is a better recommendation that the course be changed in the range of 60 to 90°. Depending on the position of the target vessel, the course alteration can be made to the port or starboard side.

The output of the decision model is the alteration of own vessel course in degrees to the port or starboard side, which is used to calculate the trajectory of the vessel in the process of Trial Manoeuvre. It is described by nine fuzzy sets with words used in maritime communication:

- Steady,
- Easy to Stbd/Port,
- Mid to Stbd/Port,
- Hard to Stbd/Port,
- Full to Stbd,
- Full Turn.

Fuzzy sets were determined subjectively by authors.

$$\mu_{hP}(deg) = (-120, -110, -90, -80) \tag{30}$$

$$\mu_{mP}(deg) = (-90, -80, -60, -50) \tag{31}$$

$$\mu_{eP}(deg) = (-60, -50, -30, -20) \tag{32}$$

$$\mu_{S}(deg) = (-30, -20, 0, 10) \tag{33}$$

$$\mu_{eS}(deg) = (0, 10, 30, 40) \tag{34}$$

$$\mu_{mS}(deg) = (30, 40, 70, 80) \tag{35}$$

$$\mu_{hS}(deg) = (70, 80, 110, 120) \tag{36}$$

$$\mu_{fS}(deg) = (110, 120, 150, 160) \tag{37}$$

$$\mu_{fT}(deg) = (150, 160, 360, 360) \tag{38}$$

The process of mapping the input parameters into the output decision is regulated by a system of fuzzy rules. Rules were created by observing the results of the database of correct solutions and the interpretation of COLREG rules 8 and 19. A database of correct solutions was created by the traditional method of calculating alteration of course, the manual radar plotting method. Manual radar plotting (Figure 4) is a method in which the azimuth and range of a signal are measured on a radar screen at time intervals, and the position of the signal is plotted on a manoeuvring board. By connecting the two points of the observed signal, the relative and true vector of the signal movement (or of the vessel) and the CPA point are obtained, which provides essential data for collision risk assessment, DCPA, and TCPA. By planning a new relative trajectory (considering a desired safety domain) of the target vessel, the resolution of the vector triangle determines the collision avoidance course (or speed change) and action point. This method assumes that the target vessel maintains its current course and speed. Database of correct solutions contains 972 decisions (course alteration) at different parameters:

- Distance to closest point of approach of the target vessel (0 M, +0.5 M, −0.5 M),
- Relative bearing of the target vessel (0–350°),
- Action point distance (2, 4, and 6 M),
- Own vessel's velocity (10, 15, and 20 kn).

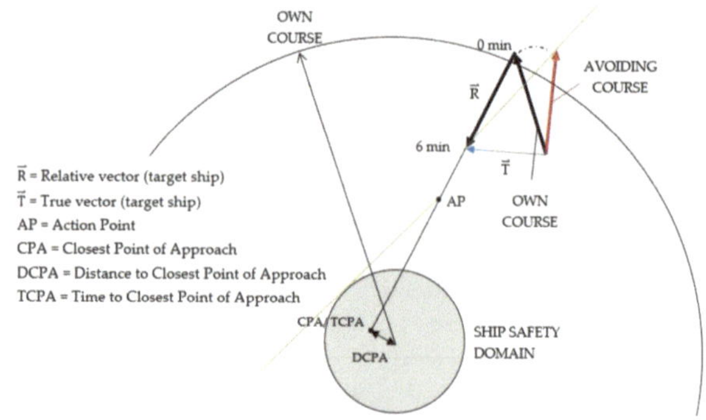

Figure 4. Manual radar plotting [26].

The safety domain for all decisions is 1 M; the relative velocity of approaching in all situations is 20 kn. Both values were chosen subjectively by the authors. COLREG rules 8 and 19 were implemented in the decisions. Figure 5 shows an excerpt from the database of correct solutions for the conditions: DCPA of the target vessel is 0 M, action point distance to target vessel is 6 M at the different relative bearing of an approaching target vessel. The decision is a course alteration for the three different velocities of each vessel.

RB [°]	AP [M]	DCPA [M]	Course deviation: Velocity own vessel: 20 kn	Velocity own vessel: 15 kn	Velocity own vessel: 10 kn
0	6	0	10	12	19
10	6	0	10	13	20
20	6	0	11	15	22
30	6	0	13	16	23
40	6	0	13	17	24
50	6	0	15	20	28
60	6	0	21	25	35
70	6	0	26	29	38
80	6	0	35	41	50
90	6	0	−25	−31	−36
100	6	0	−22	−24	−35
110	6	0	−16	−21	−30
120	6	0	−15	−18	−25
130	6	0	−12	−16	−23
140	6	0	−10	−13	−20
150	6	0	−9	−12	−18
160	6	0	−10	−14	−20
170	6	0	−10	−13	−20

RB [°]	AP [M]	DCPA [M]	Course deviation: Velocity own vessel: 20 kn	Velocity own vessel: 15 kn	Velocity own vessel: 10 kn
180	6	0	10	14	20
190	6	0	10	13	18
200	6	0	10	14	21
210	6	0	10	14	21
220	6	0	12	16	23
230	6	0	12	15	21
240	6	0	13	17	24
250	6	0	16	21	27
260	6	0	20	25	33
270	6	0	26	33	40
280	6	0	33	39	47
290	6	0	44	49	57
300	6	0	19	30	72
310	6	0	14	19	32
320	6	0	14	18	29
330	6	0	11	15	23
340	6	0	11	15	22
350	6	0	10	13	20

Figure 5. Excerpt from the database of correct solutions: green colour indicates a change of course to the starboard side, and red to the port side, at different parameters.

3.3. Decisions Validation

The decision model was tuned until the results of the fuzzy logic outlined a similar area under the curve to the results of the database of correct solutions (Figure 6). Full coverage of graphs was not expected, since fuzzy logic simulates the human way of reasoning, but it was important to match the results in the RB range from 0 to 110°, due to the rule of right (own vessel status: Underway using the engine). Figure 6 shows the results of both methods. The fuzzy logic graph covers almost the entire surface of the manual plotting graph, with minor deviations in individual RB areas.

The composition of the fuzzy rules was based on the principle of finding the maximum value of the "course alteration" for individual areas (in the range of 30 to 40 degrees) of the relative bearing. A total of 216 rules form IF–THEN statements (Table 1).

Table 1. IF-THEN statements.

IF	DCPA	Negative, Positive, Centre
AND	AP	Near, Middle, Far
AND	RB	Stbd Bow, Stbd Bow/Beam, Stbd Beam, Stbd Quarter, Stern, Port Quarter, Port Beam, Port Bow
AND	V	Low, Normal, High
THEN	Course alteration	Steady, Easy to port/starboard, Mid to port/starboard, Hard to port/starboard, Full to starboard, Full turn

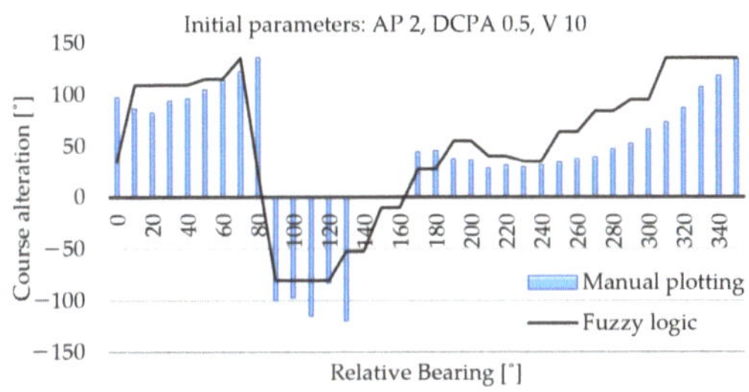

Figure 6. Comparison of the results of the database of correct solutions and the results of the fuzzy inference system.

The fuzzy inference system showed some shortcomings in areas where decisions change from positive to negative, and vice versa, as a result of COLREG rule 19, which dictates that the vessel which detects by radar alone the presence of another vessel shall avoid alteration of course towards the target vessel abeam or abaft the beam. Consequently, the coverage of the graphs in these parts is poorer. This coverage was slightly improved with the additional fuzzy set of the RB parameter »Starboard Bow/Beam«. A similar problem was addressed by Perera et al. [31], which softened the sharp boundaries between the fuzzy sets that were opposing the decisions by adding fuzzy sets that neutralised the sharp transition between the two opposing decisions.

4. Simulations

The simulations' aim is to demonstrate the quality of the operation of a multi-parametric collision avoidance decision model. There are countless possible situations of encounters at sea; Perera [6] presented, in his article, simulations of encountering a single vessel from different relative bearings: 63, 30, 106, 1, and 296° at distances between 7 and 14 M. This covered all three navigation situations according to the COLREG rules—crossing, overtaking, and head-on. A similar simulation was presented by Ni [32], using multiple generic algorithms and a linear extension algorithm for trajectory planning for different navigation situations on the open sea, and Nguyen [33], in constrained waters in a multi-encounter vessel situation. Zhuo [16] simulated an encounter with three vessels simultaneously located at relative bearings 50, 325, and 2° in a crossing and head-on situation. A similar situation is simulated by Pietrzykowski [18]. Lee et al. [21] considered crossings from the starboard side direction, concerning its own vessel (approach from the direction of relative bearing 30–50° and 90–140°), from portside direction (approach from the direction of relative bearing 300–330° and 220–270°), and overtaking situation.

Another important aspect of collision avoidance at sea is the vessel's safety domain. The authors, who solved the problem of avoiding collisions at sea comprehensively, mostly used a simple radar circle in the model for a safety domain: Zhang [34] determined the radius of the 1500 m (0.8 M) circle for the safe ship domain, while Pietrzykowski [35] and Szłapczyński [36] used a 1 M circle in the simulations. Zhuo [16] also considered the dimensions of his own and observed ship in the size of the circle. Some authors used simple ellipses, where the size of the large strip was determined by the length of the ship

(2× the length of the ship) and the small strip by the width of the ship, 2–4× the breadth of the ship [33,37]. Many authors did not specifically define domain size in decision models, among them Hwang [38], Perera [6], and Hu [39], or they chose the minimum encounter distance [8,40–42]. Based on those findings, the minimum vessel's safety domain in the presented model is assumed to be 1 nautical mile.

The validation was made using the six selected cases, which simulate encounters with the target vessel on the open sea from different angles in any visibility: head-on, crossing, and overtaking situations and different right of way. Observed criterion simulations are collision avoiding in accordance with COLREG and minimum vessel's safety domain. Moreover, since the simulations are focused on the vessel's navigation on the open sea, the influence of the navigational behaviour and environmental impacts (wind and currents) are ignored in the modelling process.

4.1. Application of the Proposed Model, Case Study 1 (Overtaking Encounter)

Following is a detailed explanation of decision model calculation (Tables 2–5). The simulation tests the fuzzy inference system for the encounter situation with target vessel. This situation is governed by COLREG rule 14, the "Head-on situation", rule 8 "Action to avoid collision", and rule 19 "Conduct of vessels in restricted visibility". Both vessels are power-driven. It is assumed that the target vessel keeps her course and speed.

Table 2. Initial parameters, 0 min.

	Own Vessel	Target Vessel
C [°]	340	161
V [kn]	16.5	13
dt [M]	-	9.5
ωt [°]	-	341

Table 3. Collision risk assessment.

DCPA [M]	0.09
TCPA [min]	19.32
CPA position	positive
RB [°]	1

Table 4. A new relative position of a target vessel for time delay 11 min.

ωt [°]	341.7
dt [M]	4.09
RB [°]	1.739

Table 5. Calculated input parameters for fuzzy inference system (FIS).

DCPA [M]	+0.09
AP [M]	4.09
RB [°]	1.739
V [kn]	16.5

Figure 7 shows a relative movement of the target vessel in a time delay. The next task is to classify the input parameters into fuzzy sets and calculate the grade of membership. This process is called fuzzification:

$$\text{DCPA} = +0.09\ \text{M}$$

$$\mu_{\text{Centre}}(\text{d}) = 1, \text{ if } -0.3 \leq \text{d} \leq 0.3$$

Explanation: DCPA +0.09 M belongs to the fuzzy set »Centre« with grade of membership 1.

$$AP = 4.09\ M$$

$$\mu_{Near}(d) = 5 - d,\ \text{if}\ 4 \le d \le 5$$

$$\mu_{Near}(d) = 5 - 4.09 = 0.91$$

$$\mu_{Middle}(d) = 1,\ \text{if}\ 4 \le d \le 7.5$$

Explanation: Distance to target vessel 4.09 M belongs to the fuzzy set »Near«, with grade of membership 0.91, and fuzzy set »Middle«, with grade of membership 1.

$$RB = 1.739^{\circ}$$

$$\mu_{S\_Bow}(deg) = 1,\ \text{if}\ deg < 60$$

Explanation: RB 1.739° belongs to the fuzzy set »Stbd Bow« with grade of membership 1.

$$V = 16.5\ kn$$

$$\mu_{Normal}(V) = 5 - V/4,\ \text{if}\ 16 \le V \le 20$$

$$\mu_{Normal}(V) = 5 - 16.5/4 = 0.875$$

$$\mu_{High}(V) = 1,\ \text{if}\ V > 16$$

Explanation: Own vessel's velocity 16.5 kn belongs to the fuzzy set »Normal« with grade of membership 0.875 and fuzzy set »High« with grade of membership 1.

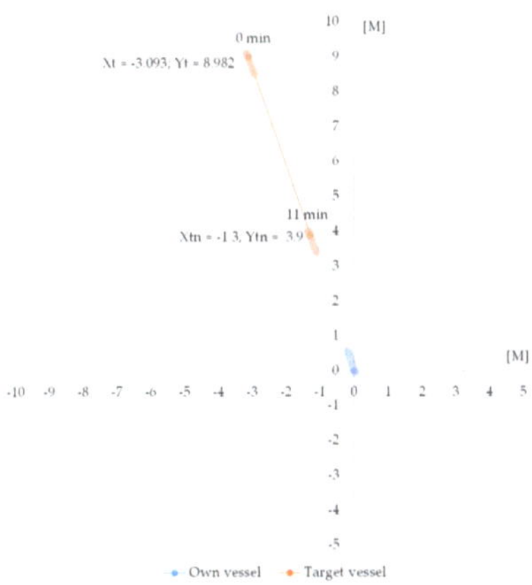

Figure 7. Graphical representation of the relative movement of the target vessel in a time delay in the polar coordinate system.

The second task is the activation of the FIS based on fuzzy rules (Table 6). Since input parameters *AP* and *V* belong to two fuzzy sets, respectively, four fuzzy rules (R_n) are activated:

Table 6. Fuzzy rules.

Rules	If DCPA	and AP	and RB	and SPEED	then Course Alteration
R1	Centre	Near	S_Bow	High	Mid to Stbd
R2	Centre	Near	S_Bow	Normal	Mid to Stbd
R3	Centre	Middle	S_Bow	High	Mid to Stbd
R4	Centre	Middle	S_Bow	Normal	Mid to Stbd

The following is the calculation of the value of each fuzzy rule output with the intersection of the fuzzy sets DCPA ∩ AP ∩ RB ∩ V:

$$\text{Rule 1: } \mu_D(x_1) = \min[1, 0.91, 1, 1] = 0.91$$

$$\text{Rule 2: } \mu_D(x_2) = \min[1, 0.91, 1, 0.875] = 0.875$$

$$\text{Rule 73: } \mu_D(x_3) = \min[1, 1, 1, 1] = 1$$

$$\text{Rule 74: } \mu_D(x_4) = \min[1, 1, 1, 0.875] = 0.875$$

The union of all outputs is

$$\mu_D(x) = \max[\mu_D(x_1), \mu_D(x_2), \ldots, \mu_D(x_n)] = \max[0.91, 0.875, 1, 0.875] = 1.$$

All outputs belong to the fuzzy set »Mid to Stbd«.

The following is the last task in the process, known as defuzzification. It is a process of calculation of the crisp output, i.e. course alteration. The defuzzification method used is a centroid method. Figure 8 shows excerpt of mapping input data into output decision using MATLAB Fuzzy Logic Toolbox.

$$\mu_{mS}(deg) = \begin{cases} 0, & deg \le 30 \\ \frac{deg}{10} - 3, & 30 \le deg \le 40 \\ 1, & 40 \le deg \le 70 \\ 8 - \frac{deg}{10}, & 70 \le deg \le 80 \\ 0, & deg \ge 80 \end{cases} \quad \mu_{mS}(deg) = 1, \text{ if } 40 \le deg \le 70$$

$$x^* = \int_0^{\infty} \frac{x - \mu_D(x)\,dx}{\mu_D(x)\,dx} = \frac{1 - 40 + 1 - 70}{2} = 55^\circ$$

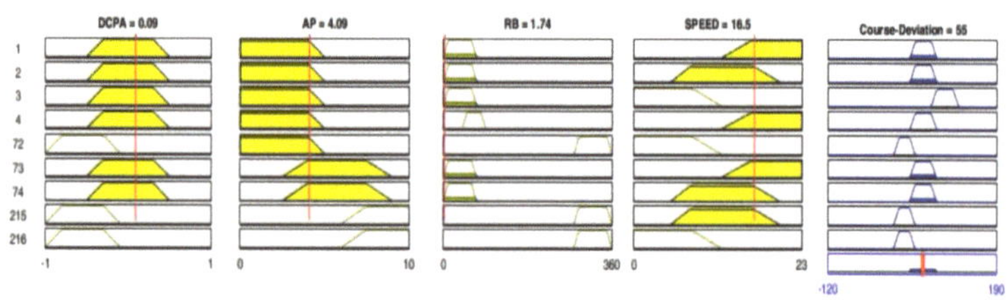

Figure 8. Fuzzy inference procedure using MATLAB Fuzzy Logic Toolbox.

Explanation: The decision model calculates course alteration x* = 55°, and the new course for collision avoidance is 35°. Figure 9 shows a relative movement of the target vessel after course alteration.

Analysis: Case study 1 shows an encounter with a target vessel approaching from opposite directions. The decision model assesses the risk of collision by calculating a DCPA value which is less than the safe ship domain of 1 M according to initial data. With a time delay of the navigational situation for 11 min, the model calculates the input parameters of the fuzzy inference system RB and AP. Based on the input parameters, the fuzzy inference system calculates the course alteration and reassess the risk of a collision. The model graphically plots the planned trajectory of the target vessel. The new DCPA of the encounter after course alteration confirms the quality of the calculated avoidance manoeuvre as it is greater than 1 M (Table 7).

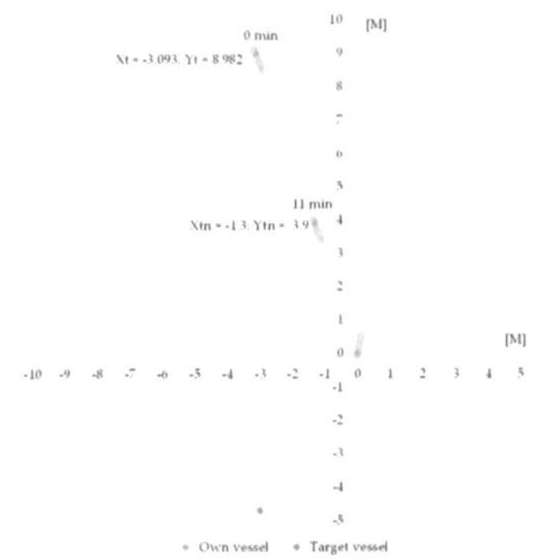

Figure 9. In the 11th minute, course alteration changes the relative path and velocity of the target.

Table 7. Reassessment of the risks of collision.

DCPA [M]	2.0
TCPA [min]	8.1
Vr [kn]	26.3
Cr [°]	191

4.2. Case Studies

The functionality of the decision model was tested on different cases; the results of the simulations are shown in the following tables (Tables 8–13). In all cases, the own vessel is in a give-way position: case 2 simulates a crossing situation with a target vessel approaching from starboard bow, case 3 simulates the situation where a target vessel acts as give-way vessel but violates rule 16 "Action by give-way vessel", and the own vessel as a stand-on vessel must perform an action to avoid the collision. Case study 4 simulates the approach situation with the target vessel from RB 274°. The target, in this case, is a stand-on vessel (constrained by the draft), the own vessel is a give-way vessel. The simulation shows the avoidance at about 4 M, the fuzzy logic with the input parameters calculates a larger course alteration, thus ensuring a meeting at 3.8 M. Case study 5 tests the response of the fuzzy inference system for the approach situation according to COLREG rule 13. Case 6 simulates the approach situation with an RB value between 90 and 112.5°. The simulation observes the compliance of the model, which considers COLREG rules 8 and 19, which require that the vessel which detects the presence of another vessel by radar alone must avoid alteration of course towards the target vessel abeam or abaft the beam [29]. The graphical representation of the relative movement of the target vessel for each case study is shown in Figure 10.

Table 8. Initial data of own vessel.

	Case 2	Case 3	Case 4	Case 5	Case 6
Vessel type	Power-driven	Power-driven	Power-driven	Power-driven	Power-driven
C [°]	225	303	134	81	359
V [kn]	20	13	10	23	17

Table 9. Initial data of target vessel.

	Case 2	Case 3	Case 4	Case 5	Case 6
Vessel type	Power-driven	Power-driven	Constrained by draft	Power-driven	Power-driven
C [°]	160	313	193	82	330
V [kn]	17	23	21	11	23
dt [M]	8	3.5	7.8	6.7	5
ωt [°]	270	146	48	80	99

Table 10. Vessel's encounter information.

	Case 2	Case 3	Case 4	Case 5	Case 6
DCPA [M]	0.73	0.03	0.9	0.01	0.49
TCPA [min]	23.85	20.11	25.8	33.49	25.8
CPA position	negative	positive	positive	negative	negative
RB [°]	45	203	274	359	100
Navigation situation	Crossing	Overtaking	Crossing	Overtaking	Crossing

Table 11. New relative position of a target vessel using time delay.

	Case 2	Case 3	Case 4	Case 5	Case 6
TD [min]	6	6	12	20	11
ω [°]	268.3	146.2	53.6	79.9	94.9
d [M]	6.01	2.46	4.24	2.70	2.9
RB [°]	43.252	203.2	279.602	358.876	95.9

Table 12. Input parameters and calculated decision.

	Case 2	Case 3	Case 4	Case 5	Case 6
DCPA [M]	−0.73	0.03	0.90	−0.01	−0.49
AP [M]	6.01	2.46	4.24	2.70	2.90
RB [°]	43.25	203.22	279.60	358.8	95.89
V [kn]	20	13.00	10.00	23	17.00
Course alteration	47.5	20	60	61.6	−41.7

Table 13. Reassessment of the collision risk.

	Case 2	Case 3	Case 4	Case 5	Case 6
DCPA [M]	3.4	1.1	2.8	2.6	2.8
TCPA [min]	9.6	12.7	17.3	−0.2	1.9

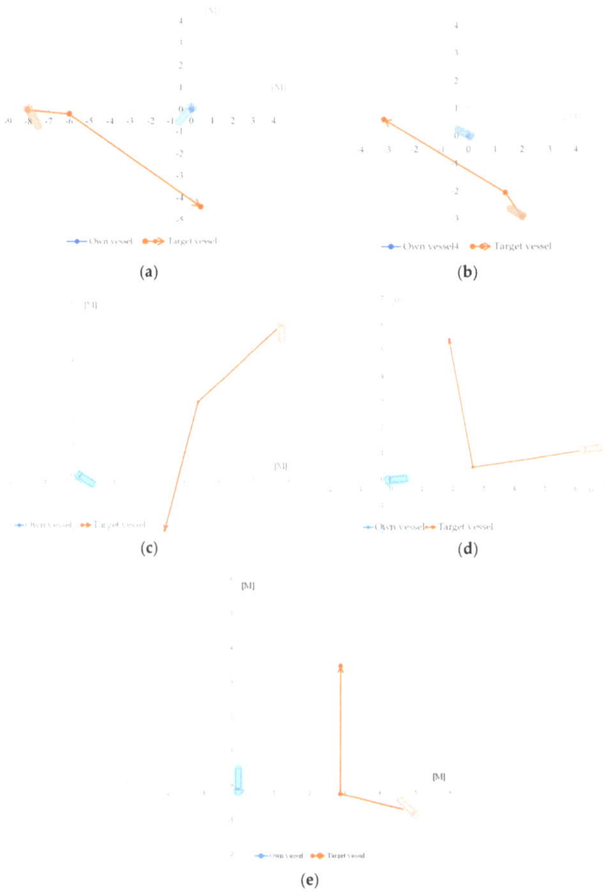

Figure 10. Planned trajectory of the target vessel: (**a**) case study 2, (**b**) case study 3, (**c**) case study 4, (**d**) case study 5, (**e**) case study 6.

5. Discussion

The simulations show the collision avoidance of a target vessel in three different navigation situations: head-on, crossing, and overtaking, whereby the own vessel has a give-way obligation. The simulations do not consider the manoeuvring characteristics of the vessels and are performed in a weather-free environment, eliding, for instance, wind and ocean currents. The results of the simulations show the decision at a subjectively determined time delay, but in real-time situations, the officer of the watch (OOW) usually has a certain timeframe to perform a collision avoidance manoeuvre that meets the desired safety criteria. For this reason, the quality of the model was also evaluated for different time delays to determine whether the decision meets the initial requirement of a DCPA value. We compared the timeframe of correct decisions with TCPA and obtained the percentage of time the model was still calculating course changes correctly and in accordance with the boundary conditions.

Case study 1: Course alteration is satisfactorily calculated in a timeframe of 15 min (see Table 14). After a 15 min time delay, the target vessel is at the distance less than 2 M, which is also the distance at which the decision model no longer operates according to the initial conditions.

Table 14. Course alteration at different time delay, case study 1.

Time Delay [min]	AP [M]	RB [°]	Course Alteration [°]
4	7.53	1.15	55.2
6	6.55	1.25	55.2
8	5.57	1.40	55.2
10	4.58	1.60	55.2
12	3.60	1.92	54.9
14	2.62	2.47	55
15	2.13	2.94	55

The graph (Figure 11) represents the DCPA value at different time delays. The figure shows that the model calculates the proper decisions in the timeframe of 15 min, which is 77% of the total time available from the beginning of the observed situation when TCPA was calculated, which in simulation case study 1, was 19 min.

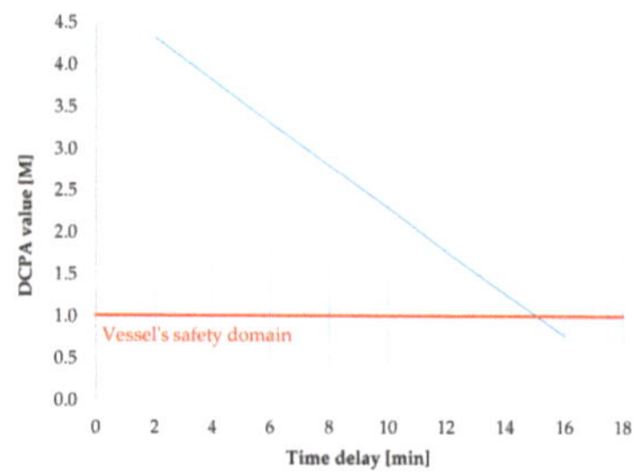

Figure 11. Calculated Distance to Closest Point of Approach (DCPA) at different time delays.

Case study 2: Table 15 shows the course alteration decision at different time delays. The safety condition DCPA (Figure 12) proves that the FIS calculates correct decisions in a timeframe between 2 and 16 min. With the time delay of 18 min, the course deviation is less than 30°, which according to Cockcroft [9], is not considered an appropriate course deviation. In 22 min, the AP of a target vessel is 1 M, and avoidance at this distance poses a high risk of vessel collision because the DCPA is less than 1 M. The model calculates the correct decisions in 67% of the total time available from the beginning of the observed situation.

Table 15. Course alteration at different time delay, case study 2.

Time Delay [min]	AP [M]	RB [°]	Course Alteration [°]
2	7.34	44.52	40.9
4	6.67	43.95	45.5
6	6.01	43.25	47.5
8	5.34	42.38	49.2
10	4.68	41.26	47.6
12	4.03	39.78	40.2
14	3.37	37.71	36.2
16	2.72	34.66	26.5
18	2.09	29.72	20
20	1.48	20.62	20
22	0.96	0.45	19.5

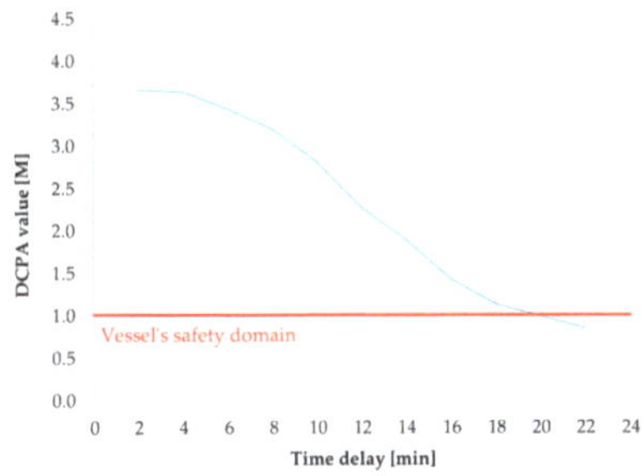

Figure 12. Calculated DCPA at different time delays.

Case study 3: In this case, the target vessel is a give-way vessel which violates the COLREG rules. The result of the avoidance manoeuvre shows that the fuzzy inference system responds well even in emergencies (Table 16). Figure 13 represents the DCPA value at different time delays. The model responds well up to 7 min of the time delay, any alteration of course afterwards would jeopardize the safety of vessels.

Table 16. Course alteration at different time delay, case study 3.

Time Delay [min]	AP [M]	RB [°]	Course Alteration [°]
2	3.15	203.06	20
4	2.80	203.13	20
6	2.46	203.22	20
8	2.11	203.34	20
10	1.76	203.51	20

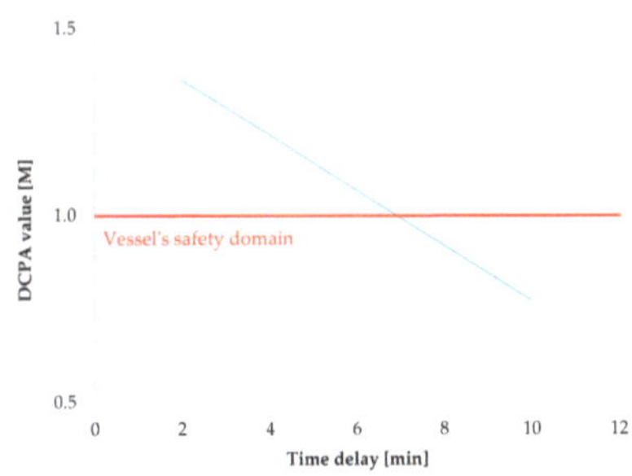

Figure 13. Calculated DCPA at different time delays.

Avoiding a dangerous vessel that violates the COLREG rules is performed in accordance with rules 8 and 19, thus changing the course to the side where there is no target vessel, in this case, the starboard side. The minimum DCPA condition is not met from 8 min on when the target vessel is at 2.1 M. The calculated change of course does not meet the criteria of minimum DCPA, so this condition is only partially confirmed. The model calculates the proper decisions in only 35% of the total time available from the beginning of the observed situation.

Case study 4: Example of a vessel encounter from the port side. As the target vessel is a vessel constrained by draft, the own vessel is obliged to avoid it. According to COLREG rule 8, the model proposed alteration of course to starboard side. The manoeuvre change is large enough to be detected on the radar. Table 17 shows the course alteration at different time delays, and Figure 14 shows the DCPA value at different time delays. The results show that the model responds well from the 2 to the 24 min. In the 24th min of observation, the distance of the observed vessel is 1.05 M. The model calculates the proper decisions in 89% of the total time available from the beginning of the observed situation.

Table 17. Course alteration at different time delay, case study 4.

Time Delay [min]	AP [M]	RB [°]	Course Alteration [°]
6	6.01	275.9	41
12	4.24	279.6	60
18	2.51	288.33	62.2
24	1.05	326.29	95

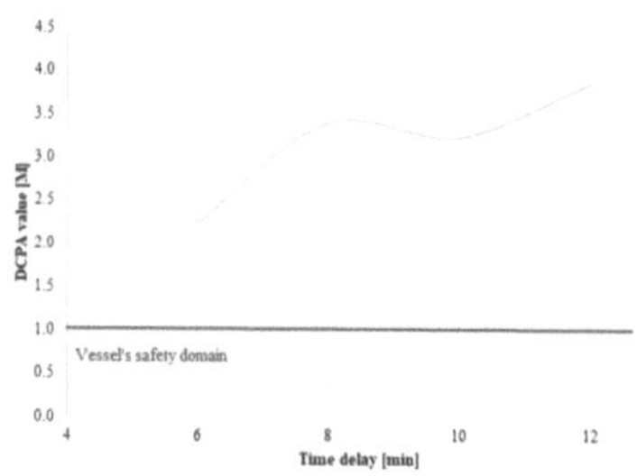

Figure 14. Calculated DCPA at different time delays.

Case study 5: In this case, the own vessel is in the position to overtake the target vessel. According to COLREG rule 13, overtaking is possible in both directions. The model offers the avoidance decision to starboard. Table 18 shows the calculated course alteration for different time delays.

Table 18. Course alteration at different time delay, case study 5.

Time Delay [min]	AP [M]	RB [°]	Course Alteration [°]
8	5.1	358.9	19.8
12	4.3	358.95	61.7
16	3.5	358.92	61.8
20	2.7	358.88	61.8
24	1.9	358.79	95.5
28	1.1	358.57	95.3

The observed DCPA condition demonstrates that fuzzy reasoning responds appropriately in the timeframe between 2 and 28 min. At 28 min (Figure 15), the vessel is at a distance of 1.1 M, and the model calculates the change of course as 95.3°, but due to the distance being too short, the manoeuvre does not reach a safe distance of approach as the calculated DCPA is 0.9372 M. The appropriate avoidance interval is between 2 and 27 min of time lag. The model calculates the proper decisions in 80% of the total time available from the beginning of the observed situation.

Figure 15. Calculated DCPA at different time delays.

Case study 6: The case deals with a meeting situation from abeam. In this case, the alteration in course towards the target vessel is dangerous. Table 19 shows the calculated course change to port side at different time delays. The decision is satisfactorily calculated within the timeframe of 2 to 17 min, as shown in Figure 16. The condition of the vessel's safety domain is met at all distances except for less than 1.7 M in the 18th min. The model calculates the proper decisions in 66% of the total time available from the beginning of the observed situation.

Table 19. Course alteration at different time delay, case study 6.

Time Delay [min]	AP [M]	RB [°]	Course Alteration [°]
2	4.62	99.53	−41.7
4	4.23	98.98	−41.7
6	3.85	98.31	−41.7
8	3.47	97.50	−41.7
10	3.09	96.49	−41.7
12	2.71	95.20	−41.7
14	2.33	93.48	−41.7
16	1.95	91.09	−41.7
17	1.76	89.52	−27.6

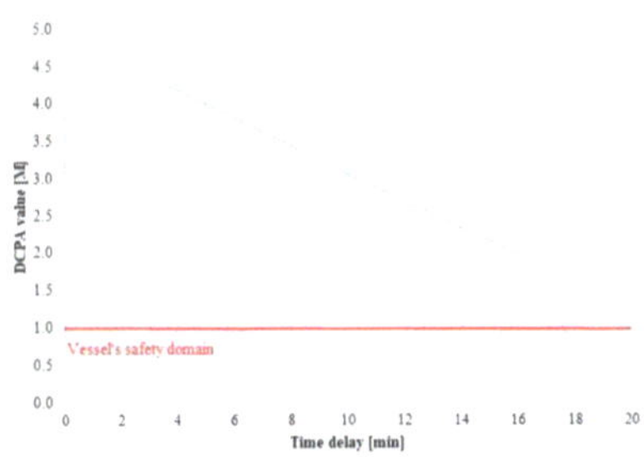

Figure 16. Calculated DCPA at different time delays.

When planning a collision avoidance manoeuvre, course alteration is the most common and effective measure to avoid a close quarters situation, especially on the high seas. According to COLREG rules, the manoeuvre must be made in good time and should be clearly visible to other vessels in the vicinity. However, the distance at which vessels should begin the avoidance manoeuvre, particularly in adverse weather conditions or in restricted areas, is not quantified, leaving the decision to each individual OOW (officer of the watch).

An important factor to consider is vessel's manoeuvrability, where the vessel type, size, and loading condition will affect its turning ability. In addition, external influences, such as the wind and local currents with their direction and force, affect the movement of the vessel and, hence, the course alteration, as they can make it difficult for the vessel to achieve the desired change of direction in the planned time. In particular, a large trim by the stern can change the vessel's wind handling characteristics quite significantly. Local currents also cause the vessel to drift or affect its speed, and their influence depends on the vessel's underwater surface exposed to the currents and their direction. In shallow waters, the turning circles become larger, but the ability to maintain course is better unless the low speed further affects the vessel's responsiveness to course alterations. Therefore, when planning a course alteration, the OOW must consider all effects on the vessel's manoeuvrability in order to make an effective and safe manoeuvre in good time and at a safe distance from the target vessel.

6. Conclusions

The article presents computer predictions of collision avoidance at sea by combining the traditional method of manual radar plotting with an artificial intelligence method—

fuzzy logic. The advantage of fuzzy logic is in generating decisions based on inaccurate data that cannot be described by mathematical notation as they are expressed linguistically. Fuzzy logic thus imitates the human way of thinking, capable of solving complex tasks, although they may also contain a great deal of uncertainty. The peculiarity of the decision model is its tuning with the results of the database of correct solutions obtained by the traditional method of manual radar plotting. Based on them, the fuzzy sets and fuzzy rules were formed.

However, there were some shortcomings in model validation: in areas where decisions change from positive to negative and vice versa, greater errors can occur, and additional fuzzy sets must be created to reduce these errors. In addition, the validation process of the fuzzy inference system (fuzzy sets and fuzzy rules) takes a long time before it is able to make useful decisions.

The multi-parametric collision avoidance decision model uses parameters obtained based on approaching vessels without considering the vessel's manoeuvrability and meteorological and oceanographic parameters that may additionally influence the choice of avoidance manoeuvre or the degree of deviation (change of course and/or speed of own vessel). The simulations also use the marine environment without vertical and/or horizontal constraints (coast, depth); the latter can further influence the choice of avoidance manoeuvre, especially the direction of avoidance. Further research will focus on avoidance using vessel speed and a combination of speed change and the course of own vessel as well as collision avoidance in a multi-vessel situation.

An important segment of further research and testing represents the integration decision model for all vessels that are included in the navigation situation and, thereby, to reduce the likelihood of violating the rules of avoidance between participating vessels.

Author Contributions: Conceptualisation, T.B. and A.A.; methodology, T.B., J.S. and R.B.; software, T.B.; validation, T.B., J.S. and R.B.; resources, A.A.; writing—original draft preparation, T.B.; writing—review and editing, T.B., A.A., and R.B. All authors have read and agreed to the published version of the manuscript.

Funding: The authors acknowledge the financial support of the Slovenian Research Agency (research core funding No. P2-0394, Modelling and Simulations in Traffic and Maritime Engineering).

Data Availability Statement: The data presented in this study are available on request from the corresponding author.

Conflicts of Interest: The authors declare no conflict of interest.

References

1. Nielsen, M.; Petersen, J. Collision avoidance at sea—Practice and problems. In Proceedings of the 20th European Annual Conference on Human Decision, Kongens Lyngby, Denmark, 1 January 2001.
2. Chauvin, C. Human factors and maritime safety. *J. Navig.* **2011**, *64*, 625–632. [CrossRef]
3. MAIB. *Accident Report MV P&O Nedlloyd Vespucci and Motor Yacht Wahkuna*; Report No. 28; Marine Accident Investigation Branch: Southhampton, UK, 2003.
4. IMO Resolution. *Strategy for the Development and Implementation of E- Navigation, MSC 85, Anex 20*; IMO Publishing: London, UK, 2009.
5. Bole, A.G.; Wall, A.D.; Norris, A. *Radar and ARPA Manual: Radar, AIS and Target Tracking for Marine Radar Users*; Butterworth-Heinemann: Oxford, UK, 2013.
6. Perera, L.; Carvalho, J.; Soares, C. Fuzzy logic based decision making system for collision avoidance of ocean navigation under critical collision conditions. *J. Mar. Sci. Technol.* **2011**, *16*, 84–99. [CrossRef]
7. Szlapczynski, R.; Szlapczynska, J. Review of ship safety domains: Models and applications. *Ocean Eng.* **2017**, *145*, 277–289. [CrossRef]
8. Wang, X.; Liu, Z.; Cai, Y. The ship manoeuvrability based collision avoidance dynamic support system in close-quarters situation. *Ocean Eng.* **2017**, *146*, 486–497. [CrossRef]
9. Cockcroft, A.N.; Lameijer, J.N.F. *Guide to the Collision Avoidance Rules*; Elsevier: Amsterdam, The Netherlands, 2003.
10. Statheros, T.; Howells, G.; McDonald-Maier, K. Autonomous Vessel Collision Avoidance Navigation Concepts, Technologies and Techniques. *J. Navig.* **2008**, *61*, 129–142. [CrossRef]

11. Perera, L.P.; Carvalho, J.P.; Soares, C.G. Intelligent Ocean Navigation and Fuzzy-Bayesian Decision/Action Formulation. *IEEE J. Ocean. Eng.* **2012**, *37*, 204–219. [CrossRef]
12. Perera, L.P.; Carvalho, J.; Soares, C.G. Solutions to the failures and limitations of Mamdani fuzzy inference in ship navigation. *IEEE Trans. Veh. Technol.* **2014**, *63*, 1539–1554. [CrossRef]
13. Wu, B.; Cheng, T.; Yip, T.L.; Wang, Y. Fuzzy logic based dynamic decision-making system for intelligent navigation strategy within inland traffic separation schemes. *Ocean Eng.* **2020**, *197*, 106909. [CrossRef]
14. Wu, B.; Yip, T.L.; Yan, X.; Soares, C.G. Fuzzy logic based approach for ship-bridge collision alert system. *Ocean Eng.* **2019**, *187*, 106152. [CrossRef]
15. Hu, Y.; Park, G.K. Collision risk assessment based on the vulnerability of marine accidents using fuzzy logic. *Int. J. Nav. Archit. Ocean Eng.* **2020**, *12*, 541–551. [CrossRef]
16. Zhuo, Y.; Tang, T. An Intelligent Decision Support System to Ship Anti-Collision in Multi-Ship Encounter. In Proceedings of the 7th World Congress on Intelligent Control and Automation, Chongqing, China, 25–27 June 2008.
17. Su, C.; Chang, K.; Cheng, C. Fuzzy Decision on Optimal Collision Avoidance Measures for Ships in Vessel Traffic Service. *J. Mar. Sci. Technol. Taiwan* **2012**, *20*, 38–48.
18. Pietrzykowski, Z.; Magaj, J.; Wołejsza, P.; Chomski, J. *Fuzzy Logic in the Navigational Decision Support Processes onboard a Sea—Going Vessel*; ICAISC Part I.; Springer: Berlin/Heidelberg, Germany, 2010.
19. Liu, Y.; Du, X.; Yang, S.; Yeung, D.; Liu, Z.; Wang, X.; Yan, H. The design of a fuzzy-neural network for ship collision avoidance. *Adv. Mach. Learn. Cybern.* **2006**, *3930*, 804–812.
20. Szłapczyński, R.; Szłapczyńska, J. COLREGS Compliance in Evolutionary Sets of Cooperating Ship Trajectories. *Electron. J. Int. Group Reliab. Reliab. Theory Appl.* **2011**, *2*, 11.
21. Lee, S.; Kwon, K.; Joh, J. A fuzzy logic for autonomous navigation of marine vehicles satisfying COLREG guidelines. *Int. J. Control Autom. Syst.* **2004**, *2*, 171–181.
22. Hu, Y.; Meng, X.; Zhang, Q.; Park, G.K. A real-time collision avoidance system for autonomous surface vessel using fuzzy logic. *IEEE Access* **2020**, *8*, 108835–108846. [CrossRef]
23. Zadeh, L.A. Outline of a new approach to the analysis of complex systems and decision processes. *IEEE Trans. Syst. Man Cybern.* **1973**, *1*, 28–44. [CrossRef]
24. Teodorović, D.; Vukadinović, K. *Traffic Control and Transport Planning: A Fuzzy Sets and Neural Networks Approach (Vol. 13)*; Springer Science & Business Media: Berlin/Heidelberg, Germany, 2012.
25. Mamdani, E.H. Application of fuzzy logic to approximate reasoning using linguistic synthesis. *IEEE Trans. Comput.* **1977**, *C-26*, 1182–1191. [CrossRef]
26. Brcko, T.; Svetak, J. Fuzzy reasoning as a base for collision avoidance decision support system. *Promet Traffic Transp.* **2013**, *25*, 555–564. [CrossRef]
27. Naik, S.L.; Vimarshini, K.R. Prudent Ship Pilotage Using Trial Maneuver. In Proceedings of the 9th International Radar Symposium India, IRSI 13, Bangalore, India, 10–14 December 2013.
28. Stateczny, A. *Radar Navigation*; GTN: Gdańsk, Poland, 2011.
29. COLREG. *Convention on the International Regulations for Preventing Collisions at Sea*; IMO Publishing: London, UK, 1973.
30. *CMA CGM Bridge Safety Manual—Safe Passing Distance*; CMA CGM Group: Marseille, France, 2015.
31. Perera, L.P.; Carvalho, J.P.; Guedes Soares, C. Mamdani type fuzzy inference failures in navigation. In Proceedings of the IEEE International Conference on Industrial Informatics (INDIN), Lisbon, Portugal, 26–29 July 2011.
32. Ni, S.; Liu, Z.; Cai, Y. Ship manoeuvrability-based simulation for ship navigation in collision situations. *J. Mar. Sci. Eng.* **2019**, *7*, 90. [CrossRef]
33. Nguyen, M.; Nguyen, V.; Tamaru, H. Automatic collision avoiding support system for ships in congested waters and at open sea. In Proceedings of the Control, Automation and Information Sciences International Conference (ICCAIS), Ho Chi Minh City, Vietnam, 26–29 November 2012.
34. Zhang, J.; Zhang, D.; Yan, X.; Haugen, S.; Soares, C.G. A distributed anti-collision decision support formulation in multi-ship encounter situations under COLREGs. *Ocean Eng.* **2015**, *105*, 336–348. [CrossRef]
35. Pietrzykowski, Z.; Wołejsza, P.; Borkowski, P. Decision support in collision situations at sea. *J. Navig.* **2017**, *70*, 447–464. [CrossRef]
36. Szłapczyński, R.; Szłapczyńska, J. A Target Information Display for Visualising Collision Avoidance Manoeuvres in Various Visibility Conditions. *J. Navig.* **2015**, *68*, 1041–1055. [CrossRef]
37. Yan, X.; Chen, X.; Sang, L.; Zhang, D. A novel approach for assistance with anti-collision decision making based on the International Regulations for Preventing Collisions at Sea. *J. Eng. Marit. Environ.* **2012**, *226*, 250–259.
38. Hwang, C.N. The integrated design of fuzzy collision—Avoidance and H∞ autopilots on vessels. *J. Navig.* **2002**, *55*, 20. [CrossRef]
39. Hu, L.; Naeem, W.; Rajabally, E.; Watson, G.; Mills, T.; Bhuiyan, Z.; Salter, I. COLREGs-Compliant Path Planning for Autonomous Surface Vehicles: A Multiobjective Optimization Approach. *IFAC-PapersOnLine* **2017**, *50*, 13662–13667. [CrossRef]
40. Fang, M.C.; Tsai, K.Y.; Fang, C.C. A Simplified Simulation Model of Ship Navigation for Safety and Collision Avoidance in Heavy Traffic Areas. *J. Navig.* **2018**, *71*, 837–860. [CrossRef]
41. Huang, Y.; van Gelder, P.; Wen, Y. Velocity obstacle algorithms for collision prevention at sea. *Ocean Eng.* **2018**, *151*, 308–321. [CrossRef]
42. Kim, D.; Hirayama, K.; Okimoto, T. Distributed Stochastic Search Algorithm for Multi-ship Encounter Situations. *J. Navig.* **2017**, *70*, 699–718. [CrossRef]

Journal of
Marine Science and Engineering

MDPI

Article

Ships Added Mass Effect on a Flexible Mooring Dolphin in Berthing Manoeuvre

Aleksander Grm

Faculty of Maritime Technology and Transport, University of Ljubljana, Pot pomorščakov 4, 6320 Portorož, Slovenia; aleksander.grm@fpp.uni-lj.si; Tel.: +386-5-6767-352

Abstract: This paper deals with the hydrodynamic effect of the ship on a flexible dolphin during a mooring manoeuvre. The hydrodynamic effect refers to the change in momentum of the surrounding fluid, which is defined by the concept of added mass. The main reason for the present study is to answer the question, "What is the effect of the added mass compared to the mass of the ship during the mooring procedure for a particular type of ship?" Measured angular frequencies of dolphin oscillations showed that the mathematical model can be approximated by the zero frequency limit. This simplifies the problem to some extent. The mooring is a pure rocking motion, and the 3D study is approximated by the strip theory approach. Moreover, the calculations were performed with conformal mapping using conformal Lewis mapping for the hull geometry. The fluid flow is assumed to be non-viscous, non-rotating and incompressible. The results showed that the additional mass effect must be taken into account when calculating the flexible dolphin loads.

Keywords: added mass; conformal mapping; lewis mapping

Citation: Grm, A. Ships Added Mass Effect on a Flexible Mooring Dolphin in Berthing Manoeuvre. *J. Mar. Sci. Eng.* **2021**, *9*, 108. https://doi.org/10.3390/jmse9020108

Received: 22 December 2020
Accepted: 18 January 2021
Published: 21 January 2021

1. Introduction

Since the beginning of naval history, ships transporting cargo or people from point A to point B have required facilities for safe berthing, loading, and unloading at both points A and B. Over time, ships have grown in size and specialised ships, terminals, and equipment have been built to handle specific types of cargo, such as liquid bulk, dry bulk, and containers. For liquid bulk terminals, a jetty is the typical berthing facility. The ship is usually moored at berths to dedicated breasting dolphins, which may be single-pile flexible dolphins or multi-pile rigid dolphins with fenders.

The primary objective of this work is to estimate the ship added mass. A typical situation of this research geometry and motion is shown in Figures 1 and 2. A ship is moving in a pure sway direction with a constant speed towards the pier. To avoid direct contact with the infrastructure of the liquid cargo terminal, two flexible dolphins reduce the speed of the ship and act as two huge shock absorbers. The current cargo terminal was designed for smaller types of ships, but now larger ships also call at the Port of Koper. As far as safety is concerned, it is also about the safety of the docking process. In the safety analysis of the docking manoeuvre, many different factors need to be analysed in order to get a complete picture of the ship dynamics and the response of the port infrastructure. In this article, we focus exclusively on the estimation of the added mass for such an operation.

Hydrodynamic modelling of added mass phenomena goes way back to names such as Green, Stokes, etc. The influence of added mass has been expressed mathematically and accurately by the expression of the added mass of a sphere. The influence of a free surface on the added mass for surface piercing bodies began many years later. For a given ship, it can be determined by an experimental method. However, the experimental method is limited to a certain condition. To simulate the ship motion, especially in the initial stage of design, the added mass must be calculated by a theoretical method.

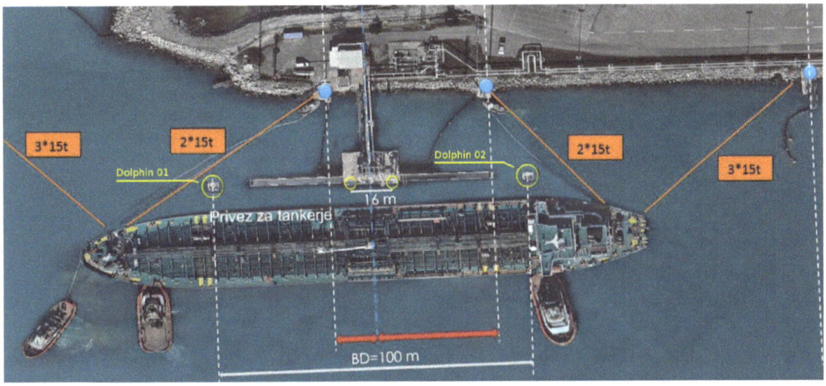

Figure 1. View of the berth in Port of Koper. The dolphins are to the right and the left from the central pier-yellow circles on the sea (photo M.Perkovic).

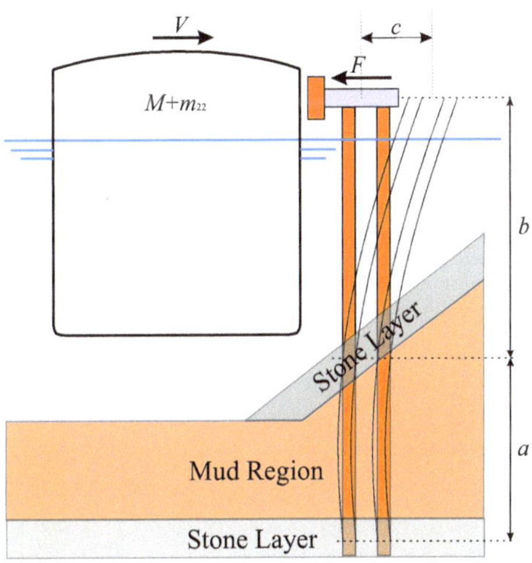

Figure 2. Flexible dolphin mooring with all dimensions. The bottom structure consists of different layers of material: Stones and mud. A ship with mass $M + m_{22}$ and velocity V approaches the mooring. The dolphin is curved by c and the force acting at this moment is F.

The principal for calculating the added mass for surface piercing bodies began with the work of Ursell [1,2] for a cylindrical cross-section. The mathematical model is based on the multipole expansion approach and is in some sense restricted to simple cross-sectional geometries and infinite water depth. The extension of the model to shallow water goes back to Thorne [3]. An important work by Ursell and co-authors can be found in [4]. The multipole expansion method was later used by many researchers, in particular it is very attractive for those working in theoretical hydrodynamics. The completely different approach began with Frank [5], who developed a method for arbitrary cross-sections based on the integral equation approach. The problem can be solved in the frequency domain, introducing a linear consideration of all quantities involved. However, the mean drift forces of order 2nd can only be obtained with the linear solution, e.g., [6]. In addition, Inglis and Price [7], Newman and Sclavounos [8], and Nakos and Sclavounos [9] are among the most important studies of this type.

All of the above methods implement the potential flow assumption and completely neglect viscous effects. The added mass can typically be approximated as not depending on viscosity for the particular case of sinusoidal relative motion between the flow and the object [10]. Similarly, viscous effects are negligible for radiated gravity waves due to body motion, but the same is not always true for damping. It is known that viscous damping during roll is typically the most significant viscous effect on the motion of a ship. Lavrov et al. [11] performed CFD calculations using the Navier-Stokes equations with the $k-\omega$ turbulence model to study the flow in the vicinity of 2D ship sections subjected to forced rolling motions. They concluded that for the same shapes, a 10–20% difference in added mass was observed over the entire frequency range compared to results from using a linear frequency domain potential flow code.

The approach taken in the present study is more in line with the Ursell method, combined with the Conformal Mapping approach. Lewis [12] proposed the classical extended Joukowski transformation method, creating the two-parameter Lewis family of ship-like sections. The family was extended by Landweber and Macagno [13,14] to include an additional parameter, the second moment of the cross-sectional area about the horizontal x-axis. Ursell's approach was used extensively in ship hydrodynamics by Grim [15], Tasai [16], Porter [17], De Jong [18] and others. Later, Athanassoulis and co-authors [19–21] extended this approach to unsymmetric sections as well. It should be noted that the use of only three parameters leads to a quite satisfactory description of ship sections of conventional hull shapes, as is the case here. This property was exploited, for example, by Grigoropoulos and Loukakis [22,23] to optimize the hull shape in terms of the seakeeping.

The problem of determining added mass traditionally falls within the scope of ship manoeuvrability analysis [24–26]. The manoeuvrability of a ship under various conditions has been studied by several authors [27–31] and many others. The most complex theories of manoeuvring and seakeeping involve nonlinear wave loads with higher-order effects [25]. In our case, it is possible to simplify most of the complex theory. Incoming waves are neglected since the ships sail in mostly closed waters. The measured periods of ship motion are very small [32], so a common approach is to further simplify the motion at a zero frequency limit. In this case, only radiated terms are relevant. A similar approach with experimental setup was also studied in [33,34].

The underlying fluid model is nonviscous, nonrotating, and incompressible to simulate flow around the hull. The ideal flow is represented by a complex velocity potential for the channel geometry (the bottom boundary is included in the geometry—Figure 3). Using the theory of complex functions with conformal mapping, it is possible to solve the flow problem of a complex geometry in a simplified geometry [35–37]. In this study, a cylindrical geometry is mapped to a hull geometry using Lewis mapping [12]. The complex velocity potential is integrated over the simplified geometry to obtain the added mass coefficient. The strip theory approach [38] simplifies the 3D problem to a set of 2D problems. The added mass is calculated for three representative ships: Middle Range oil tanker (MR) with range 25,000 t–55,000 t, Long Range type one oil tanker (LR1) with range 55,000 t–80,000 t and Long Range type 2 oil tanker (LR2) with range 80,000 t–160,000 t. The analysis of the under kill clearance (UKC) effect is also studied. For each type of ship, the velocity field is calculated for 20 different drafts from the summer waterline at the intervals of 0.1 m.

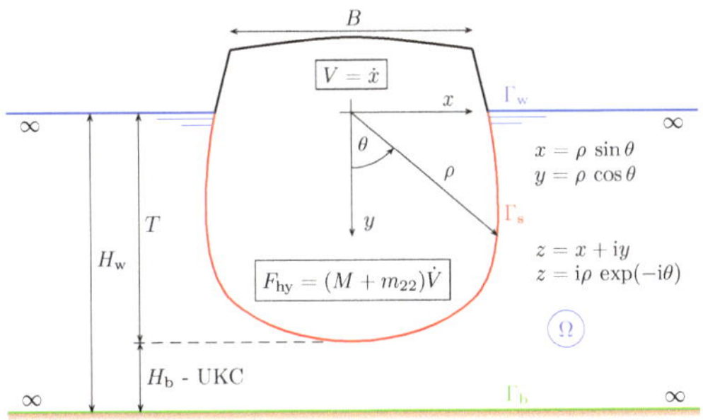

Figure 3. Description of the computational domain.

2. Formulation of the Problem

The added mass is associated with the change in momentum of the surrounding fluid over time [24]. If the fluid is ideal (non-viscous and irrotational) and incompressible, then the fluid is completely described by the complex velocity potential Φ in 2D [39]. Consider a two-dimensional ideal and incompressible fluid in a bounded geometry Ω bounded by the water surface (Γ_w), the bottom (Γ_b) and the hull (Γ_s), as shown in Figure 3. In the (x, y) coordinate system (Figure 3), the velocity potential $\Phi(\boldsymbol{x}, t)$, where $\boldsymbol{x} = (x, y)$ is a point in domain Ω, for a moving body in an otherwise still fluid can be given by the differential equation

$$\frac{\partial^2 \Phi}{\partial x^2} + \frac{\partial^2 \Phi}{\partial y^2} = 0, \quad \boldsymbol{x} \in \Omega \tag{1}$$

and the boundary conditions

$$\frac{\partial \Phi}{\partial y} + k\,\Phi = 0, \qquad \boldsymbol{x} \in \Gamma_w \tag{2a}$$

$$\boldsymbol{n} \cdot \nabla \Phi = 0, \qquad \boldsymbol{x} \in \Gamma_b \tag{2b}$$

$$\boldsymbol{n} \cdot \nabla \Phi = \boldsymbol{n} \cdot \boldsymbol{V}, \qquad \boldsymbol{x} \in \Gamma_s \tag{2c}$$

where $\boldsymbol{n}$ is the normal unit vector always pointing out of domain Ω and k is a wavenumber defined by the relation $k = \omega^2/g$ (infinite depth [26]), where ω is the frequency of the oscillating body, g is the acceleration due to gravity, and $\boldsymbol{V}$ is the velocity of the body. Furthermore, the velocity potential for the oscillatory phenomena can be written in the form

$$\Phi(\boldsymbol{x}, t) = \Re\Big(\phi(\boldsymbol{x})e^{-i\omega t}\Big), \tag{3}$$

where the potential Φ is split into the temporal ($e^{-i\omega t}$) and spatial components ($\phi(\boldsymbol{x})$). It can be shown that the system (1)–(2) is also valid for ϕ [26].

In the case we study, the oscillations are very slow ($\omega \ll 1$), so the boundary condition (2a) can be simplified to

$$\frac{\partial \Phi}{\partial y} = \boldsymbol{n} \cdot \nabla \Phi = 0, \quad \boldsymbol{x} \in \Gamma_w \;\; (\text{for } k \to 0). \tag{4}$$

Now, the solution ϕ must satisfy the following system

$$\frac{\partial^2 \phi}{\partial x^2} + \frac{\partial^2 \phi}{\partial y^2} = 0, \qquad \boldsymbol{x} \in \Omega \tag{5a}$$

$$\boldsymbol{n} \cdot \nabla \phi = 0, \qquad \boldsymbol{x} \in \Gamma_w, \Gamma_b \tag{5b}$$

$$\boldsymbol{n} \cdot \nabla \phi = \boldsymbol{n} \cdot \boldsymbol{V}, \qquad \boldsymbol{x} \in \Gamma_s \tag{5c}$$

where the boundary notations are shown in Figure 3. The ship moves with velocity V in x (sway) direction according to the orientation of the coordinate system shown in Figure 3. The fluid flow can be represented by the potential ϕ as a moving dipole potential for a body described by a cylindrical shape [24].

Let us convert the (x, y) coordinate system into the complex notation

$$z = x + \mathrm{i}y, \quad x, y \in \mathbb{R}, \quad z \in \mathbb{C}. \tag{6}$$

Such a representation simplifies the solution procedure. It is always possible to write the complex velocity potential as the sum of two real-valued functions

$$\Phi(z) = \phi(x, y) + \mathrm{i}\psi(x, y), \tag{7}$$

where we have the fluid velocity defined as the gradient of the real part of the complex potential [26]

$$\boldsymbol{v} := \nabla\Re(\Phi(z)) = \nabla\phi(x, y). \tag{8}$$

The imaginary part of the complex potential $\psi(x, y) = \Im(\Phi(z))$ is known in the literature as streamlines [24].

The motion of the ship in the sway direction can be represented by a moving complex dipole velocity potential oriented in the x direction and defined as

$$\Phi(z) := \frac{A}{z}, \tag{9}$$

where the constant A is the dipole strength that opposes the fluid at the body boundary and satisfies the nonpenetration boundary condition. The constant A has the unit $[\mathrm{m}^3/\mathrm{s}]$ while the potential in a dimensional for has unit $[\mathrm{m}^2/\mathrm{s}]$. The potential (9) does not satisfy the boundary condition at the bottom Γ_b (2b) and must be corrected somehow. The potential correction is done by using dipole images on both sides of the dipole center in y direction at different positions, which are summed up in an infinite series (method of images [37]). The resulting series converges to a new dipole potential

$$\Phi(z) = \frac{A}{2h}\coth\left(\frac{\pi z}{2h}\right), \tag{10}$$

which also satisfies the missing boundary condition at Γ_b (2b), where the distance between the Γ_w and Γ_b equals to $h = H_\mathrm{w}$ (Figure 3).

Proposition 1. *The real part of the potential (10) is the solution of the system (5). The constant A is the strength of the dipole potential $\Phi(z)$ and is obtained from the body boundary condition (5c)*

$$\boldsymbol{n} \cdot \boldsymbol{V} = \boldsymbol{n} \cdot \nabla\phi, \quad \boldsymbol{x} \in \Gamma_\mathrm{b}.$$

Proof. The proof of the proposition 1 is trivial. We need to start with $\phi = \Re(\Phi)$ and substitute this into the system (5). Since $\Phi(z)$ is holomorphic, its real part automatically satisfies Laplace's equation. Write the flow velocity $\boldsymbol{v} = (v_x, v_y)$, then the constant A follows from the body boundary condition (5c) at the point $\boldsymbol{x} = (1, 0)$ where the velocity equals $\boldsymbol{v} = (V, 0)$ and the constant A equals

$$A = V\frac{(2h)^2}{\pi}\sinh\left(\frac{\pi}{2h}\right)^2, \tag{11}$$

where h is a dimensionless water column height

$$h = \frac{H_\mathrm{w} - T}{T} + 1, \tag{12}$$

where discussed parameters are shown in Figure 3. Velocity is a time-dependent quantity and the potential can be decomposed as

$$\phi = V\tilde{\phi} \quad \rightarrow \quad \frac{\mathrm{d}\phi}{\mathrm{d}t} = \dot{V}\tilde{\phi}, \tag{13}$$

assuming that the velocity V and potential $\tilde{\phi}$ are related to the velocity and potential in the sway direction [26]. □

Typical solutions of Equation (10) for variables ϕ and ψ can be seen in Figures 4–9, for $V = 1$ and various h in the case of a cylindrical body geometry. Let us further rewrite the coordinate system into a more natural one for cylindrical geometry. The transformation from Cartesian coordinates (x, y) to polar coordinates (ρ, θ) with the notation of the complex plane is

$$x = \rho \sin\theta, \quad y = \rho\cos\theta, \quad x, y, \rho, \theta \in \mathbb{R} \tag{14a}$$

$$z = x + \mathrm{i}y = \mathrm{i}\rho\exp(-\mathrm{i}\theta), \quad z \in \mathbb{C} \tag{14b}$$

as can be seen in Figure 3. The geometry of the cylindrical body can be transformed into a shape similar to the ship-like shape using the conformal mapping $w = f(z)$, preserving the shape of the complex velocity potential $\Phi(z)$ [35]. This fact is used to compute the hydrodynamic force in the cylindrical geometry Ω_c of the flow generated by the ship geometry Ω_s Figure 10.

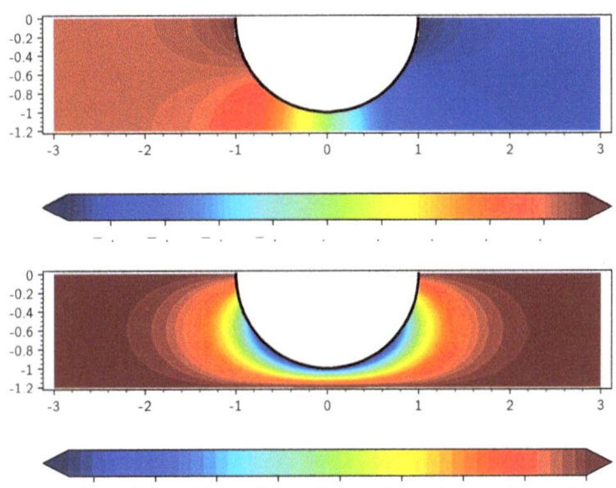

Figure 4. Plot of Equation (10) in the form (7). The top plot shows the real part of the complex potential $\phi(x, y)$, the bottom plot shows the imaginary part of the complex potential $\psi(x, y)$ for velocity $V = 1$ and channel gap width $h = 1.2$ for cylindrical geometry Γ_s with $\rho = 1$.

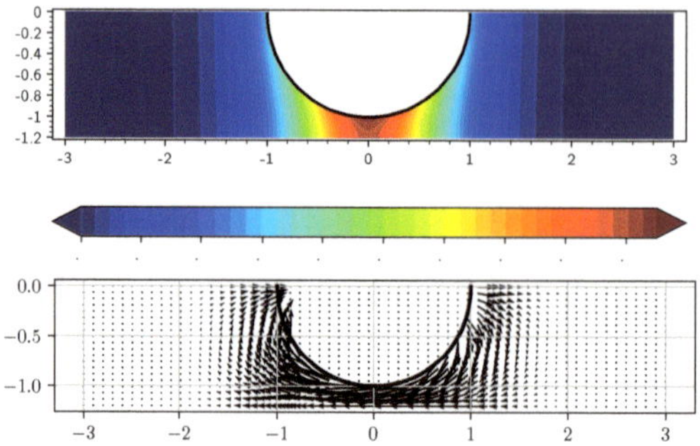

Figure 5. Plot of the Equation (8). The top plot shows the velocity amplitude $\|\boldsymbol{v}\|$, the bottom plot shows the velocity vector field for velocity $V = 1$ and the channel gap of width $h = 1.2$ for cylindrical geometry Γ_s with $\rho = 1$.

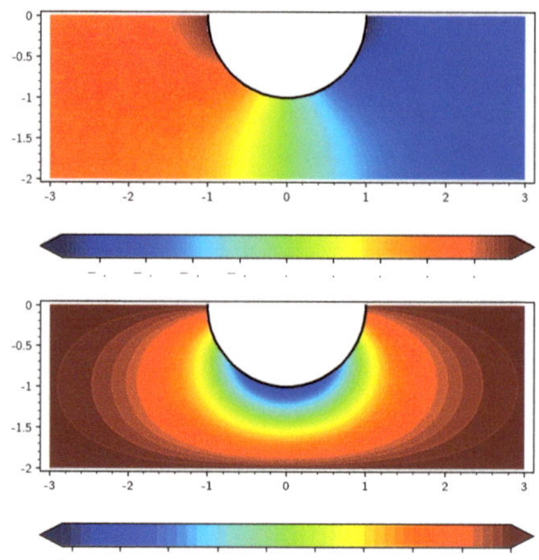

Figure 6. Plot of the Equation (10) in the form (7). The top plot shows the real part of the complex potential $\phi(x,y)$, the bottom plot shows the imaginary part of the complex potential $\psi(x,y)$ for velocity $V = 1$ and channel gap width $h = 2.0$ for cylindrical geometry Γ_s with $\rho = 1$.

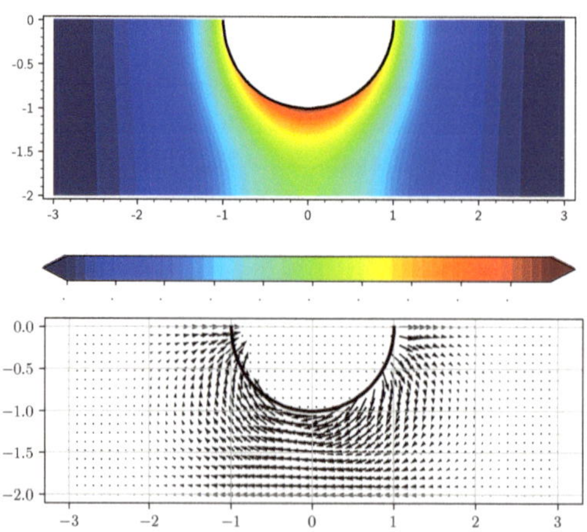

Figure 7. Plot of the Equation (8). The top plot shows the velocity amplitude $\|\boldsymbol{v}\|$, the bottom plot shows the velocity vector field for velocity $V = 1$ and the channel gap of width $h = 2.0$ for cylindrical geometry Γ_s with $\rho = 1$.

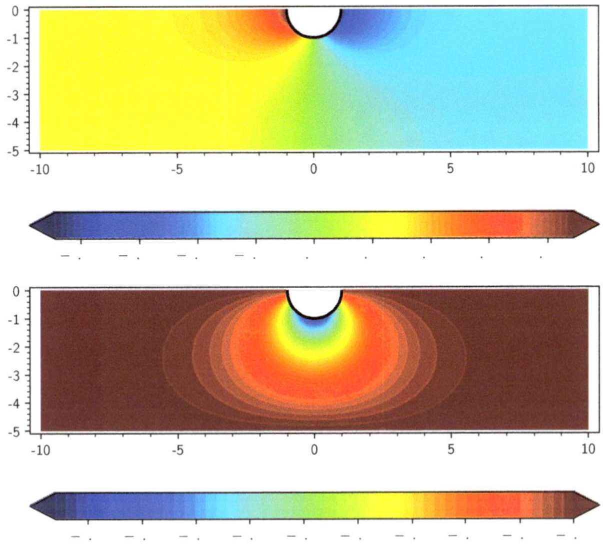

Figure 8. Plot of the Equation (10) in the form (7). The top plot shows the real part of the complex potential $\phi(x,y)$, the bottom plot shows the imaginary part of the complex potential $\psi(x,y)$ for velocity $V=1$ and channel gap width $h=5.0$ for cylindrical geometry Γ_s with $\rho=1$.

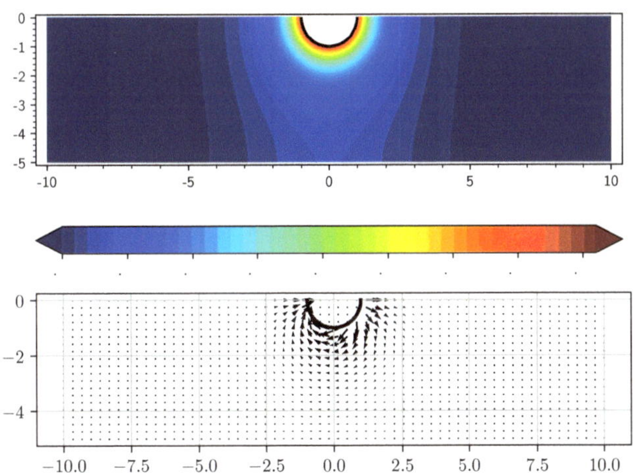

Figure 9. Plot of the Equation (8). The top plot shows the velocity amplitude $\|\boldsymbol{v}\|$, the bottom plot shows the velocity vector field for velocity $V=1$ and the channel gap of width $h=5.0$ for cylindrical geometry Γ_s with $\rho=1$.

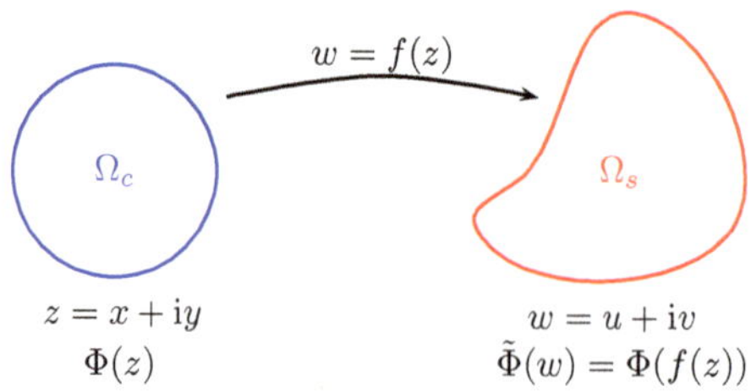

Figure 10. Conformal mapping $w = f(z)$ of a circular domain Ω_c onto a ship-like domain Ω_s with coordinates and velocity complex potentials preserved by the mapping.

One of the most commonly used conformal mappings for ship-like forms is the Lewis transformation [12], which uses only 3 free parameters a, a_1 and a_3

$$w = a\left(z + \frac{a_1}{z} + \frac{a_3}{z^3}\right), \quad z \in \mathbb{C}, \quad a, a_1, a_3 \in \mathbb{R}, \tag{15}$$

where a only causes the shape to expand/compress, but does not affect the appearance of the shape. The free parameters are determined with the basic parameters of the specific ship cross-section: B—maximal breadth, T—draft and S—area

$$\sigma_s = \frac{S}{BT}, \tag{16a}$$

$$H = \frac{B}{2T}, \tag{16b}$$

$$C_1 = \left(3 + \frac{4\sigma_s}{\pi}\right) + \left(1 - \frac{4\sigma_s}{\pi}\right)\left(\frac{H-1}{H+1}\right)^2, \tag{16c}$$

$$a = \frac{B}{2}(1 + a_1 + a_3), \tag{16d}$$

$$a_1 = (1 + a_3)\frac{H-1}{H+1}, \tag{16e}$$

$$a_3 = \frac{-C_1 + 3 + \sqrt{9 - 2C_1}}{C_1}. \tag{16f}$$

Figure 11 and Table 1 show the data used in the present calculations. The ship constructed from these cross-sections is referred to as the *Lewis ship*. The hydrodynamic properties of sway motion for the Lewis ship are shown in Figure 12 for infinite depth and in Figure 13 for finite depth.

Table 1. Lewis mapping coefficients for MR, LR1 and LR2 oil tanker type with $C_b = 0.78$ producing shapes in Figure 11. Only B_k needs to be scaled with $\beta = B/T$ ratio for different draft calculations.

Section (k)	B_k	T_k	σ_k	$\tilde{L}_k$
0	0.8	0.2	0.60	0.05
1	1.2	0.9	0.50	0.05
2	1.6	1.0	0.68	0.05
3	2.0	1.0	0.93	0.05
4	2.0	1.0	0.99	0.60
5	2.0	1.0	0.93	0.05
6	1.8	1.0	0.68	0.05
7	1.2	1.0	0.56	0.05
8	0.3	0.7	0.56	0.05

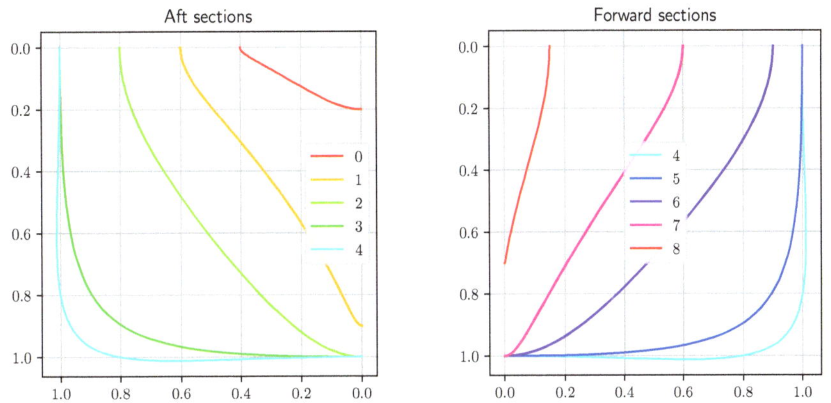

Figure 11. Ship cross-sections used in the calculation. The parameters of the cross-section are shown in Table 1.

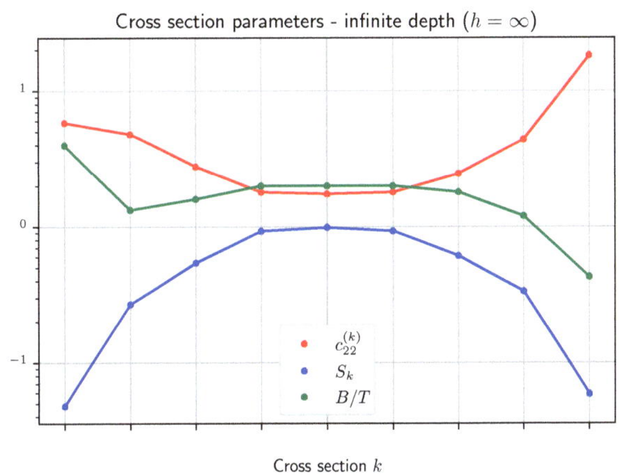

Figure 12. Results for Lewis cross-sections k from Table 1 (Figure 11) for infinite water depth. $c_{22}^{(k)}$ is the added mass coefficient, B/T is the ratio of beam/draft cross-section, and S_k is the cross-section area. The scales on the ordinate are logarithmic for better result representation.

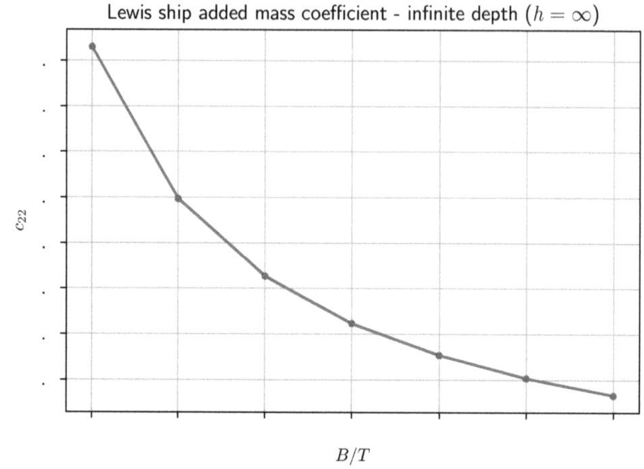

Figure 13. Results for Lewis ship defined in Table 1 (Figure 11) for infinite water depth. c_{22} is the added mass coefficient for the Lewis ship, and B/T is the ratio between beam and draft.

The hydrodynamic force resulting from the time variation of the surrounding fluid is defined in [24] and is equal to

$$\begin{aligned} \boldsymbol{F} = &-\rho \frac{\mathrm{d}}{\mathrm{d}t} \int_{\Gamma_s} \phi \, \boldsymbol{n} \, \mathrm{d}S \\ &- \rho \int_{\Gamma_w \cup \Gamma_b} (\boldsymbol{n} \cdot \nabla\phi) \, \nabla\phi \, \mathrm{d}S \\ &+ \rho \frac{1}{2} \int_{\Gamma_w \cup \Gamma_b} (\nabla\phi \cdot \nabla\phi) \, \boldsymbol{n} \end{aligned} \tag{17}$$

In the present study, the integrals over the boundary $\Gamma_w \cup \Gamma_b$ are zero, since we are only interested in the sway component of the motion. Splitting the potential ϕ into a velocity part and a space part (13) gives the final form of the hydrodynamic force

$$\begin{aligned} \boldsymbol{F} &= -\rho \, \dot{V} \, \mathcal{V} \int_{\Gamma_s} \tilde{\phi} \, \boldsymbol{n} \, \mathrm{d}S \\ &= -\dot{V} \, \rho \, \mathcal{V} \, c_{22} \\ &= -\dot{V} \, m_{22}, \end{aligned} \tag{18}$$

where $\mathcal{V}$ is the displacement of the body, ρ is the fluid density, $\dot{V}$ is the acceleration of the body, c_{22} is added mass coefficient in sway mode, and m_{22} is the added mass in sway mode. To calculate the integral over the body surface Γ_s, we perform the integration for each cross-section k according to Table 1 and add their contribution to the total added mass. The coefficient of the added mass for each cross-section k is calculated in the circular cross-section in space Ω_c and transferred to the ship cross-section in space Ω_s using the conformal mapping (15). The integral in (18) is transformed from Ω_c to Ω_s

$$c_{22}^{(k)} = \frac{2}{S} \int_0^{\pi/2} \tilde{\phi}(w) \, (\boldsymbol{n}(w) \cdot \boldsymbol{e}_x) \left| \frac{\mathrm{d}w}{\mathrm{d}z} \frac{\mathrm{d}z}{\mathrm{d}\theta} \right| \mathrm{d}\theta, \tag{19}$$

where $w = f(z)$ is a conformal mapping (15), $z = z(\rho, \theta)$ is defined in (14b), $\boldsymbol{e}_x$ is a unit vector in x (sway) direction in Ω_s (Figure 3), and S is the area of the cross-section k. The integral (19) is computed as a contour integral over the cylinder in the polar coordinates with radius $\rho = 1$ and $\theta \in [0, \pi/2]$. The term $\mathrm{d}w/\mathrm{d}z$ is the Jacobian of the conformal mapping and $\mathrm{d}z/\mathrm{d}\theta$ follows from the chain rule in the derivative of conformal mapping.

The potential in (19) is written in dimensionless form. The length scale is scaled by T_i (specific draft configuration) and the velocity by the ship velocity V according to the following scheme

$$x = \tilde{x}T_i, \qquad y = \tilde{y}T_i, \tag{20a}$$
$$\dot{x} = \tilde{v}_x V, \qquad \dot{y} = \tilde{v}_y V, \tag{20b}$$
$$\ddot{x} = \tilde{a}_x V^2/T_i, \qquad \ddot{y} = \tilde{a}_y V^2/T_i. \tag{20c}$$

The added mass of a cross-section k is given by the cross-section k added mass coefficient (19) multiplied by the respective water density ρ and volume V_k

$$m_{22}^{(k)} = c_{22}^{(k)}\,\rho\,V_k = c_{22}^{(k)}\,\rho\,(\tilde{S}_k B T_i)\,(\tilde{L}_k L), \tag{21}$$

where $\tilde{S}_k$ is the dimensionless cross-sectional area and $\tilde{L}_k$ is the dimensionless cross-sectional length. For each cross-section k the values for $\tilde{S}_k$ and $\tilde{L}_k$ are taken from Table 1, and for each ship type, the constants B, T_i and L are taken from Table 2. The final added mass of the ship for the slow sway motion is the sum of all added mass contributions of the cross-sections k

$$m_{22} = \sum_{k=0}^{8} m_{22}^{(k)}. \tag{22}$$

Detailed description of added mass calculation procedure is described in next section.

Table 2. Oil tanker types used in simulation: $L = L_{\mathrm{bp}}$—length between perpendiculars, B—maximal breadth, $T_{\max}$—draft at summer line, $T_{\min}$—minimal draft in simulation, C_b block coefficient. Specific draft T_i is in the interval $[T_{\min}, T_{\max}]$.

Type	L [m]	B [m]	$T_{\min}$ [m]	$T_{\max}$ [m]	C_b
MR	185.0	29.1	8.50	10.50	0.78
LR1	220.0	36.3	10.50	12.50	0.78
LR2	238.0	41.3	12.20	14.20	0.78

3. Results

The ship moves at a relatively slow speed when docked. In this study, the problem's formulation contains many reasonable simplifications to obtain results based only on symbolic derivations. The further simplification of the full 3D problem is based on the strip theory approach. The first step was to decompose the representative geometry of the oil tanker into some cross-sections to obtain relevant shape differences. The Levis map (15) is used to describe different cross-sections. The generated data for each cross-section describing the shape of the oil tanker are shown in Table 1. The results can be seen in Figure 11.

In Figures 4–9 are plots of the complex dipole potential (10) for different values of the water height h, where $\rho = 1$ and $V = 1$. The sequence of images for different h shows the difference between the deep water solution ($h >> 1$) and the shallow water solution ($(h-1) < 1$. The gap effect can be well observed from the intensity of the velocity potential ϕ. The maximum value is in the range from 2.5 to 1.2, for water heights from $h = 1.2$ to $h = 5.0$. The magnitude of the velocity in the gap increases with smaller h. The higher values of ϕ at the cylinder boudary result in a larger additional mass. The magnitude of the velocity in the gap is related to the viscous damping. The larger the magnitude of the velocity in the gap, the smaller the gap width and the stronger the viscous forces act.

Three different representative oil tanker types are studied for the selection of ship types. The different types show the difference in the added mass in terms of ship size, their particulars, and UKC distance. The influence of UKC on the added mass was determined with 20 different ship drafts T_i. In this case, the number of draft subdivisions is not a limit, since the calculations for a single geometry are very fast (order of magnitude of a few seconds). Table 2 gives the main specifications for the different oil tanker types used in the simulation. All three types have the same block coefficient $C_b = 0.78$ with the cross-sectional shapes defined in Table 1 and their particulars defined in Table 2.

To obtain the Lewis cross-sectional forms for various drafts T_i, we only need to multiply the coefficient B_k in Table 1 by the constant β_i

$$B_k \to \beta_i B_k, \quad \beta_i = \frac{B/T_i}{2}. \tag{23}$$

The ratio β_i is defined as the ratio between the ship's beam B and the current ship's draft T_i and the constant ratio $B/T = 2$ for the Lewis ship (Table 1). The values of a given ship configuration "i" (B/T_i) are calculated from Table 2. The cross-sectional area S_k for a given configuration i is determined as

$$S_k = \sigma_k \, \beta_i \, B_k \, T_k,$$

where $k = \{0, 1, \ldots, 8\}$ is the cross-section number and $i = \{1, 2, 3, 4, \ldots, N\}$ is the specific draft configuration, where N is the number of different draft scenarios for a given tanker type. In the present case, N was set to 20 to get nice continuous plots. The calculations are very fast, and it takes about SI 1s to calculate a single draft configuration. One of the main considerations in the present work was also the speed of the computation, and it could only be achieved with a semi-analytical approach.

In the previous section, a complete model for calculating the added mass in slow sway motion was formulated. The model is based on a potential flow theory with linear boundary conditions (5). For simple geometries, such as the circular one, the solution ϕ of (5) is a pulsating dipole with origin at the free surface (10) with constant A defined in Equation (11). The solution (10) satisfies the PDE system (5) only for a circular body geometry. The added mass coefficient c_{22} of a circular geometry can be easily obtained using the integral (18) for different water heights h. Figure 14 shows the solution for the added mass coefficient as a function of different dimensionless gap widths (UKC/R). For this particular case, one obtains the explicit expression for the added mass coefficient

$$\begin{aligned} c_{22}(h) &= \frac{2}{S}\int_0^{\pi/2} \tilde{\phi} \sin\theta \, \mathrm{d}\theta \\ &= \frac{4}{\pi}\int_0^{\pi/2} \Re\left[\frac{2h}{\pi}\sinh^2\left(\frac{\pi}{2h}\right)\coth\left(\frac{\pi}{2h}z\right)\right]\sin\theta \, \mathrm{d}\theta, \\ & z \to \mathrm{i}\exp(-\mathrm{i}\theta) \end{aligned} \tag{24}$$

$$\begin{aligned} c_{22}(h) &\approx \left[\frac{1}{3} + \left(\frac{2h}{\pi}\right)^2\right]\sinh^2\frac{\pi}{2h}, \quad h > 1, \\ h &= 1 + \mathrm{UKC}/R, \end{aligned} \tag{25}$$

where the term $\coth(x)$ in the integral function (24) has been expanded into Taylor series (see [40]). For $|z| = 1$ the series converges very quickly. Already the first three terms yield the solution error below 10^{-3}. To obtain the added mass coefficient, the value of the integral must be divided by the area of the cross-section. In this particular case for a circular cross-section with unit radius, the value of the area is $S = (\pi R^2)/2 = \pi/2$. The result shown in Figure 14 will be used later when verifying the results of the proposed method for calculating the added mass of a tanker-type ship.

The average water depth at the liquid terminal in the Port of Koper is approximately $H_w = 14.5$ m. The variable h is calculated using Equation (12) for different Lewis shapes (Table 1) and ship particulars (Table 2) for each draft configuration T_i. If h is known, the coefficient of added mass coefficient c_{22}, as defined in Equation (19), can be calculated for each cross-section k

$$c_{22}^{(k)} = \frac{2}{S} \int_0^{\pi/2} \tilde{\phi}(w) \left(\boldsymbol{n}(w) \cdot \boldsymbol{e}_x\right) \left| \frac{\mathrm{d}w}{\mathrm{d}z} \frac{\mathrm{d}z}{\mathrm{d}\theta} \right| \mathrm{d}\theta.$$

Now each term of the integral is explained in detail. Let us begin with the velocity potential

$$\begin{aligned}
\tilde{\phi}(w) &= \Re\left[\frac{2h}{\pi} \sinh^2 \frac{\pi}{2h} \coth\left(\frac{\pi}{2h} w\right)\right] \\
&= \Re\left[\frac{2h}{\pi} \sinh^2 \frac{\pi}{2h} \coth\left(\frac{\pi}{2h} a\left(z + \frac{a_1}{z} + \frac{a_3}{z^3}\right)\right)\right], \quad z \to \mathrm{i}\exp(-\mathrm{i}\theta), \\
&= \frac{2h}{\pi} \frac{\sinh^2(\frac{\pi}{2h}) \sinh(\frac{\pi}{2h} \sin\theta) \cosh(\frac{\pi}{2h} \sin\theta)}{\sin^2(\frac{\pi}{2h} \cos\theta) + \sinh^2(\frac{\pi}{2h} \sin\theta)}
\end{aligned}$$

Next is the Jacobian of the transformation

$$\begin{aligned}
\frac{\mathrm{d}w}{\mathrm{d}z} \frac{\mathrm{d}z}{\mathrm{d}\theta} &= a[a_1 \exp(\mathrm{i}2\theta) - 3a_3 \exp(\mathrm{i}4\theta) + 1] \exp(-\mathrm{i}\theta) \\
&= a[a_1 \exp(\mathrm{i}\theta) - 3a_3 \exp(\mathrm{i}3\theta) + \exp(-\mathrm{i}\theta)] \\
&= a[(a_1 + 1) \cos\theta - 3a_3 \cos 3\theta] + \mathrm{i}a[(a_1 - 1) \sin\theta - 3a_3 \sin 3\theta].
\end{aligned}$$

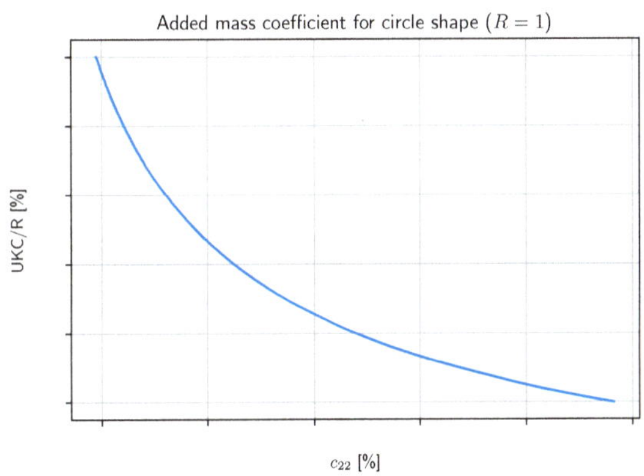

Figure 14. Plot of the solution (25) for the added mass coefficient c_{22} with respect to the dimensionless UKC for the solution with circular (Figures 4–9). For the larger UKC, the typical result for the solution with infinite depth ($c_{22} = 100\%$) can be seen. UKC is scaled in dimensionless form with the radius of the circle R and is related to h defined in Equation (25).

The absolute value of the Jacobian is found using the relation $|z| = \sqrt{\Re(z)^2 + \Im(z)^2}$ The normal vector is found by

$$\boldsymbol{r}(w) = \frac{\mathrm{d}w}{\mathrm{d}z}\frac{\mathrm{d}z}{\mathrm{d}\theta} = (r_x, r_y) = (\Re(\boldsymbol{r}), \Im(\boldsymbol{r})),$$

$$\boldsymbol{t}(w) = \frac{\boldsymbol{r}(w)}{|\boldsymbol{r}(w)|} = \frac{(r_x, r_y)}{\sqrt{r_x^2 + r_y^2}},$$

$$\boldsymbol{n}(w) = \mathrm{i}\,\boldsymbol{t}(w),$$

$$n_x = \boldsymbol{n}(w) \cdot \boldsymbol{e}_x = \Re[\boldsymbol{n}(w)],$$

$$n_y = \boldsymbol{n}(w) \cdot \boldsymbol{e}_y = \Im[\boldsymbol{n}(w)],$$

where the vector $\boldsymbol{n}$ is written in complex notation, where the $x$component is equal to the real part ($n_x = \Re(\boldsymbol{n})$) and the $y$component is equal to the imaginary part ($n_y = \Im(\boldsymbol{n})$). The last one is the explanation of the cross-sectional area

$$S = \sigma_k\, \beta_i\, B_k\, T_k,$$

where all the coefficients are taken/calculated form Table 1. The integral is evaluated numerically using the Gaussian quadrature rule for each cross-section k for a single draft and tanker-type configuration with arbitrary accuracy.

The cross-sectional added mass coefficient $c_{22}^{(k)}$ is then multiplied by the corresponding cross-sectional volume to obtain the cross-sectional added mass $m_{22}^{(k)}$ for a given ship type under various draft conditions. Finally, all cross-sectional masses are summed to obtain the ship added mass m_{22} for the sway motion for a given ship type and draft.

Figures 15–17 show the results of calculated ship added mass m_{22} (Equation (22)) for all three tanker types MR, LR1 and LR2 for different drafts. The maximum draft is the draft at the summer load line as given in Table 2. The results show that the added mass increases with ship draft T (green line) resulting in smaller UKC (blue line). Smaller UKC causes higher velocity magnitudes in the hull neighbourhood and higher values of the potential ϕ at the ship boundary. The effect appears weakly nonlinear and could not be predicted using crude approximation methods, especially if one is interested in fairly good estimates of the added mass for a given ship type. In contrast, the added mass relative to displacement increases almost linearly (red line). The difference in added mass relative to draft is about 30–45% per 1 meter draft change. Assuming that it is constant over the entire draft range is not good practice in this case, and the effect of draft should always be considered in calculations for the flexible mooring problem.

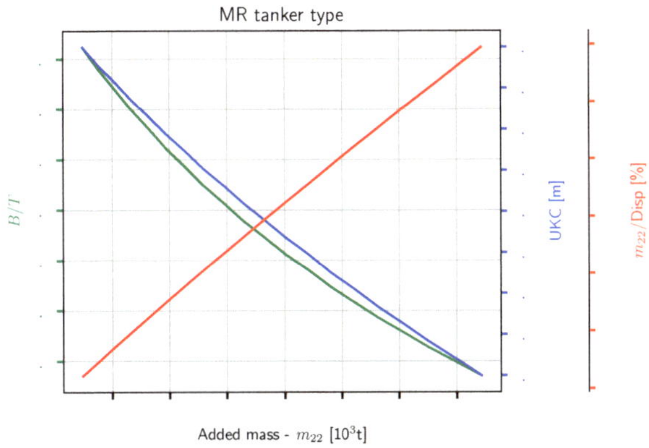

Figure 15. Results for the added mass m_{22} of the **MR** tanker type for different drafts. Labeled variables are: B/T (green line—left side scale), UKC (blue line—first right side scale), and the ratio between the added mass and the displacement in percent $m_{22}/Disp$ (red line—second right side scale).

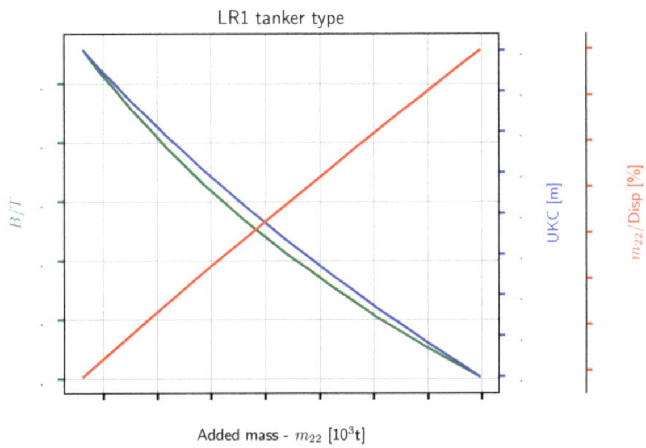

Figure 16. Results for the added mass m_{22} of the **LR1** tanker type for different drafts. Labeled variables are: B/T (green line—left side scale), UKC (blue line—first right side scale), and the ratio between the added mass and the displacement in percent $m_{22}/Disp$ (red line—second right side scale).

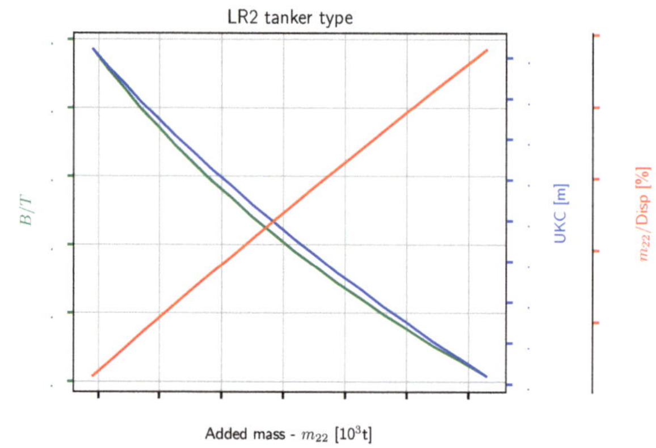

Figure 17. Results for the added mass m_{22} of the **LR2** tanker type for different drafts. Labeled variables are: B/T (green line— left side scale), UKC (blue line—first right side scale), and the ratio between the added mass and the displacement in percent $m_{22}/Disp$ (red line—second right side scale).

Figures 18–20 show the same result as in Figures 15–17, but are composed in a different way. Figure 18 shows the added mass as a function of the B/T ratio. The effect of smaller UKC is seen in a faster increase of the added mass. The same phenomenon is observed in Figures 19 and 20. The result shown in Figure 21 is very revealing. The plot shows the added mass coefficient with respect to the dimensionless UKC. Compared with Figure 14 (dash-dot line), the same trend is observed. There is a difference in the added mass coefficient c_{22} between the circular cross-section and the ship-shaped geometry. The difference is due to the different cross-section shapes. Figure 22 is from Vugts research published in [33] and clearly shows the dependence on the B/T ratio with respect to the added mass coefficient c_{22} for the square cross-section. The larger the B/T ratio is, the smaller the added coefficient is. In our case, the B/T ratio is in the interval between 2.8 and 3.4 (Figure 18). The results in Figure 22 were obtained for infinite water depth. To obtain a clear validation of the present results, the same experiment is performed for the Lewis ship (Table 1) for different B/T ratios. The results are shown in Figure 13 and show the decay of c_{22} of the ship-like shape versus the B/T ratio. Comparing the range of the B/T ratio and the data from Figure 13, the estimate of the added mass coefficient for the Lewis-type shape lies in the interval $c_{22} \in [0.6, 0.73]$. For each cross-section k, the results for the infinitely deep water are shown in Figure 12 for $c_{22}^{(k)}$, S_k and B/T ($\beta_i = 1$). The added mass coefficients of the ship-shaped cross-section are always smaller than the added mass coefficients of the circular cross-section. This fact is mostly related to the B/T ratio.

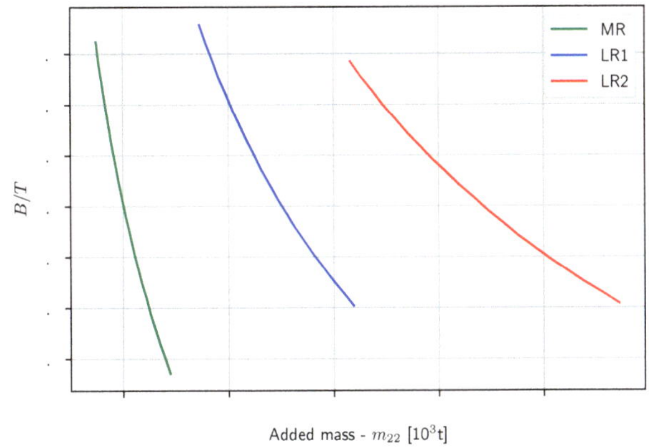

Figure 18. Results for the added mass m_{22} of the **MR**, **LR1** and **LR2** tanker type with respect to B/T ratio.

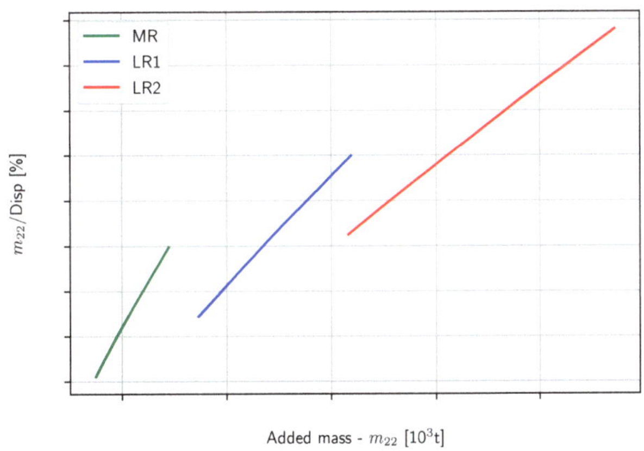

Figure 19. Results for the added mass m_{22} of the **MR**, **LR1** and **LR2** tanker type with respect to m_{22}/Disp ratio.

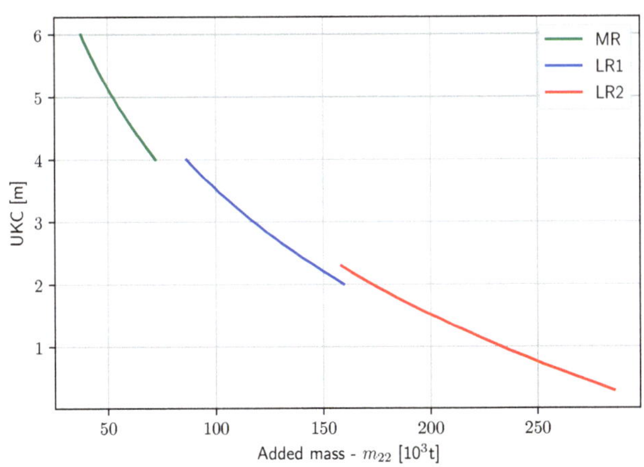

Figure 20. Results for the added mass m_{22} of the **MR**, **LR1** and **LR2** tanker type with respect to UKC.

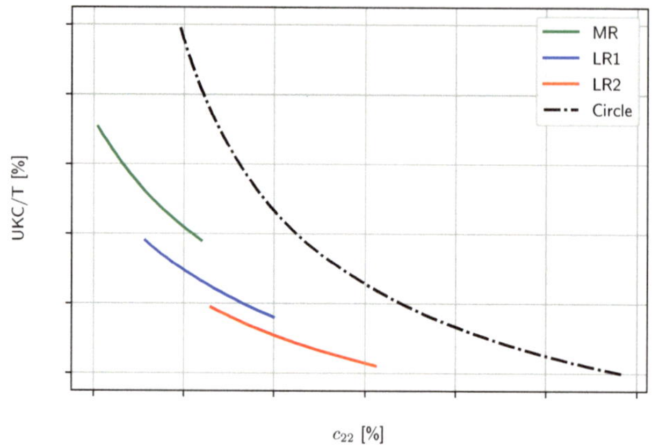

Figure 21. Results for the added mass coefficient $c_{22} = m_{22}/\text{Disp}$ of the **MR**, **LR1** and **LR2** tanker type with respect to UKC/T ratio. Dashed line is the same as in Figure 14.

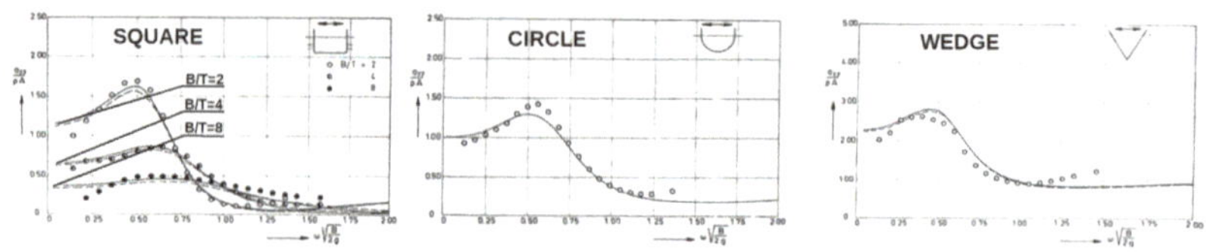

Figure 22. Results for the added mass coefficient c_{22} obtained from Vugts [33]. Comparison with present results can be made with the results of zero frequency case $\omega = 0$.

4. Discussion and Conclusions

The effect of added mass during the berthing manoeuvre was analysed at the liquid berth in the port of Koper for different types of oil tankers. The formulation of the problem is based on the theory of ideal incompressible fluid so that the velocity of the surrounding fluid can be expressed as a complex velocity potential. Measured ship oscillation times under dolphin loading are long, and the simplification of the zero-frequency limit leads to the simplification of the free surface boundary condition (longwave approximation). The described simplifications and the use of complex analysis methods facilitate the calculation of added mass. One of the missing effects is the viscosity effect. If viscosity were included, it would complicate the system of equations to such an extent that a symbolic solution would not be possible, which was the motivation of this study to avoid numerical calculations as much as possible.

In the present case, the complex velocity potential represents the finite depth situation to include the effect of under keel clearance (UKC) in the calculations of added mass. The simplification of 3D calculations into 2D calculations is applied with the strip theory approach for the zero head velocity. All the described simplifications resulted in a system of equations that can be solved symbolically. The rather complicated system of equations is described in *Python* [41] environment with *SymPy* [42] module for the symbolic calculations and can be found in the *Zenodo* repository [43].

Conformal maps as Lewis map [12] defines a simplified ship geometry with only three parameters. The geometry is simplified, but the overall shape is very close to that of an oil tanker. A similar system is discussed in [34]. The results in [34] are very similar to those in this study for the larger values of UKC/R. The sway motion was also analysed in [33] and the results are comparable. The computational system is written in complex Python language form and it is very easy to manipulate with it for a variety of different parameters, cross-section geometries, ship details, UKC etc.

The main objective of this study was to accurately estimate the amount of added mass for certain types of ships docking at the liquid jetty where flexible dolphins are installed. The information of added mass can now be used in future fatigue analyses of flexible dolphins. To support a broader analysis, three different ship types are identified as the representative fleet: MR oil tanker, LR1 oil tanker and LR2 oil tanker. Each class is analysed under different draft conditions with a constant water height of the port basin in the full simulation procedure. In the port of Koper, the average tidal range is about 0.5 m. In this case, the minimum mooring UKC at low tide should always be 10 cm. All these aspects were included in the analyses to obtain accurate data for the ship added mass.

One of the general aspects of added mass in relation to UKC can be reduced from the results shown in Figure 21. With a fair degree of confidence, it can be extrapolated to similar scenarios for different ports and a variety of ships with $Cb \approx 0.8$.

The observed added mass is in the range of 100–160% of displacement for MR oil tanker type, 130–200% of displacement for LR1 oil tanker type and 170–260% of displacement for LR2 oil tanker type. As observed, the values of added mass are very high and must always be considered in the loading analysis of flexible dolphins.

Funding: This research was co-funded by ARRS grant number P2-0394.

Data Availability Statement: Complete code project available at Zenodo [43].

Conflicts of Interest: The author declares no conflict of interest.

References

1. Ursell, F. On the heaving motion of a circular cylinder on the surface of a fluid. *Q. J. Mech. Appl. Math.* **1949**, *2*, 218–231. [CrossRef]
2. Ursell, F. On the rolling motion of cylinders in the surface of a fluid. *Q. J. Mech. Appl. Math.* **1949**, *2*, 335–353. [CrossRef]
3. Thorne, R. Multipole expansions in the theory of surface waves. In *Mathematical Proceedings of the Cambridge Philosophical Society*; Cambridge University Press: Cambridge, UK, 1953; Volume 49, pp. 707–716.
4. Ursell, F. *Ship Hydrodynamics, Water Waves, and Asymptotics: Collected Papers of F. Ursell, 1946–1992*; Number v. 1 in Advanced Series on Fluid Mechanics; World Scientific: Singapore, 1994.
5. Frank, W. Oscillation of Cylinders in Or Below the Free Surface of Deep Fluids; Technical Report; David W Taylor Naval Ship Research and Development Center Bethesda Md Dept. 1967. Available online: https://ci.nii.ac.jp/naid/10008438661/ (accessed on 19 December 2020).
6. Newman, J.N. The drift force and moment on ships in waves. *J. Ship Res.* **1967**, *11*, 51–60. [CrossRef]
7. Inglis, R.; Price, W. The influence of speed dependent boundary conditions in three dimensional ship motion problems. *Int. Shipbuild. Prog.* **1981**, *28*, 22–29. [CrossRef]
8. Newman, J.; Sclavounos, P. The Computation of Wave Loads on Large Offshore Structures. The BOSS'88. 1988; pp. 605–622. Available online: http://salsahpc.indiana.edu/dlib/articles/00000714/ (accessed on 19 December 2020).
9. Nakos, D.; Sclavounos, P. Ship Motions by a Three Dimensional Rankine Panel Method. The 18th International Symposium Naval Hydrodynamics. 1990; pp. 21–40. Available online: https://trid.trb.org/view/439613 (accessed on 19 December 2020).
10. Fackrell, S. Study of the Added Mass of Cylinders and Spheres. Ph.D. Thesis, University of Windsor, Windsor, ON, USA, 2011. Available online: https://scholar.uwindsor.ca/cgi/viewcontent.cgi?article=1457&context=etd (accessed on 19 December 2020).
11. Lavrov, A.; Rodrigues, J.; Gadelho, J.; Soares, C.G. Calculation of hydrodynamic coefficients of ship sections in roll motion using Navier-Stokes equations. *Ocean. Eng.* **2017**, *133*, 36–46. [CrossRef]
12. Lewis, F.M. The inertia of the water surrounding a vibrating ship. *SNAME* **1929**, *37*, 1–20.
13. Landweber, L.; de Metcagno, M. Added mass of two-dimensional forms oscillating in a free surface. *J. Ship Res.* **1957**, *1*, 20–30. [CrossRef]
14. Landweber, L.; Macagno, M. Added mass of a three-parameter family of two-dimensional forces oscillating in a free surface. *J. Ship Res.* **1959**, *3*, 36–48. [CrossRef]
15. Grim, O. Berechnung der durch Schwingungen eines Schiffskörpers erzeugten hydrodynamischen Kräfte. *Jahrb. Schiffsbautechnischen Ges.* **1953**, *47*, 277–299.
16. Tasai, F. On the damping force and added mass of ships heaving and pitching. *J. Zosen Kiokai* **1959**, *1959*, 47–56. [CrossRef]
17. Porter, W.R. Pressure Distributions, Added-Mass, and Damping Coefficients for Cylinders Oscillating in a Free Surface. Ph.D. Thesis, University of California, Berkeley, CA, USA, 1961.
18. De Jong, B. Computation of the Hydrodynamic Coefficients of Oscillating Cylinders. TUDelft, Faculty of Marine Technology, Ship Hydromechanics Laboratory, Nederlands Scheeps Studiecentrum TNO, Delft, Report 145S. 1973. Available online: https://repository.tudelft.nl(accessed on 19 December 2020).
19. Athanassoulis, G. An expansion theorem for water-wave potentials. *J. Eng. Math.* **1984**, *18*, 181–194. [CrossRef]

20. Athanassoulis, G.; Loukakis, T.; Lyberopoulos, A. Exciting Forces on Two-Dimensional Bodies of Arbitrary Shape. In *Ocean Space Utilization'85: Proceedings of the International Symposium*; Nihon University: Tokyo, Japan, 1985; Volume 1, p. 95.
21. Athanassoulis, G.; Loukakis, T. An extended-Lewis form family of ship-sections and its applications to seakeeping calculations. *Int. Shipbuild. Prog.* **1985**, 32, 33–43. [CrossRef]
22. Grigoropoulos, G.J.; Loukakis, T. A new Method for Developing Hull Forms with Superior Seakeeping Qualities. Proceedings of CADMO. 1988, p. 1. Available online: https://trid.trb.org/view/434165 (accessed on 19 December 2020).
23. Grigoropoulos, G.; Loukakis, T. On the optimization of hull forms with respect to seakeeping. In Proceedings of the 5th IMAEM Congress, Athens, Greece, 16–17 May 1990.
24. Newman, J.N. *Marine Hydrodynamics*; MIT Press: Cambridge, MA, USA, 1978.
25. Faltinsen, O. *Sea Loads on Ships and Offshore Structures*; Cambridge University Press: Cambridge, UK, 1990.
26. Linton, C.M.; McIver, P. *Handbook of Mathematical Techniques for Wave/Structure Interactions*; Chapman & Hall/CRC: Boca Raton, FL, USA, 2001.
27. Cummins, W.E. The impulse response function and ship motions. *Schiffstechnik* **1962**, 9, 101–109.
28. Ogilvie, T.F. First- and second-order forces on a cylinder submerged under a free surface. *J. Fluid Mech.* **1963**, 16, 451–472. [CrossRef]
29. Salvesen, N.; Tuck, E.; Faltinsen, O. Ship motions and sea loads. *Trans. SNAME* **1970**, 78. Available online: https://trid.trb.org/view/495 (accessed on 19 December 2020).
30. Bailey, P. A unified mathematical model describing the maneuvering of a ship travelling in a seaway. *Trans RINA* **1997**, 140, 131–149.
31. Fossen, T.I. A nonlinear unified state-space model for ship maneuvering and control in a seaway. *Int. J. Bifurc. Chaos* **2005**, 15, 2717–2746. [CrossRef]
32. Batista, M.; Grm, A. Experimental Estimation of Material and Support Properties for Flexible Dolphin Structures. *Our Sea Int. J. Marit. Sci. Technol.* **2017**, 64, 50–53.
33. Vugts, J.H. The hydrodynamic coefficients for swaying, heaving and rolling cylinders in a free surface. *Int. Shipbuild. Prog.* **1968**, 15, 251–276. [CrossRef]
34. Clarke, D. A Two-Dimensional Strip Method for Surface Ship Hull Derivatives: Comparison of theory with Experiments on a Segmented Tanker Model. *J. Mech. Eng. Sci.* **1972**, 14, 53–61. [CrossRef]
35. Needham, T. *Visual Complex Analysis*; Oxford University Press: Oxford, UK, 1998.
36. Cohen, H. *Complex Analysis with Applications in Science and Engineering*; Springer Science & Business Media: Berlin/Heidelberg, Germany, 2007.
37. Da Costa Campos, L.M.B. *Complex Analysis with Applications to Flows and Fields*; CRC Press: Boca Raton, FL, USA, 2010.
38. Ogilvie, T.F.; Tuck, E.O. *A Rational Strip Theory of Ship Motions: Part I*; Technical Report; University of Michigan: Ann Arbor, MI, USA, 1969.
39. Batchelor, G. *An Introduction to Fluid Dynamics*; Cambridge University Press: Cambridge, UK, 2000.
40. Gradstein, Y.; Ryzhik, Y. *Tables of Integrals, Summs, Series and Products*; Elsevier: Amsterdam, The Netherlands, 2015.
41. Foundation, P.S. Python Language. 2020. Available online: https://www.python.org (accessed on 19 December 2020).
42. Team, S.D. SymPy—Python Library for Symbolic Mathematics. 2020. Available online: https://www.sympy.org (accessed on 19 December 2020).
43. Grm, A. Ship Added Mass Calculations Available online: https://zenodo.org/record/4452633#.YAjhHhYRVPY (accessed on 19 December 2020).

Journal of
Marine Science and Engineering

MDPI

Article

A Fuzzy Optimization Model for the Berth Allocation Problem and Quay Crane Allocation Problem (BAP + QCAP) with *n* Quays

Edwar Lujan [1], Edmundo Vergara [1], Jose Rodriguez-Melquiades [1], Miguel Jiménez-Carrión [2], Carlos Sabino-Escobar [3] and Flabio Gutierrez [2,*]

[1] Faculty of Physical and Mathematical Sciences, Universidad Nacional de Trujillo, Trujillo 13011, Peru; elujans@unitru.edu.pe (E.L.); evergara@unitru.edu.pe (E.V.); jrodriguez@unitru.edu.pe (J.R.-M.)
[2] Science Faculty and Faculty of Industrial Engineering, Universidad Nacional de Piura, Piura 20002, Peru; mjimenezc@unp.edu.pe
[3] Faculty of Economic Sciences, Universidad Nacional de Tumbes, Tumbes 24001, Peru; csabinoe@untumbes.edu.pe
* Correspondence: flabio@unp.edu.pe

Abstract: This work introduces a fuzzy optimization model, which solves in an integrated way the berth allocation problem (BAP) and the quay crane allocation problem (QCAP). The problem is solved for multiple quays, considering vessels' imprecise arrival times. The model optimizes the use of the quays. The BAP + QCAP, is a NP-hard (Non-deterministic polynomial-time hardness) combinatorial optimization problem, where the decision to assign available quays for each vessel adds more complexity. The imprecise vessel arrival times and the decision variables—berth and departure times—are represented by triangular fuzzy numbers. The model obtains a robust berthing plan that supports early and late arrivals and also assigns cranes to each berth vessel. The model was implemented in the CPLEX solver (IBM ILOG CPLEX Optimization Studio); obtaining in a short time an optimal solution for very small instances. For medium instances, an undefined behavior was found, where a solution (optimal or not) may be found. For large instances, no solutions were found during the assigned processing time (60 min). Although the model was applied for $n = 2$ quays, it can be adapted to "n" quays. For medium and large instances, the model must be solved with metaheuristics.

Keywords: berth allocation problem; BAP; quay cranes assignment problem; QCAP; fuzzy optimization; fully fuzzy linear programing; combinatorial optimization; robust optimization

Citation: Lujan, E.; Vergara, E.; Rodriguez-Melquiades, J.; Jiménez-Carrión, M.; Sabino-Escobar, C.; Gutierrez, F. A Fuzzy Optimization Model for the Berth Allocation Problem and Quay Crane Allocation Problem (BAP + QCAP) with *n* Quays. *J. Mar. Sci. Eng.* **2021**, *9*, 152. https://doi.org/10.3390/jmse9020152

Academic Editors: Marko Perkovic and Ferrari Claudio
Received: 16 December 2020
Accepted: 27 January 2021
Published: 2 February 2021

1. Introduction

Maritime container terminals (MCTs) are vital elements in the global supply chain. The essential objective of an MCT is to provide the resources and organization to the transport of containers between the landside and maritime mediums. In this process, the MCT must guarantee the best conditions of speed, efficiency, and safety, in accordance with the environment [1].

Due to the current rise in the global maritime trade, many ports have suffered from resource constraints, such as space, time, quay cranes, etc. The problems that exist in an MCT are different. This work addresses the berth allocation problem (BAP) and the quay crane allocation problem (QCAP). The BAP is a NP-hard (Non-deterministic polynomial-time hardness) combinatorial optimization problem [2], which involves assigning each incoming vessel a berth position and arrival time at the quay. On the other hand, the QCAP tries to assign a number of quay cranes (QC) to each berth vessel, aiming to perform all the necessary unload or load movements of the containers in the vehicles. The QC are giant cranes that are mounted on rails, therefore, they cannot break through to each other.

The vessel arrival times are very uncertain, i.e., they can arrive earlier or later than their expected arrival time. This imprecision depends on several factors, such as: technical problems, weather conditions (winds, storms), additional terminal visits, or other reasons [3,4]. This has effects over other MCT activities, such as load and unload operations, negatively impacting the services required by customers. This work assumes the imprecise arrival times of the vessels, where the arrival is known but the exact time is uncertain, as this kind of uncertainty (ambiguity) can be modeled with fuzzy logic [5].

In [6], an exhaustive literature review which show a very limited number of reports, the authors deal with the imprecise times in the BAP + QCAP.

The authors of [7] use discrete event simulation to show that the collaboration between two MCTs (sharing resources such as berths, quay cranes, and the container yard) can help reduce uncertainty in arrival time and the number of containers brought in by vessel, although the authors do not present any mathematical models.

A reactive strategy for the BAP + QCAP with discrete berths is proposed in [8], a mixed integer linear programming (MILP) model with practical constraints, is formulated to obtain a basic planning when uncertainty problems appear (deviation of vessels' arrival times, deviation of vessels' loading and unloading operation times, unscheduled vessel calls, quay breakdowns, etc.), and a moving horizon heuristic is used to obtain good feasible solutions. Another reactive strategy for the BAP + QCAP is proposed in [9], where basic planning is obtained through the use of a multi-objective optimization model that penalizes: the deviations of the vessels from their preferred berthing positions, the delay in the berthing time in comparison to estimated arrival times, and the delay in departure times compared to estimated departure times; then, with movable time windows the berthing plans are updated.

Given that the model we propose is proactive, a review of works on this approach is made below.

Regarding discrete and dynamic BAP, where a single berth quay is available with some kind of imprecision present, a fuzzy mixed integer linear programming (MILP) model was proposed in [10], the imprecise arrival times were represented by triangular fuzzy numbers; however, this model does not addressed continuous BAP. According to [6], the design of a continuous model would have a more complicated berthing plan than a discrete one. Nevertheless, a continuous model has the advantage of better using the quay space. In [11], a fully fuzzy linear programing (FFLP) model was developed for dynamic and continuous BAP. The model obtained a robust berthing plan, which supports incidents in the vessel arrivals. In [12], a MILP model and a genetic algorithm (GA) were created for dynamic and continuous BAP+QCAP. In the modeling methodology, spaces of times were added in the vessel departure, which mitigates the effects of imprecision and strengthens the model's accuracy.

In the case of BAPs, where vessels can berth in two quays; [13] proposed a MILP model and a GA. However, imprecision was not included in the problem parameters. On the other hand, a FFLP model for BAP which addressed the imprecision in the vessel arrivals was presented in [14]. A MILP model for integrated laycan and berth allocation and quay crane assignment problem (LBACAP) that considers multiple quays is proposed in [15].

As far as we know, no models have been developed for the BAP + QCAP which considers "n" quays and the uncertainty in the arrival times.

In this work, we present a fuzzy optimization model for the BAP + QCAP with two quays, continuous and dynamic, which considers the vessels' imprecise arrival times, meaning they can arrive early or late of an allowed time. The vessels' imprecise arrival times are represented by triangular fuzzy numbers. The model optimizes the use of the quays. In order to analyze the behavior and efficiency, the model is applied to a small, medium, and large instances respectively.

2. Materials and Methods

2.1. Fuzzy Arithmetic

Definition 1. *Let X be the discourse universe, and a fuzzy set $\widetilde{A}$ in X is a set of ordered pairs.*

$$\widetilde{A} = \{(x, \mu_{\widetilde{A}}(x)), x \in X\} \tag{1}$$

where $\mu_{\widetilde{A}} : X \to [0,1]$ is the membership function, which represents the belonging degree of x with the set $\widetilde{A}$.

In this work, fuzzy sets are defined over the real numbers $\mathbb{R}$. A membership function can be triangular, trapezoidal, sigmoidal, quadratic, etc.

Definition 2. *A fuzzy singleton is a fuzzy set with a membership function.*

$$\mu_{\widetilde{A}}(x) = \begin{cases} 1, & x \in [a-\varepsilon, a+\varepsilon] \\ 0, & x \notin [a-\varepsilon, a+\varepsilon] \end{cases} \tag{2}$$

where ε is a margin of error. The fuzzy singleton allows the expression of a real number as a fuzzy set.

Definition 3. *A fuzzy number is a normal and convex fuzzy set in $\mathbb{R}$.*

Definition 4. *A triangular fuzzy number is represented by $\widetilde{a} = (a1, a2, a3)$.*

Definition 5. *A triangular fuzzy number $\widetilde{a} = (a1, a2, a3)$ is positive if and only if $a1 > 0$.*

Definition 6. *A real number a can be represented by a fuzzy singleton in the form of a triangular fuzzy number $\widetilde{a} = (a - \varepsilon, a, a + \varepsilon)$*, where ε is a margin of error.

Definition 7. *For two non-negative triangular fuzzy numbers $\widetilde{a} = (a1, a2, a3)$ and $\widetilde{b} = (b1, b2, b3)$*, the sum difference and multiplication operations are defined as follows:

$$\begin{aligned} \widetilde{a} + \widetilde{b} &= (a1 + b1, a2 + b2, a3 + b3) \\ \widetilde{a} - \widetilde{b} &= (a1 - b3, a2 - b2, a3 - b1) \\ \widetilde{a}.\widetilde{b} &= (a1.b1, a2.b2, a3.b3). \end{aligned} \tag{3}$$

Ordering methods allows us to decide the greater number between two fuzzy numbers $\widetilde{a}$ and $\widetilde{b}$. Fuzzy numbers do not always provide an ordered set as found in real numbers. Ordering methods in fuzzy numbers have advantages and disadvantages. There are different ordering methods, depending on the representation: preference, rationality, and robustness, etc. [16].

In this work, the objective of the fuzzy model optimization is to achieve an ordering (planning) for the berthing of vessels, therefore, any ordering method that orders fuzzy numbers could be used. Ordering methods that use intervals to order fuzzy numbers are not recommended, as they could add more imprecision to the model.

This work utilizes the Yagger First Index [17] ordering method, which is defined below.

Definition 8. *Given two triangular fuzzy numbers $\widetilde{a}$ and $\widetilde{b}$ alongside the ordering function,*

$$\Re(\widetilde{a}) = \frac{a1 + a2 + a3}{3} \tag{4}$$

$\widetilde{a} \leq \widetilde{b}$ when, $\Re(\widetilde{a}) \leq \Re(\widetilde{b})$, Meaning, $\widetilde{a} \leq \widetilde{b}$ when $a1 + a2 + a3 \leq b1 + b2 + b3$.

2.2. Fully Fuzzy Linear Programming (FFLP)

There are different approaches in fuzzy mathematical programming, the classical fuzzy linear programming methods are used when the parameters are fuzzy; in this work, some of the decision variables are fuzzy, for this case, the FFLP approach is the most appropriate.

In the FFLP approach, the decision parameters and variables are fuzzy and linear, respectively. Several solution methods can be applied to the FFLP model [18]. Most of them convert the fuzzy model into a classic linear programming one. In this work, we use the method proposed by Nasseri [19].

Given the FFLP problem:

$$max \sum_{j=1}^{n} \widetilde{C}_j \widetilde{X}_j \tag{5}$$

Subject to:

$$\sum_{j=1}^{n} \widetilde{a}_{ij}\widetilde{x}_j \leq \widetilde{b}_i \, , \, \forall i = 1 \ldots m \tag{6}$$

where the parameters $\widetilde{C}_j$, $\widetilde{a}_{ij}$, $\widetilde{b}_j$ and the decision variable $\widetilde{x}_j$ are non-negative triangular fuzzy numbers, $\widetilde{C}_j = (c1_j, c2_j, c3_j)$, $\widetilde{a}_{ij} = (a1_{ij}, a2_{ij}, a3_{ij})$, $\widetilde{b}_j = (b1_i, b2_i, b3_i)$ y $\widetilde{x}_j = (x1_j, x2_j, x3_j)$.

The Nasseri method transforms the previous model into a classic linear programming problem,

$$\max \Re(\sum_{j=1}^{n} (c1_j, c2_j, c3_j) \; (x1_j, x2_j, x3_j)) \tag{7}$$

Subject to:

$$\sum_{j=1}^{n} a1_{ij}x1_j \leq b1_i \, , \, \forall i = 1 \ldots m \tag{8}$$

$$\sum_{j=1}^{n} a2_{ij}x2_j \leq b2_i \, , \, \forall i = 1 \ldots m \tag{9}$$

$$\sum_{j=1}^{n} a3_{ij}x3_j \leq b3_i \, , \, \forall i = 1 \ldots m \tag{10}$$

where $\Re$ is an ordering function (see Definition 8).

2.3. Problem Description

According to [20], an MCT is a composition of several subsystems integrated into a single one. This system has physical and information connections with the transport networks (landside and maritime). The subsystems (see Figure 1) are:

1. The loading–unloading container subsystem, which is responsible for resolving the maritime interface.
2. The storage container subsystem, which occupies most of the MCT surface.
3. The landside reception and delivery subsystem, which act as gates in the landside for trucks and/or railways.
4. The internal connection subsystem. To the previous three subsystems, which address the basic terminal functions, a fourth subsystem must be added, this ensures the horizontal transport of containers between the previous subsystems.

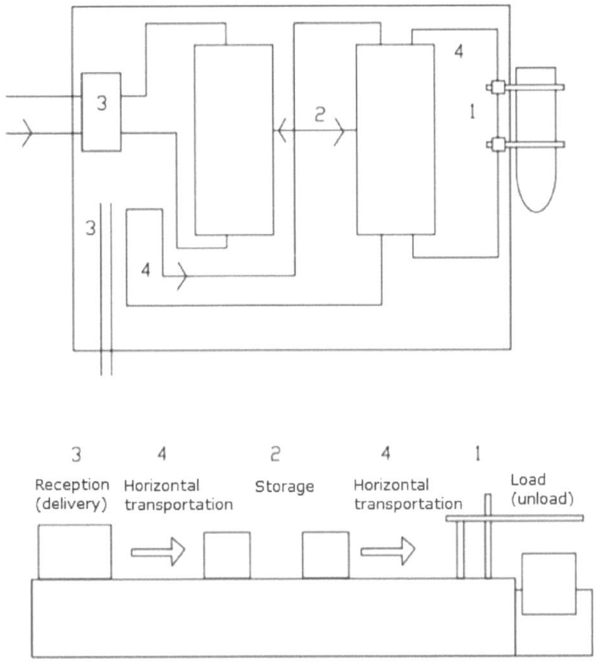

Figure 1. MCT subsystems in plant (**top** image) and elevation (**bottom** image) [20].

The BAP + QCAP discussed in this work occur in subsystem 1, however any incident will affect the other subsystems. For example, if a vessel berths outside their scheduled time, it will cause problems to the assigned cranes, affecting the prepared warehouses for the containers and the trucks waiting to receive them.

The existence of n quays in the port is assumed, where vessels can berth in any of them. Arrival time is imprecise, i.e., vessels are allowed to be early or late until a predetermined time. The BAP focuses on choosing a quay (if any) and the arrival time and position where each arriving vessel at the MCT must berth. The other problem is the QCAP, which involves the assigning of a number of quay cranes to each vessel to be handled.

The BAP can be represented in a two-dimensional form (see Figure 2), where the horizontal axis represents the time and the vertical axis represents the berth length.

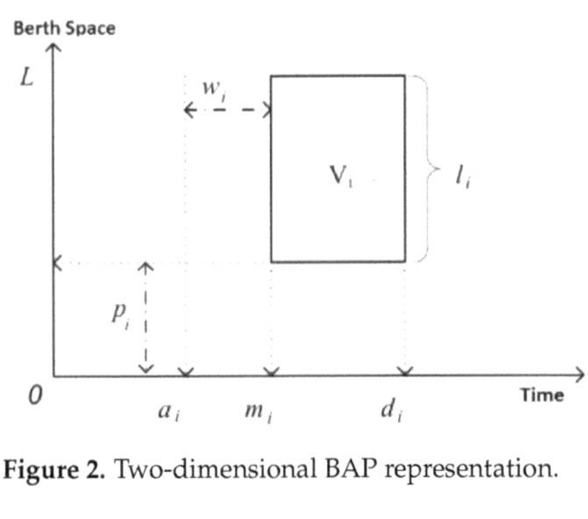

Figure 2. Two-dimensional BAP representation.

We consider the following assumptions and limitations:

Assumptions: The vessel information is previously known, each vessel has a draft less or equal than the quay, and that the berthing and departure do not consume much time. Simultaneous berthing is allowed and the safety distance between vessels is not considered. The number of cranes assigned to the vessels does not change during the berthing time, once a QC starts a task on a vessel they must complete it without any pause, all QCs assigned to vessel i have the same working time ($t_{ik} = h_i,\ \forall k \in QC,\ u_{ik} = 1$). The model assigns the berthing place to each vessel, that is, preference zones are not allowed for berthing a vessel.

Limitations: The quay length must be limited, the available crane number in each quay is five. There is a safety distance between cranes which must to be maintained (35 m). The maximum crane number that can be assigned to a vessel is four. The number of crane movements performed in a given time is 2.5. There must be at least one QC assigned to each vessel.

In this work we use the following notation, which was taken from [12]:

H: Planning Horizon.

Let be Q the set of existent quays in an MCT, where $q \in Q$ is a quay.

QC_q: Available cranes at quay q. All QCs perform the same number of movements per unit time ($movsQC$), data provided by the MCT.
QC^+_{iq} : The maximum number of QCs assigned to each vessel i at quay q
L_q: Length of quay q.

Let be V a set of vessels which arrive at the port, the data for each vessel $i \in V$ is given by:

a_i: Vessel arrival time at port.
l_i: Vessel length.
c_i: Required number of movements to load or unload the containers from the vessel.

The decision variables are:

BM_{iq}: Takes the value of 1 when vessel i berths at quay q, and takes 0 otherwise.
m_{iq}: Berthing time of vessel i at quay q. The waiting time (w_i) is computed as $(w_i = m_{iq} - a_i)$
p_{iq}: Position at quay q, where vessel i must berth.
n_{iq}: Number of QCs in quay q assigned to vessel i.
u_{ik}: Indicates whether k, (which belonging to QC_q) works (1) or not (0) on vessel i.

The variables that are deduced from above are:

h_i: Vessel handling time i. $h_i = c_i/(n_{iq} * movsQC)$.
t_{ik}: Working time of k (belonging to QC_q) which is assigned to vessel i.
d_i: Vessel departure time $(d_i = m_{iq} + h_i)$.
s_{iq} ; e_{iq}: Indices for the first and last QC, on quay q, used in vessel i, respectively.
M: Is a sufficiently large number.

2.4. Fuzzy Optimization Model for the BAP + QCAP with Two Quays

The arrival, berthing, handling, waiting and departure time of each vessel are considered imprecise (fuzzy), they are denoted by: $\widetilde{a}$, $\widetilde{m}$, $\widetilde{h}$, $\widetilde{w}$ and $\widetilde{d}$ respectively.

Assuming the presence of imprecision in some parameters and decision variables, the fuzzy optimization model for BAP + QCAP is introduced. This model is based on the deterministic model developed in [12]. The objective function minimizes the waiting and handling time for all vessels.

$$min \sum_{i \in V} \left(\widetilde{w}_i + \widetilde{h}_i \right) \tag{11}$$

Subject to:

$$\textstyle\sum_{q \in Q} BM_{iq} = 1 \qquad , \ \forall i \in V , \forall q \in Q \tag{12}$$

$$\widetilde{m}_{iq} \geq \widetilde{a}_i \quad , \forall i \in V, \forall q \in Q \tag{13}$$

$$\widetilde{w}_i + \widetilde{a}_i = \widetilde{m}_{iq} \quad , \forall i \in V, \forall q \in Q \tag{14}$$

$$\widetilde{d}_i = \widetilde{m}_{iq} + \widetilde{h}_i \quad , \forall i \in V, \forall q \in Q \tag{15}$$

$$p_{iq} + l_i \leq L_q \quad , \forall i \in V, \forall q \in Q \tag{16}$$

$$n_{iq} = \sum_{k \in QC_q} u_{ik} \quad , \forall i \in V, \forall k \in QC_q, \forall q \in Q \tag{17}$$

$$1 \leq n_{iq} \leq QC_{iq}^{+} \quad , \forall i \in V, \forall q \in Q \tag{18}$$

$$1 \leq s_{iq}, e_{iq} \leq |QC_q| \quad , \forall i \in V, \forall q \in Q \tag{19}$$

$$s_{iq} \geq e_{iq} \quad , \forall i \in V, \forall q \in Q \tag{20}$$

$$n_{iq} = e_{iq} - s_{iq} + 1 \quad , \forall i \in V, \forall q \in Q \tag{21}$$

$$\sum_{k \in QC_q} t_{ink} * movQC \geq c_i \quad , \forall i \in V, \forall q \in Q \tag{22}$$

$$\widetilde{h}_i = max_{k \in QC_q} t_{ik} \quad , \forall i \in V, \forall q \in Q \tag{23}$$

$$t_{ik} - M * u_{ik} \leq 0 \quad , \forall i \in V, \forall k \in QC_q, \forall q \in Q \tag{24}$$

$$\widetilde{h}_i - M(1 - u_{ik}) - t_{ik} \leq 0 \ , \forall i \in V, \forall k \in QC_q, \forall q \in Q \tag{25}$$

$$u_{ik} + u_{jk} + z^x_{ijq} < 2 \quad , \forall i,j \in V, i \neq j, \forall k \in QC_q, \forall q \in Q \tag{26}$$

$$M(1 - u_{ik}) + (e_{iq} - k) \geq 0 \quad , \forall i \in V, \forall k \in QC_q, \forall q \in Q \tag{27}$$

$$M(1 - u_{ik}) + (k - s_{iq}) \geq 0 \quad , \forall i \in V, \forall k \in QC_q, \forall q \in Q \tag{28}$$

$$p_{iq} + l_i \leq p_{jq} + M\left(1 - z^x_{ij}\right) \quad , \forall i,j \in V, i \neq j, \forall q \in Q \tag{29}$$

$$e_{iq} + 1 \leq s_{jq} + M\left(1 - z^x_{ijq}\right) \quad , \forall i \in V, i \neq j, \forall q \in Q \tag{30}$$

$$\widetilde{d}_i \leq \widetilde{m}_{jq} + M\left(1 - z^y_{ijq}\right) \quad , \forall i \in V, i \neq j, \forall q \in Q \tag{31}$$

$$\widetilde{m}_{iq} + \widetilde{h}_i \leq H \quad , \forall i \in V, \forall q \in Q \tag{32}$$

$$z^x_{ijq} + z^x_{jiq} + z^y_{ijq} + z^y_{jiq} \geq 1 \quad , \forall i,j \in V, i \neq j, \forall q \in Q \tag{33}$$

$$z^x_{ijq}, z^y_{ijq} \in \{0,1\} \quad , \forall i,j \in V, i \neq j, \forall q \in Q \tag{34}$$

There are two auxiliary variables: z^x_{ijq} is a decision variable that indicates whether the vessel i is located to the left of vessel j at quay ($z^x_{ijq} = 1$), while z^y_{ijq} indicates that vessel i berths before vessel j in time, at quay q ($z^x_{ijq} = 1$) (see restriction 34).

Constraint (12), assigns each vessel to a quay q. Constrain (13), indicates that all vessels can berth once they arrive at the port. Constraints (14) and (15) set the waiting and departure times for vessel i, according with the berthing time. Constraint (16) ensures the berth position of vessel i does not exceed the length of the quay q.

Constraints (17)–(21) assign a number of working QCs at quay q for vessel i. Constraint (22) sets the minimum handling time required to load or unload containers, according with the assigned number of QCs to vessel i. Constraint (23) assigns the handling time to vessel i. Constraint (24) ensures that unassigned QCs have a value of $t_{ik} = 0$. Constraint (25) forces the assigned QCs in vessel i to work the same number of hours. Constraint (26) prevents a QC from being assigned to different vessels at the same time. Constraints (27) and (28) force QCs to be contiguously assigned (from s_i to e_i) at quay q. Constraint (29) takes into account the safety distance. Constraint (30) prevents a vessel from using a QC which breaks through other QCs. Constraint (31) prevents vessel j from berthing at the quay, while vessel i is still berthed at the same quay. Constraint (32) indicates that the berthing and handling time for vessel i must not exceed the planning horizon (H). Finally, constraint (33) establishes the relationship between each pair of vessels.

2.5. Fuzzy Optimization Model Solution

It is assumed that, in the fuzzy optimization model from Section 2.4, the imprecision times related with the vessel operations: arrival, berthing, handling, and departure, are represented by the triangular fuzzy numbers $\widetilde{a} = (a1, a2, a3)$, $\widetilde{m} = (m1,\ m2,\ m3)$, $\widetilde{h} = (h1, h2, h3)$, and $\widetilde{d} = (d1, d2, d3)$, respectively. $\widetilde{h}$, is considered a fuzzy singleton; also, if the model parameters are linear, the model is a FFLP.

The method used to transform the FFLP model into a classic linear programming model requires the application of fuzzy arithmetic operations; restrictions (25), (31), and (32) show such operations between fuzzy and real numbers, in order to perform such operations, real numbers are considered to be fuzzy singletons (See Definition 6), for example, in constraint (31),

$$\widetilde{d}_i \le \widetilde{m}_{jq} + M\left(1 - z^y_{ijq}\right)$$

Is transformed into,

$$(d1_i, d2_i, d3_i) \le \left(m1_{jq},\ m2_{jq}, m3_{jq}\right) + M\left((1,1,1) - \left(z^y_{ijq},\ z^y_{ijq},\ z^y_{ijq}\right)\right)$$

Simplifying,

$$(d1_i, d2_i, d3_i) \le \left(m1_{jq} + M\left(1 - z^y_{ijq}\right),\ m2_{jq} + \left(1 - z^y_{ijq}\right),\ m3_{jq} + \left(1 - z^y_{ijq}\right)\right)$$

As indicated in the Nasseri method (see Section 2.2), operations between fuzzy numbers which involve sum, difference, and multiplication operations, are performed in the FFLP model; an ordering method is applied to the objective function, in this case, the First Yagger Index is used (see Definition 8), obtaining the following MILP model.

$$min \sum_{i\in V} \frac{1}{3}\left(\left(m1_{iq} - a3_i + h1_i\right) + \left(m2_{iq} - a2_i + h2_i\right) + \left(m3_{iq} - a1_i + h3_i\right)\right) \quad (35)$$

Subject to:

$$\textstyle\sum_{q\in Q} BM_{iq} = 1 \quad , \forall i \in V , \forall q \in Q \quad (36)$$

$$m1_{iq} \ge a1_i\ ,\ m2_{iq} \ge a2_i\ , m3_{iq} \ge a3_i \quad , \forall i \in V,\ \forall q \in Q \quad (37)$$

$$m3_{iq} \ge m2_{iq}\ ,\ m2_{iq} \ge m1_{iq} \quad , \forall i \in V,\ \forall q \in Q \quad (38)$$

$$w1_i + a1_i = m1_{iq},\ w2_i + a2_i = m2_{iq}\ , \quad w3_i + a3_i = m3_{iq} \quad , \forall i \in V,\ \forall q \in Q \quad (39)$$

$$d1_i = m1_{iq} + h1_i\ ,\ d2_i = m2_{iq} + h2_i\ ,\ d3_i = m3_{iq} + h3_i \quad , \forall i \in V,\ \forall q \in Q \quad (40)$$

$$p_{iq} + l_i \le L_q \quad , \forall i \in V,\ \forall q \in Q \quad (41)$$

$$n_{iq} = \sum_{k\in QC_q} \widetilde{u}_{ik} \quad , \forall i \in V,\ \forall k \in QC_q,\ \forall q \in Q \quad (42)$$

$$1 \le n_{iq} \le QC^+_{iq} \quad , \forall i \in V,\ \forall q \in Q \quad (43)$$

$$1 \le s_{iq}\ ,\ e_{iq} \le \left|QC_q\right| \quad , \forall i \in V,\ \forall q \in Q \quad (44)$$

$$s_{iq} \ge e_{iq} \quad , \forall i \in V,\ \forall q \in Q \quad (45)$$

$$n_{iq} = \widetilde{e}_{iq} - \widetilde{s}_{iq} + 1 \quad , \forall i \in V,\ \forall q \in Q \quad (46)$$

$$\sum_{k\in QC_q} t_{ink} * movQC \ge c_i \quad , \forall i \in V,\ \forall q \in Q \quad (47)$$

$$h1_i = max_{k \in QC_q} t_{ik} \ , \ h2_i = max_{k \in QC_q} t_{ik} \ , h3_i = max_{k \in QC_q} t_{ik} \quad , \forall i \in V, \ \forall q \in Q \tag{48}$$

$$t_{ik} - M * u_{ik} \le 0 \qquad , \ \forall i \in V, \ \forall k \in QC_q, \ \forall q \in Q \tag{49}$$

$$h1_i - M(1 - u_{ik}) - t_{ik} \le 0 \quad , \ \forall i \in V, \ \forall k \in QC_q, \ \forall q \in Q \tag{50}$$

$$h2_i - M(1 - u_{ik}) - t_{ik} \le 0 \quad , \ \forall i \in V, \ \forall k \in QC_q, \ \forall q \in Q \tag{51}$$

$$h3_i - M(1 - u_{ik}) - t_{ik} \le 0 \quad , \ \forall i \in V, \ \forall k \in QC_q, \ \forall q \in Q \tag{52}$$

$$u_{ik} + u_{jk} + z^x_{ijq} < 2 \qquad , \ \forall i,j \in V \ , \ i \neq j, \ \forall k \in QC_q, \ \forall q \in Q \tag{53}$$

$$M(1 - u_{ik}) + (e_{iq} - k) \ge 0 \quad , \ \forall i \in V, \ \forall k \in QC_q, \ \forall q \in Q \tag{54}$$

$$M(1 - u_{ik}) + (k - s_{iq}) \ge 0 \quad , \ \forall i \in V, \ \forall k \in QC_q, \ \forall q \in Q \tag{55}$$

$$p_{iq} + l_i \le p_{jq} + M\left(1 - z^x_{ij}\right) \quad , \ \forall i,j \in V, \ i \neq j, \ \forall q \in Q \tag{56}$$

$$e_{iq} + 1 \le s_{jq} + M\left(1 - z^x_{ijq}\right) \quad , \ \forall i \in V, \ i \neq j, \ \forall q \in Q \tag{57}$$

$$d1_i \le m1_{jq} + M\left(1 - z^y_{ijq}\right) \quad , \ \forall i \in V, \ i \neq j, \ \forall q \in Q \tag{58}$$

$$d2_i \le m2_{jq} + M\left(1 - z^y_{ijq}\right) \quad , \ \forall i \in V, \ i \neq j, \ \forall q \in Q \tag{59}$$

$$d3_i \le m3_{jq} + M\left(1 - z^y_{ijq}\right) \quad , \ \forall i \in V, \ i \neq j, \ \forall q \in Q \tag{60}$$

$$m1_{iq} + h1_i \le H \ , \ m2_{iq} + h2_i \le H \ , \ m3_{iq} + h3_i \le H \quad , \forall i \in V, \ \forall q \in Q \tag{61}$$

$$z^x_{ijq} + z^x_{jiq} + z^y_{ijq} + z^y_{jiq} \ge 1 \quad , \ \forall i,j \in V, \ i \neq j, \ \forall q \in Q \tag{62}$$

$$z^x_{ijq}, \ z^y_{ijq} \in \{0,1\} \qquad , \ \forall i,j \in V, \ i \neq j, \ \forall q \in Q \tag{63}$$

3. Results

For both the case study and the model evaluation efficiency a set of uniform distributed instances were used. The model was implemented with the IBM ILOG CPLEX Optimization Studio (CPLEX) solver and was programmed on a personal computer: Intel(R) Core (TM) i7-8550U CPU, 1.80 GHz processor and 8 GB of RAM. The experiments were conducted within a 60 min timeout.

3.1. Study Case

To evaluate the model, an instance of ten vessels was used as a case study (see Table 1); the input data are the vessel imprecise arrival times (early, exact, late), its length, and the number of crane movements required to handle it. Figure 3 displays the vessels' uncertain arrivals from Table 1, represented as triangular fuzzy numbers.

Table 1. Instance of ten vessels with imprecise arrival times.

Vessel	a1	a2	a3	l (m)	QC Mov.
V1	14	16	20	260	4160
V2	18	31	48	232	9680
V3	56	68	86	139	3640
V4	81	82	97	193	7610
V5	92	105	119	287	6860
V6	106	116	133	318	6300
V7	126	138	147	366	8110
V8	155	167	176	166	1560
V9	159	163	164	109	7830
V10	162	179	186	251	2220

Table 2. The fuzzy berthing plan and crane allocation obtained by the model.

Vessel	m1	m 2	m 3	h	d1	d2	d3	p	Cranes	Q
V1	14	16	20	694	708	710	714	440	2	0
V2	927	928	943	1076	2003	2004	2019	0	3	1
V3	56	68	86	607	663	675	693	193	2	1
V4	81	82	97	846	927	928	943	0	3	1
V5	583	600	607	763	1346	1363	1370	0	3	0
V6	708	710	714	1050	1758	1760	1764	382	2	0
V7	663	675	693	1352	2015	2027	2045	334	2	1
V8	409	426	433	174	583	600	607	0	3	0
V9	1346	1363	1370	870	2216	2233	2240	0	3	0
V10	162	179	186	247	409	426	433	189	3	0

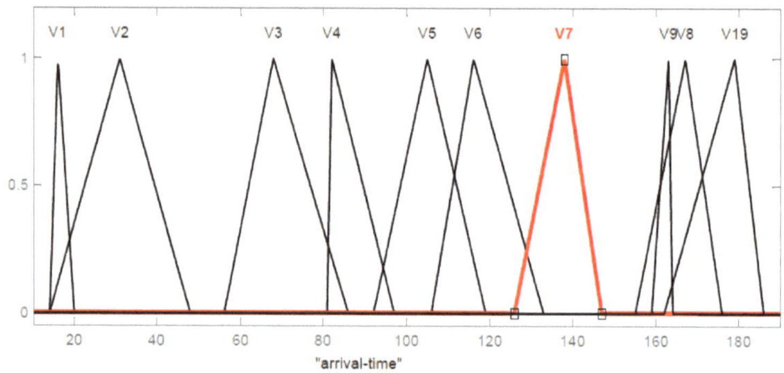

Figure 3. Imprecise vessel arrivals (See Table 2).

For example, for vessel V7, the arrival time with all its possibilities gives us 138 time units, but it can also arrive early or late by up to 126 and 147 time units, respectively. The vessel length is 366 m and 8110 crane movements are required to handling it.

The model assumes that: two quays are used; each having five cranes, with a minimum of one crane operating on each vessel to a maximum of four cranes per vessel. Each quay has a length of 700 m.

The obtained results are shown in Table 2. Within an hour of computing time, an objective value of 15,570 was obtained, this is not the optimal value; but the best value obtained within that time.

Berthing time (m1, m2, m3) and output time (d1, d2, d3) are triangular fuzzy numbers. Q = 1 refers to quay 1, and Q = 0 to quay 2.

For example, vessel V7 can berth between the time units 663 and 693, with more possibility at time unit 667. It can depart between units 2015 and 2045, with more possibility at unit 2027. Additionally, it must berth at position 334 of the quay. Finally, two cranes are assigned to this vessel, and they must berth at quay 1.

The fuzzy berthing plan is displayed in Figure 4. Vessels are represented as polygons (not as rectangles as in the deterministic problem). The red polygon line represents the allowed early time which can be tolerated for the vessel to berth; meanwhile the green line is the tolerated late arrival time. The small triangle represents the berthing with more possibility.

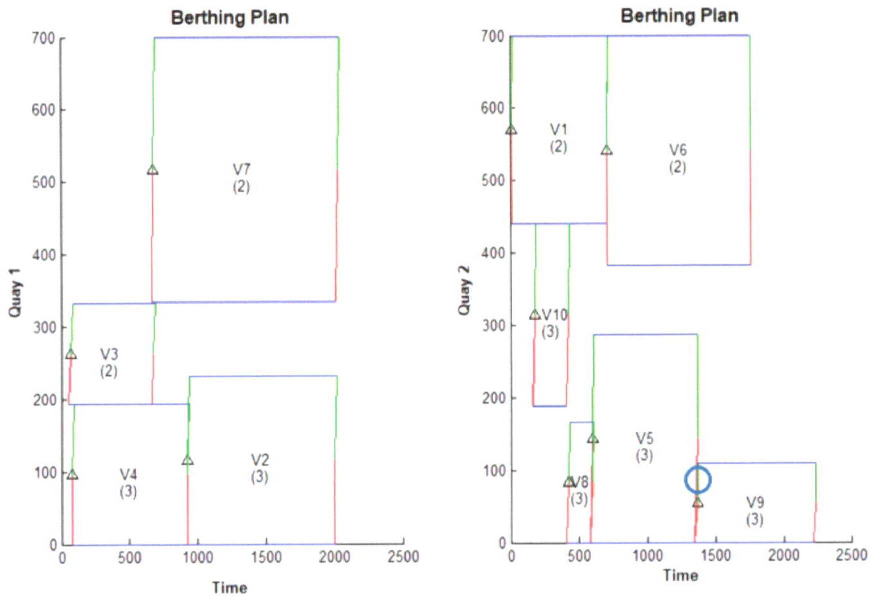

Figure 4. Fuzzy berthing plan for two quays (See Table 2).

In Figure 4, in quay 2, the blue circle suggests a conflict area between the departure and berthing of vessels V5 and V9. However as explained in Figure 5, such conflict is not real. For example, if vessel V5 departures late from the quay at time unit 1365, vessel V9 could berthing between the 1365 and 1370 units.

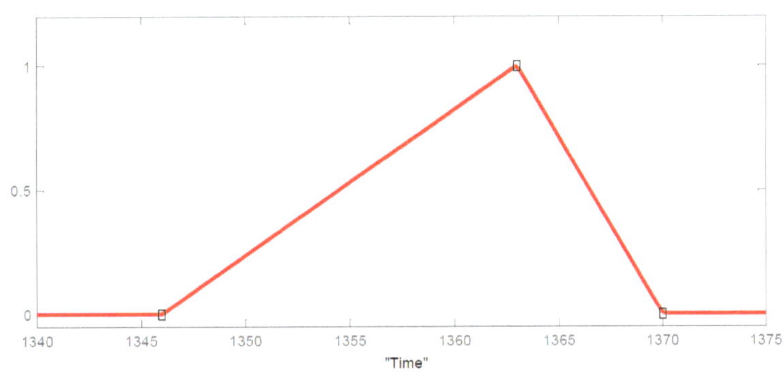

Figure 5. Fuzzy triangular number representing the imprecise departure and berthing times of vessels V5 and V9 (See Table 2).

In order to verify the fuzzy model robustness, incidences were simulated in the vessel arrivals (see Table 3). With these incidences, the final berthing plan is obtained (see Table 4). Figure 6 shows that the final berthing plan is part of the fuzzy berthing plan.

Table 3. Arrival incidence in vessels.

Vessel	Incidence	Time
V1	Exactly	0
V2	Exactly	0
V3	Early	6
V4	Late	10
V5	Exactly	0
V6	Late	15
V7	Exactly	0
V8	Late	8
V9	Early	3
V10	Exactly	0

Table 4. Final berthing plan.

Vessel	m	l	h	d	p	Cranes	Q
V1	16	260	694	710	440	2	0
V2	938	232	1076	2014	0	3	1
V3	62	139	607	669	193	2	1
V4	92	193	846	938	0	3	1
V5	583	287	763	1346	0	3	0
V6	710	318	1050	1760	382	2	0
V7	669	366	1352	2021	334	2	1
V8	409	166	174	583	0	3	0
V9	1346	109	870	2216	0	3	0
V10	179	251	247	426	189	3	0

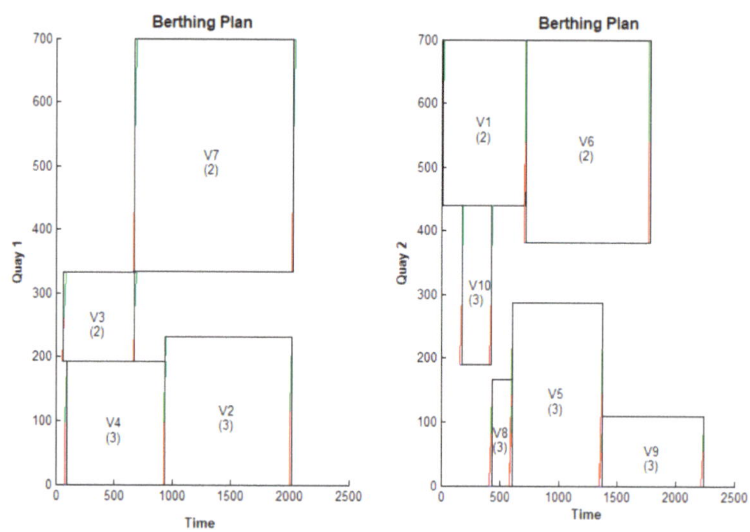

Figure 6. Final berthing plan included in the fuzzy plan (See Table 4).

3.2. Model Efficiency Analysis

In order to analyze the model efficiency, data from vessels 5 to 35 with 100 instances each were used. The results are shown in Table 5.

Table 5. Instances evaluation results.

Vessels	Average Objective Value	Average Processing Time (Minutes)	Optimal	No Optimal
5	4922.0	0.2	100	0
6	6655.3	1.3	47	53
7	9401.3	60.0	0	1
8	10,499.7	60.0	0	1
9	13,724.7	60.0	0	1
10	15,522.3	60.0	0	1
11	19,741.3	60.0	0	1
12	23,714.3	60.0	0	1
13	28,762.3	28.9	2	98
14	36,194.7	19.0	3	97
15	39,153.0	38.8	1	99
16	42,786.3	30.8	2	98
17	49,753.0	25.0	2	98
18	57,623.7	28.5	2	98
19	68,661.0	24.7	2	98
20	71,727.3	22.8	2	98
21	80,968.3	21.2	2	98
22	92,723.3	22.1	2	98
23	88,050.0	60.0	0	1
24	96,369.0	60.0	0	1
25	110,842.3	60.0	0	1
26	108,655.3	60.0	0	1
27	128,984.3	60.0	0	1
28	116,706.3	60.0	0	1
29	172,058.3	39.7	1	99
30	142,758.3	60.0	0	1
31	158,178.0	60.0	0	1
32	177,955.3	60.0	0	1
33	156,638.7	60.0	0	1
34	200,806.0	60.0	0	1
35	-	60.0	0	0

Optimal values were found for five vessels, with an average objective value and processing time of 4922 and 0.2 min, respectively; this was the unique number of vessels for which an optimal solution was obtained in all its instances. For six vessels, an average objective value of 6655.3 was obtained within an average processing time of 1.3 min, where a total of 47 instances were optimally solved. No optimal solutions were found for 7 to 12 vessels. Instead, a single non-optimal solution was found in just one instance. For 13 to 22 vessels there were between one and three optimal solutions. No optimal solutions were found for 23 to 34 vessels, rather, just a best solution in the given processing time. However, for 29 vessels an optimal solution was obtained. For 35 vessels onwards, no solution was found.

Regarding the processing time (see Figure 7), is noted that for six vessels, the average time it took to find a solution was 1.3 min. Meanwhile, for seven vessels the average processing time drastically increases until it matches the allowed processing time (60 min). For 13 to 22 vessels the time varies between 19 and 38.8 min, respectively. For 23 vessels onwards, the time processing was 60 min, except for 29 vessels, which takes 39.7 min.

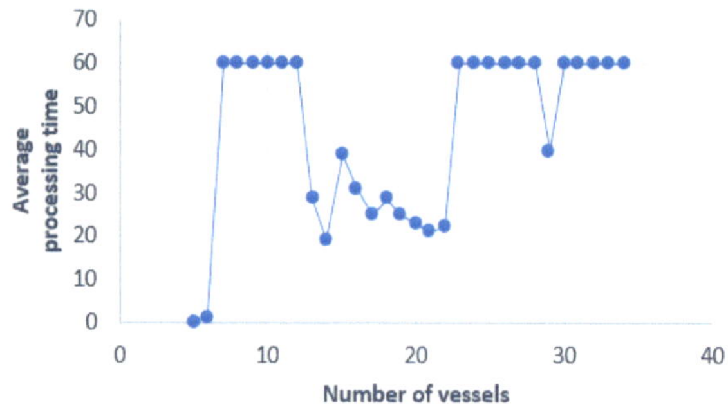

Figure 7. Average processing time trend.

Figure 8 shows a trend of increase in the target value, alongside, the polynomial curve which best adjusts the data; this is a quadratic curve, with a determination coefficient R^2 of 0.9775.

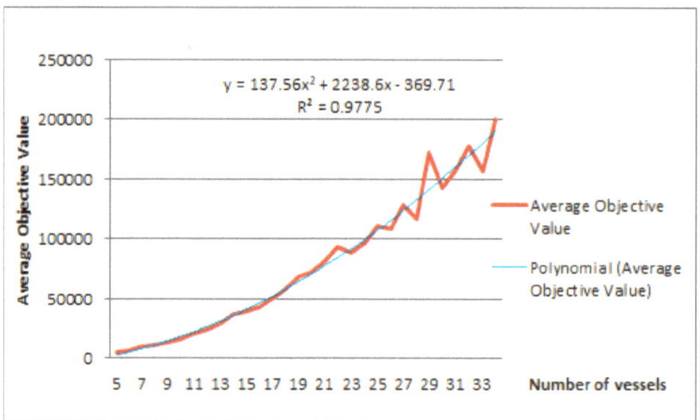

Figure 8. Average objective value trend.

4. Discussion

A fuzzy optimization model for the BAP + QCAP with two quays was developed, considering the vessels' imprecise arrivals. The model was implemented in the CPLEX solver. For previously known input data from a set of vessels, the model obtained a fuzzy berthing and crane allocation plan, which supported incidences of vessels being early or late in their arrival times.

The model efficiency was evaluated with a benchmark of 100 instances for each number of vessels, with one hour of computing processing. Only for a very small number of instances (five vessels) were optimal solutions obtained in all instances. For 6 to 34 vessels, only some optimal and non-optimal solutions were obtained: these do not follow a defined behavior regarding the processing time, this is because the problem is NP-hard and solutions will only be obtained in medium instances that have a structure that allows the algorithm used by the Solver CPLEX to find a solution. For 35 vessels onwards, the CPLEX solver was unable to find solutions within the allowed processing time. The same results applied for large instances.

The model was designed for "n" quays, but in this work is applied to only $n = 2$. Each time the quay number increases, the complexity will increase as well.

For medium and large instances, the model must be solved with metaheuristics, because it is a highly complex combinatorial optimization problem.

Author Contributions: Conceptualization, E.L. and F.G.; methodology, C.S.-E.; software, E.L.; validation, M.J.-C., J.R.-M. and C.S.-E.; formal analysis, F.G.; investigation, E.L.; resources, M.J.-C.; data curation, F.G.; writing—original draft preparation, C.S.-E.; writing—review and editing, F.G.; visualization, J.R.-M.; supervision, E.V.; project administration, E.V.; funding acquisition, E.V. All authors have read and agreed to the published version of the manuscript.

Funding: This research was funded by INNOVATE-PERU, grant number Project N PIBA-2-P-069-14.

Conflicts of Interest: Declare conflicts of interest or state—The authors declare no conflict of interest.

References

1. CCIM (Maritime Industry Knowledge Center) "PORTS AND TERMINALS". Available online: https://www.maritimeinfo.org/es/Maritime-Directory/ports-and-terminals-es-21f71f7e802b11e2bf310013721274c6 (accessed on 17 June 2019).
2. Lim, A. The berth planning problem. *Oper. Res. Lett.* **1998**, *22*, 105–110. [CrossRef]
3. Bruggeling, M.; Verbraeck, A.; Honig, H. Decision support for container terminal berth planning: Integration and visualization of terminal information. In *Proceedings of the Transport Logistics Working Days (VLW2011)*; University Press: Zelzate, Belgium, 2011; pp. 263–283.
4. Laumanns, M. Robust adaptive resource allocation in container terminals. In Proceedings of the 25th Mini-EURO Conference Uncertainty and Robustness in Planning and Decision Making, Coimbra, Portugal, 15–17 April 2010; pp. 501–517.
5. Zadeh, L.A. Fuzzy sets. *Inf. Control* **1965**, *8*, 338–353. [CrossRef]
6. Bierwirth, C.; Meisel, F. A survey of berth allocation and quay crane scheduling problems in container terminals. *Eur. J. Oper. Res.* **2010**, *202*, 615–627. [CrossRef]
7. Budipriyanto, A.; Wirjodirdjo, B.; Pujawan, I.N.; Gurning, S. A Simulation Study of Collaborative Approach to Berth Allocation Problem under Uncertainty. *Asian J. Shipp. Logist.* **2017**, *33*, 127–139. [CrossRef]
8. Xiang, X.; Liu, C.; Miao, L. Reactive strategy for discrete berth allocation and quay crane assignment problems under uncertainty. *Comput. Ind. Eng.* **2018**, *126*, 196–216. [CrossRef]
9. Xiao, L.; Hu, Z.-H. Berth Allocation Problem with Quay Crane Assignment for Container Terminals Based on Rolling-Horizon Strategy. *Math. Probl. Eng.* **2014**, *1*. [CrossRef]
10. Exposito, C.; Lalla, E.; Melian, B.; Moreno, J. Fuzzy optimization models for seaside port logistics: Berthing and quay crane scheduling. *Comput. Intell.* **2016**, 323–343. [CrossRef]
11. Gutierrez, F.; Lujan, E.; Vergara, E.; Asmat, F. Fuzziness in the Berth Allocation Problem. *Recent Adv. Comput. Optim. Stud. Comput. Intell.* **2019**, *795*, 149–174. [CrossRef]
12. Rodriguez, M.; Ingolotti, L.; Barber, F.; Salido, M.; Puente, J. A genetic algorithm for robust berth allocation and quay crane assignment. *Prog. Artif. Intell.* **2014**, *2*, 177–192. [CrossRef]
13. Frojan, P.; Correcher, J.; Alvarez, R.; Koulouris, G.; Tamarit, J. The continuous Berth Allocation Problem in a container terminal with multiple quays. *Expert Syst. Appl.* **2015**, *42*, 7356–7366. [CrossRef]
14. Gutierrez, F.; Lujan, E.; Vergara, E.; Asmat, R. Fully Fuzzy Linear Programming Model for the Berth Allocation Problem with Two Quays. *Uncertain. Manag. Fuzzy Rough Sets Recent Adv. Appl. Stud. Fuzziness Soft Comput.* **2019**, *377*, 87–113. [CrossRef]
15. Bouzekri, H.; Alpan, G.; Giard, V. Integrated Laycan and Berth Allocation and time-invariant Quay Crane Assignment Problem in tidal ports with multiple quays. *Eur. J. Oper. Res.* **2021**, in press. [CrossRef]
16. Young-Jou, L.; Hwang, C. Fuzzy mathematical programming: Methods and applications. In *Lecture Notes in Economics and Mathematical Systems*; Springer: Berlin/Heidelberg, Germany, 1992; Volume 394, pp. 74–186. [CrossRef]
17. Yager, R. A procedure for ordering fuzzy subsets of the unit interval. *Inf. Sci.* **1981**, *24*, 143–161. [CrossRef]
18. Das, S.K.; Mandal, T.; Edalatpanah, S.A. A mathematical model for solving fully fuzzy linear programming problem with trapezoidal fuzzy numbers. *Appl. Intell.* **2017**, *46*, 509–519. [CrossRef]
19. Nasseri, S.H.; Behmanesh, E.; Taleshian, F.; Abdolalipoor, M.; Taghi-Nezhad, N.A. Fully fuzzy linear programming with inequality constraints. *Int. J. Ind. Math.* **2013**, *5*, 309–316.
20. Saurí, S. Operations and Tails of Vessels in Ports. Available online: https://upcommons.upc.edu/handle/2099.1/6271 (accessed on 16 June 2019).

Journal of
*Marine Science
and Engineering*

MDPI

Article

Determining Residual Deviation and Analysis of the Current Use of the Magnetic Compass

Andrej Androjna [1,*], Blagovest Belev [2], Ivica Pavic [3] and Marko Perkovič [1]

1 Faculty of Maritime Studies and Transport, University of Ljubljana, 6320 Portorož, Slovenia; marko.perkovic@fpp.uni-lj.si
2 Nikola Vaptsarov Naval Academy, Black Sea University, 9002 Varna, Bulgaria; bl.belev@naval-acad.bg
3 Faculty of Maritime Studies, University of Split, 21000 Split, Croatia; ipavic71@pfst.hr
* Correspondence: andrej.androjna@fpp.uni-lj.si

Abstract: The use of electronic compasses and satellite systems has led to the magnetic compass becoming a secondary means of navigation. Yet this means of navigating is not only not obsolete, it is a necessary backup device: the construction simplicity of the magnetic compass, without electrical windings, rotating elements, and control units, remains resistant to power losses, hardware malfunction, and thus is reliable under the harshest conditions. This durability, however, comes at some cost; the magnetic compass is influenced by ships' permanent and transient magnetism, cargo gears. For the proper use of a magnetic compass, it is necessary to perform an adjustment to determine the residual deviation at regular intervals. The paper analyses selected methods to manage this, and to identify the main features of classical methods. The research was supplemented by a study carried out during the practical compensations of the magnetic compass at sea and by a survey among navigation officers on its basic requirements for proper use. The results indicate insufficient inspection of the magnetic compass. Further, an investigation into the causes of deviation delivers information regarding the causes under varying conditions including type of ship and latitudinal circumstances. This paper presents findings and recommendations to improve the compensation and use of the ships magnetic compasses.

Keywords: magnetic compass; compass error; residual deviation; magnetic compass compensation

Citation: Androjna, A.; Belev, B.; Pavic, I.; Perkovič, M. Determining Residual Deviation and Analysis of the Current Use of the Magnetic Compass. *J. Mar. Sci. Eng.* **2021**, *9*, 204. https://doi.org/10.3390/jmse9020204

Academic Editor: Dracos Vassalos
Received: 17 January 2021
Accepted: 12 February 2021
Published: 16 February 2021

1. Introduction

The introduction of contemporary technologies in design and marine equipment has led to significant improvement and development in maritime navigation, including the development of unmanned and autonomous ships and e-navigation. The most recent required system is the Electronic Chart Display and Information System (ECDIS). The use of ECDIS with the Global Navigation Satellite System (GNSS) and other mandatory navigation devices enable the permanent display of a ship's position and heading and other relevant navigational information. Apparently, the problem of determining and continuously displaying the ship's position at any location in the world's oceans has been solved.

Keeping the vessel on a given course is usually carried out with the aid of a gyrocompass [1], which is the master compass on board. The reliability of the compass plays a central role in any steering mode. Its accuracy is much greater than that of magnetic compasses; however, even so the gyrocompass is still not accurate enough to ensure safe berthing of large ships in many ports with narrow basins and restricted approaches. In fact, in practice, the gyrocompass often does not meet the standards established by the International Maritime Organization (IMO) [2]. Late or missed detection of a gyrocompass malfunction can lead to accidents at sea. In recent years, there have been some occasions when a sudden, unexpected loss of power triggered the undetected inaccuracy of electronic instruments, which then developed into a serious crisis. To support this contention, one need only recall the 2014 accident involving the m/v "Atlantic Erie", a bulk carrier which ran aground in Ontario. The crew was unaware of a gyrocompass defect. If they had the ability and readiness to switch to "old fashioned" navigation using a reliable magnetic compass, that should have prevented the grounding [3].

The magnetic compass's advantage is entirely related to its simplicity, durability, and autonomy [4]. A transmitting magnetic compass with integrated correction of magnetic variation and deviation curve can be used as a source of heading information. Other heading sensors are the free-directional gyro updated by a Kalman filter, rate-of-turn gyro, GNSS compass filtered and stabilized by the inertia of the free-directional gyro, and a vessel's course-over-ground (CoG) received from differential global positioning system (DGPS). With these aids to navigation, the officer of the watch (OOW) has in his hands many reliable sources to identify ships' heading and CoG, but the only independent one—is the magnetic compass.

As with all technological developments, an adjustment period has followed that indicates that the improvements bring with them new maritime navigation problems, such as over-reliance on satellite and computer-based navigation systems and their vulnerability to cyber threats. Today's global maritime sector depends on digitalization, integration of operations, and automation [5]. While automation offers excellent benefits, it also introduces a set of corresponding cybersecurity-related risks [6,7]. According to the International Convention for Safety of Life at Sea (SOLAS) regulation V/19.2.1.1, all ships, regardless of size, shall have a properly adjusted standard magnetic compass, or other means, independent of any power supply, to determine the ship's heading and display the reading at the main steering position [8]. The standard magnetic compass (and spare magnetic compass) must be appropriately compensated, and its table or curve of residual deviation must be available on board in the vicinity of the compass at all times [9].

The IMO also requires that masters and officers in charge of navigational watch know, understand, and have proficiency in regard to the principle of operation and error determination of the magnetic compass [10]. The usual method used on board merchant vessels is comparison between bearing, measured by magnetic compass, and the azimuth of a celestial body. Another method uses bearing to a distant object on shore in the vicinity and comparison in the same way. Thus, despite the view that the magnetic compass belongs to the past of navigation, it retains value as a backup instrument [11]. The International Convention on Standards of Training, Certification and Watchkeeping for Seafarers (STCW) requires that the officer in charge of the navigational watch shall conduct regular checks to ensure that the standard compass error is determined at least once a watch and, when possible, after any significant alteration of course, and that the standard and gyrocompasses are frequently compared and repeaters are synchronized with their master compasses [10]. According to a procedure in the ship's safety management system, it is mandatory to record this in the compass observation (deviation) book [12]. Masters and officers of the watch do not have in their hands options to mitigate compass deviation at the time of the ship's voyage. According to the STCW Convention standards, they do not have such obligations. It is one of the reasons why they must check compasses and record their errors. Noncompliance with these obligations is one reason for remarks in port state control (PSC) inspections. Clearly, the mandates regarding the magnetic compass are no longer being taken quite seriously. The analysis of the annual reports of the regional memorandum of understanding (MoU) on PSC shows that deficiencies related to SOLAS Convention Chapter V (Safety of Navigation) are among the five main reasons for PSC remarks in virtually all regions (ranging from 11 to 15%) [13–16]. Further, the analysis of the top 20 deficiencies (from the Safety of Navigation group) shows extensive compass deficiencies recorded, resulting even in "detention of the ship" [17]. To provide some illustration, problems serious enough to record regarding the magnetic compass include air bubbles, lack of spare magnets, missing bearing devices, rust, and—this is key—irregular inspections by crew members [18].

This article aims to underscore the importance of the magnetic compass and exhort navigation officers to properly use and maintain this historic yet still necessary navigation aid. Further advances in navigation are required before the magnetic compass can be abandoned. For instance, a truly redundant safety measure is not the addition of a second gyrocompass, comforting as that may be. In a sense, this paper is about securing the advances in navigation. Furthermore, classical methods of magnetic compass compensation are discussed, results of adjustments summarized, and findings on the influence of ship types and age on deviation are shown. This article presents the results of a survey of navigation officers regarding the current use and recommendations provided on means to improve the compensation.

This article consists of four sections. Section 2 provides the literature review, Section 3 describes the methodology and presents the results, and Section 4 provides discussion and conclusions.

2. Literature Review

The ship's magnetic compass is a classic navigational instrument that has been thoroughly researched. Noteworthy are the achievements of Airy and Kolong that laid the foundation for the compensation and determination of the residual deviation of a compass onboard a ship with a metal hull. In late 1835, Airy conducted a series of studies, as a result of which he developed a method for compensating and determining the residual deviation of a magnetic compass [19]. The deviation device of modern compasses and the additional soft iron of various shapes attached to the compass allows for the application of this method. Compass adjustment is a job that can take from one to several hours if well organized, experienced, and under favorable conditions. Usually, it takes between two to four hours [20].

Kolong's method requires a more complicated sequence of execution, so it is applicable only if the compass adjuster has the necessary equipment. The method is described in detail by Kozhukov [21], and the author also presents the construction of the device invented by Kolong, called a "deflector". The Kolong method is not widely used because of the need for additional instruments, the time required to prepare the compass on shore, and to perform the maneuver of the ship to determine and compensate for the residual deviation.

Airy's method puts into practice the theoretical developments of Smith and Evans, published in 1863 in their work "Admiralty Manual for Ascertaining and Applying the Deviations of the Compass Caused by the Iron in a Ship" [22]. Evans–Smith's formulae, for which the magnetic course is plotted in the Fourier series, is the basis for the entire mathematical means for calculating the deviation table of any magnetic compass [23]:

$$\tan\delta = \frac{A' + B'\cdot\sin\zeta + C'\cdot\cos\zeta + D'\cdot\sin 2\zeta + E'\cdot\cos 2\zeta}{1 + B'\cdot\cos\zeta - C'\cdot\sin\zeta + D'\cdot\cos 2\zeta - E'\cdot\sin 2\zeta} \tag{1}$$

The symbols ζ and ζ' denote the magnetic and compass course, whereas A', B', C', D', and E' indicate the exact coefficients that were expressed originally by Archibald–Smith as A, B, C, D, and E [24].

Another formula widely used in practice for determining the residual deviation (2) allows the calculation of the deviation in degrees simply, although various trigonometric estimations turned it into a rough equation [25–27]. This approximate deviation (Δ) is normally used to adjust the magnetic compass [28–30]:

$$\Delta = A + B\cdot\sin\zeta' + C\cdot\cos\zeta' + D\cdot\cos 2\zeta' + E\cdot\cos 2\zeta' \tag{2}$$

The symbols A, B, C, D, and E are approximate deviation coefficients and their values match up with the sine of the exact coefficients. Both coefficients may be considered constant for a long time [2,24]. Nevertheless, this may not always be the case, since a bolt of lightning or a shipment of steel cargo may affect the ship's magnetism [31].

After transformations (1), formulas are derived for the exact coefficients of the deviation of the magnetic compass (3):

$$\begin{aligned} A &= \frac{\delta_N+\delta_S+\delta_E+\delta_W+\delta_{NE}+\delta_{SW}+\delta_{SE}+\delta_{NW}}{8} \\ B &= \frac{\delta_E-\delta_W}{4} + \frac{\delta_{NE}-\delta_{SW}+\delta_{SE}-\delta_{NW}}{4}.\boldsymbol{sin45^\circ} \\ C &= \frac{\delta_N-\delta_S}{4}\frac{\delta_{NE}-\delta_{SW}+\delta_{SE}-\delta_{NW}}{4}.\boldsymbol{cos45^\circ} \\ D &= \frac{\delta_{NE}+\delta_{SW}-\delta_{SE}-\delta_{NW}}{4} \\ E &= \frac{\delta_N+\delta_S-\delta_E-\delta_W}{4} \end{aligned} \quad (3)$$

It is important to note that magnetic compass deviation calculations start with coefficient calculations. Because of this, the choice of the deviator between Formulae (3) and (4) for coefficient calculations is essential for the accuracy of his work.

The practical application of the Formula (3) requires the ship to steer successively on eight compass courses: 0°, 45°, 90°, 135°, 180°, 225°, 270°, and 315° [32]. The magnetic compass deviation table is computed for every 10° or 15°, replacing the compass course in Formula (1) with the corresponding value of ζ.

If the ship turns 360°, but only the deviation of the main magnetic courses 0°, 90°, 180° and 270° is measured, the abbreviated Formula (4) can be used to calculate the coefficients:

$$\begin{aligned} A &= \frac{\delta_N+\delta_S+\delta_E+\delta_W}{4} \\ B &= \frac{\delta_E-\delta_W}{2} \\ C &= \frac{\delta_N-\delta_S}{2} \\ D &= \frac{\delta_{NE}+\delta_{SW}-\delta_{SE}-\delta_{NW}}{4} \\ E &= \frac{\delta_N+\delta_S-\delta_E-\delta_W}{4} \end{aligned} \quad (4)$$

Lushnikov and Pleskacz argue that the method for determining the residual deviation required by the regulations is overly time-consuming and costly [20]. They propose using a more straightforward method based on the compass' directional force (λH). By increasing the coefficient of the guiding force (λ), there will be a reduction in all types of deviation. Thus, one does not have to deal with individual types of magnetic compass deviation before, but with all deviations simultaneously [21]. This method is based on the installation of a standard suspension device that compensates for the semicircular deviation during the initial installation of the compass and increases the directional force at the same time, or simply, after that, it is not necessary to perform compensation. The disadvantage of this method is that the existing magnetic compasses must be reconstructed, while newly built compasses must be equipped with this device, of course, after thorough testing and evaluation in practice.

Kozhukov also proposes abbreviated formulas for calculating the coefficients. His proposal is based on the permissible value for the magnetic compass deviation and sets ±4° as the limit. If the compass deviation is above this value, the full formulae and the corresponding ship maneuvering method should be used [21]. This theory is also found in other studies [24,33,34]. To be widely accepted, these methods need some additional empirical evaluations and comparisons with the "classical" methods of compensation in the cardinal and intercardinal courses.

It is essential to note that the magnetic field of a ship is variable and depends on several factors, such as the sailing area, the type of cargo the ship is carrying, the type of repairs carried out on board, the installation of additional equipment, the age of the ship, [35,36]. Based on these factors, applying one or another type of formula in the calculation process should be done after a critical analysis of the circumstances.

The STCW-78 Convention does not require deck officers and masters to have the knowledge and skills to adjust the magnetic compass. Deck officers and masters must be able to use the information on magnetic compass readings. Compensation and determination of residual compass error is the responsibility of persons certified by the competent maritime authorities [37]. Attention should be paid to the maritime authorities' provisions, which also allow masters to compensate the deviation and compile a new table with the values of the residual deviation [38–41]. In such a manner, the master needs to be aware that any changes to the ships' hull, to loading/discharging equipment, to machinery, and other mechanisms that lead to changes in the ship's magnetic field and magnetic compass deviation.

Another serious problem affecting the accuracy of the magnetic compass is the latitude error caused by a significant change in the ship's latitude [24,36], especially near the magnetic poles [31]. Any ship's trading operation requires it to travel thousands of miles across different latitudes. The variable nature of the Earth's magnetic field causes the ship's magnetic field to change due to the intersection of the Earth's magnetic lines of force at different angles, as illustrated in Figure 1a. That accumulates an error, as per Figure 1b, in the deviation table, and eventually, the actual deviation of the magnetic compass does not correspond with the deviation table data.

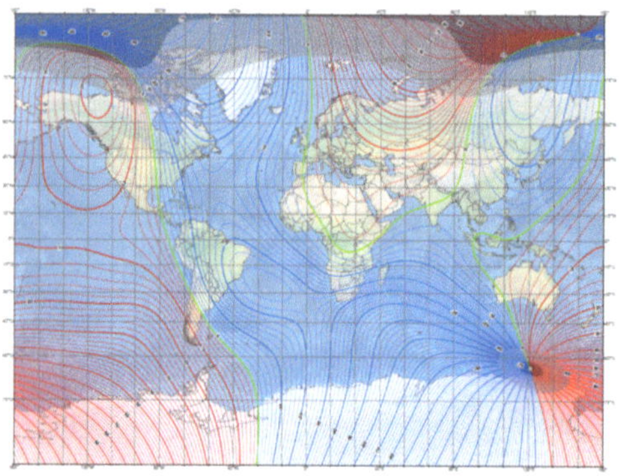

(a)

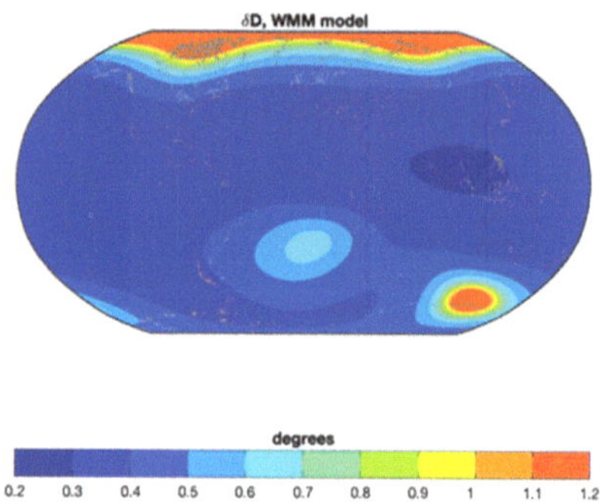

(b)

Figure 1. (**a**) World magnetic model—main field declination [42]; (**b**) global distribution of the declination error model. Color scale limited to 1.2 degrees [43]. Available online: https://www.ngdc.noaa.gov/geomag/WMM/image.shtml (accessed on 27 January 2021).

In this context, studies are being conducted to apply the least-squares method to determine the deviation of any ship's course [44], although the author has not yet tested the algorithm in practice. Felski [45] has proposed a device that automatically collects information about the deviation and continuously compares it with the deviation table values. Other authors propose an algorithm to compute the latitude error while the ship is in service somewhere in the world [33]. However, all the listed improvements in compass adjustment activity are based on the ship's existing deviation table. That means that at least once the residual error's compensation and determination must be performed by a highly qualified and certified compass adjuster.

3. Methodology and Results

The study was conducted using two approaches: by analyzing the results of magnetic compass compensation and surveying the current use of compasses on ships.

The compensation of the deviation of a ship's magnetic compass and the subsequent determination of the residual deviation were carried out according to Airy classical method's rules with a full swing of the ship on eight magnetic courses by in situ measurements. The method for determination used by the authors is applicable in the area at sea specially designated for magnetic compass adjustment or in an area with sufficient space for maneuver and accurately distinct onshore markers. Using the gyrocompass course as reference direction is incorrect because of gyro inertia.

Deviation activities begin with turning the vessel on every cardinal course: 000°–180° and 090°–270°. If the compass deviation is within the limit, the vessel is turned to the quadrantal courses: 045°–225° and 135°–315°. If the deviation is within the limit, one full circle from 000° to 360° is necessary to maintain the ship on every cardinal and quadrantal course for at least one minute. The deviation activities are correct if the weather and sea are calm; otherwise, the ship's deviation is affected by rolling and pitching.

For this study, ships were grouped by type and age (years of service). The plots show calculated deviation values based on Formulae (3) and (4) for the selected ship types and ages. Deviation and specific differences that may indicate the need to perform compass compensation using the Airy method were noted.

The second part of the research methodology in this article was a survey. The objectives (targets) of this survey are to review the level of knowledge and current use of magnetic compasses on board. The research instrument was a questionnaire consisting of an introduction, general questions, and specific questions. The introductory part of the questionnaire contains general remarks (i.e., research objectives, instructions for respondents, importance of the survey, and a statement about voluntary and anonymous participation in the survey). The questionnaire contains nine general and specific questions. The questions were closed-ended with predefined single-choice responses. The general questions aimed to categorize the profile of the respondents. These included questions on certificates of competency (CoC), seafaring experience, and assignments on board. The group of specific questions related to the use and ability to set the magnetic compass. This group of questions was divided into two categories. The first category includes questions related to regular deviation checks and intervals to perform this task, use of the deviation table (curve), and experience with PSC inspections of the magnetic compass. The purpose of these questions was to collect data on knowledge of this specific related topic. The second category includes questions about the intensity of the importance and frequency of deviation control and compensation. This category consisted of numerical values ranging from one (lowest level of importance and the lowest level of frequency of performing the task) to five (highest level of importance and the highest frequency of performing the task). Thus, the relationship between the answers was established. After developing the questionnaire, a test of the survey was conducted by the professors of navigation and experienced experts (retired shipmasters). The questionnaire was then corrected and distributed online to the target group of 320 respondents. The survey was conducted from 9 December 2020 to 9 January 2021. During this period, 123 responses were collected from navigation officers. The target group consisted of deck officers and captains. To verify the results and their clear interpretation, a face-to-face interview or correspondence via email was used for a selected 10% of the respondents according to their CoC and assignment on board.

3.1. Results of the Magnetic Compass Compensation Study

The compensation and determination of the residual deviation were carried out in the area of Varna Bay in the western part of the Black Sea with approximate coordinates latitude $\varphi = 43°11.0' $ N and longitude $\lambda = 28°55.0'$ E. The total number of ships included in the study was 252. The distribution by type is shown in Table 1.

Table 1. The number of ships included in the study according to type.

Type of Ships	Number of Ships
General Cargo	83
Tankers	60
Bulk carrier	57
Reefer	25
Offshore supply ship	13
Container ship	13
Ro-Ro Cargo	1

One of the theses put forward by the authors is that the type of ship is influences the change in the magnetic field. The transported cargo over time and while the ship is sailing at different latitudes changes the magnetic compass' deviation. Experiments have shown that the loading and unloading activities of metal-bearing cargoes and liquid bulk cargoes affect the compass deviation differently. In addition, the transport of goods over long distances in different latitudes of the world's oceans influences the magnetic compass error. In this context, it is important to reiterate the importance of regularly determining compass errors during the voyage.

Another hypothesis that we support and for which we provide evidence is the dependence of the change in magnetic compass deviation on the ship's age. The accumulation of changes, including design changes, alters the ship's magnetic field, and these changes directly affect the deviation. Therefore, the studied ships are divided into five groups according to their age into deviation activities. The cutoff point is five years, the time for class repairs where structural changes are made if necessary. The percentage of the different ship groups by their age is shown in Table 2.

Table 2. Distribution of the examined ships depending on their age.

Group of Ships (Per Their Age)	%
0–5	20
6–10	15
11–15	17
16–20	16
>20	32

The data processing involves calculating the coefficients of deviation of the magnetic compass of each ship using the Formulae (3) and (4). The empirical values of the deviation were calculated when the ships maneuvered to eight compass courses: $0°$, $45°$, $90°$, $135°$, $180°$, $225°$, $270°$, and $315°$. The results show a graph of the deviation constructed with coefficients from the full formulas and one constructed with coefficients from the short formulas. For the calculations using Formula (4), only the deviations on the four main magnetic courses: $0°$, $90°$, $180°$ and $270°$, are considered. The analyses were also performed considering the ship type.

3.1.1. Compasses on Newer Vessels

Barring force majeure, ships five years old and less have not undergone any hull and machinery alterations. Thus, the conditions for constructive preservation of the ship's magnetic field are met. Any alterations would result from the nature of the cargo carried. Figure 2a show the deviation curves of a bulk carrier that was two years old at the time of the deviation activities. The graph shows the deviation curve after calculated coefficients according to Formula (3) (red line), and the deviation curve after calculated coefficients according to Formula (4) (blue line).

Figure 2a shows that the deviation of the magnetic compass is mainly caused by the magnetic field of the solid ship's iron and has a semicircular character. The values are approximately the same with respect to maximum and minimum and are within $\pm 5°$. In this case, it may be necessary to reduce the deviation values, but such measures are not mandatory.

Figure 2b shows graphs of a five-year-old tanker. The lines show the deviation curve after calculated coefficients according to Formula (3) (red line), and the deviation curve after calculated coefficients according to Formula (4) (blue line).

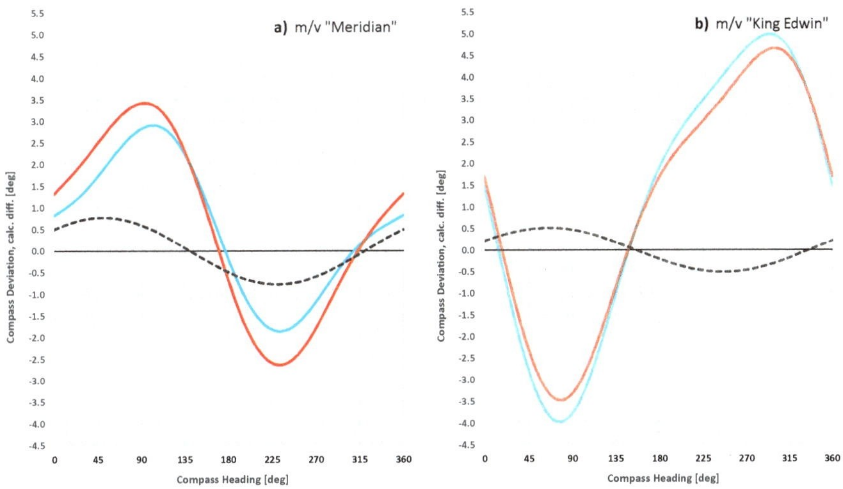

Figure 2. Graph of the deviation, according to Formulae (3) and (4): (**a**) m/v "Meridian", (**b**) m/v "King Edwin".

In this case, the same results are observed. The following conclusion can be drawn:

- The type of deviation of a newer ship (age 0 to 5 years) is mainly semicircular;
- Clearly, it takes time for cargo type to affect the magnetism of a vessel—changes are generally a result of latitudinal effects.

3.1.2. Compasses on Vessels from Five to Ten Years of Age

The deviations were measured after a class repair of 39 ships 5–10 years old. Figure 3a shows the diagram of a ten-year-old ship carrying petroleum products and chemicals. The magnetic compass deviation is semicircular, but the graphs are shifted along the abscissa. In this case, applying the short formulas to determine only the semicircular deviation leads to a calculated value of coefficient A equal to zero. Although the two figures' deviation is within the allowable values, complete measurements and application of the formulae are necessary.

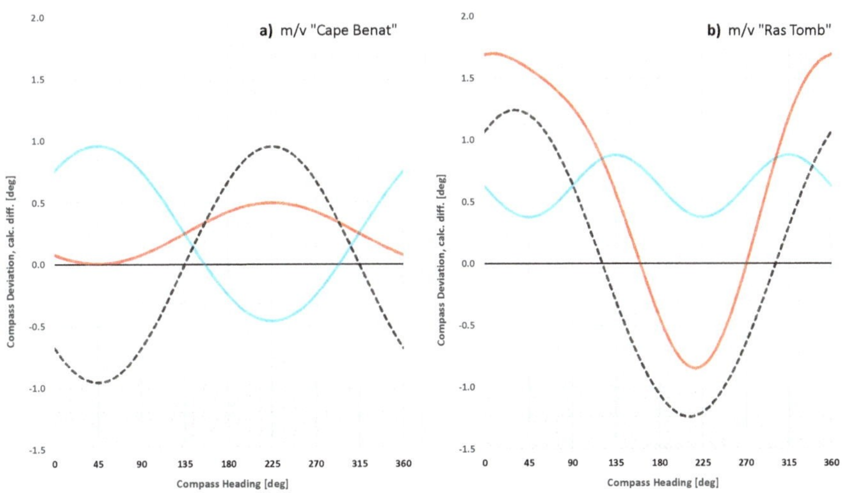

Figure 3. Graph of the deviation, according to Formulae (3) and (4): (**a**) m/v "Cape Benat", (**b**) m/v "Ras Tomb".

The other example, in Figure 3b, shows the deviation of a nine-year-old general cargo ship. The lines present the deviation curve as in the previous figures. The ship's sailing area was the Mediterranean, and the North and Baltic Seas—i.e., latitudes no more than 15 degrees above those at which the deviation was determined in 2012. Figure 3b shows that the nature of the deviation is beginning to change to quadrantal. Although the values are still within acceptable limits, this is a sign that ignoring the determination of the deviation of the magnetic quarter courses 0°, 90°, 180°, and 270° will result in erroneous values of the magnetic compass.

3.2. *Analysis of the Results of the Survey*

A survey method was used to collect the data on the use and knowledge level of magnetic compasses. The responses obtained were processed using statistical, descriptive, and comparative methods. Of the total number of respondents, 36 officers (29.3%) hold the CoC for Officer in charge of a navigational watch on ships of 500 GT or more, 20 officers (16.3%) hold the CoC for Chief mate on ships of 3000 GT or more, and 57 officers (46.3%) hold Master mariner unlimited certificates. Other navigation officers who responded to the survey (10 officers) hold the CoC for Master on ships between 500 and 3000 GT, or naval and other certificates issued in accordance with national regulations. In terms of seafarering experience, 24.4% of respondents have 1–4 years of sea service, 25.2% have 4–10 years of sea service, 21.1% have 10–15 years of sea service, and 29.3% have more than 15 years of sea service. The current (or last) assignment onboard ships is the third officer for 16.3% of the respondents, second officer for 28.5%, Chief mate for 23.6%, while 31.7% are Masters. The number, qualifications, seagoing service, and duties of the respondents constitute a relevant sample to draw reasonable conclusions about the knowledge and use of a magnetic compass on ships.

The respondents were asked about conducting a regular deviation check (error determination) of the magnetic compass; 82.9% of the respondents answered positively, while 17.1% answered negatively. A regular deviation check may indicate the need for repair, testing, or adjustment of the magnetic compass [46]. Table 3 presents the response.

Table 3. Distribution of responses to the regular deviation check of the magnetic compass.

Regular Deviation Check of the Magnetic Compass	%
At least once a month	7.3
At least once a week	11.4
At least once a day	17.9
At least once a watch	17.9
At least once a watch and, when possible, after any major alteration of course	31.7
Other	13.8

Only 49.6% (31.7% + 17.9%; as per Table 3) of the navigation officers perform a regular magnetic compass deviation check following the provisions of the STCW Convention (STCW, 2017, Section A- VIII/2, Part 4-1). Pleskacz obtained almost identical results (53%) in a survey conducted in 2017 among a sample of 212 navigation officers [34]. These responses indicate an inadequate level of knowledge or negligence in applying the relevant provisions of the STCW Convention relating to watchkeeping or the improper use of a magnetic compass on board. Comparing these results suggests that navigation officers generally consider that compass inspection requirements are overly stressed in the STCW.

Clarification was provided, as navigation officers were asked to estimate the importance and frequency of magnetic compass deviation checks. The responses are shown in Figure 4.

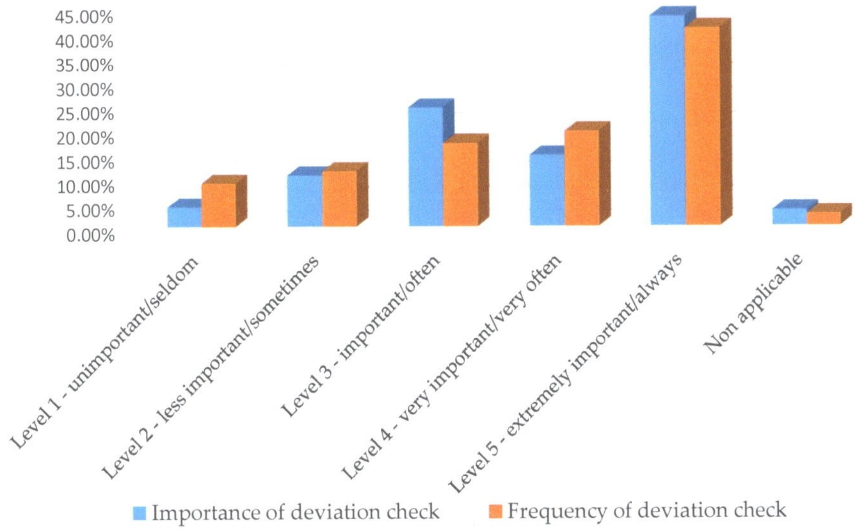

Figure 4. The perceived importance and frequency of deviation check of the magnetic compass.

Comparing the data on the regular deviation check (Table 3) with the data on its importance and frequency (Figure 4), we can see that an almost identical percentage of navigation officers perform the regular deviation check of the magnetic compass and at the same time give it the highest level of importance (43.1%) and frequency (40.7%). That suggests that there is disagreement with the required procedures, or a lack of knowledge.

Another critical factor affecting the use of a magnetic compass is the validity of the deviation table (curve). When asked about the significance of the deviation table (curve) in navigation, only 55.3% of the respondents answered positively. This relatively high percentage of responses further suggests the failure of navigation officers to implement regulations regarding the magnetic compass.

The authors compared the responses regarding the regular deviation check and the use of the magnetic compass deviation table with the respondents' current (or last) assignment. The results presented in Table 4 are as expected in that they show equal or very high agreement between the application of STCW regulations and the correct use of the magnetic compass in navigation. Yet the results are somewhat surprising as they show that the highest percentage of responses with positive answers to this question are given by second officers and not by masters and chief mates.

Table 4. Responses to the regular deviation check and use of the deviation table by rank.

Current (Last) Assignment (Nr.)	Deviation Check		Deviation Table (Curve) Use	
	Regular (Nr./%)	Irregular (Nr./%)	Yes (Nr./%)	No (Nr./%)
Master (39)	19/48.7	20/51.3	19/48.7	20/51.3
Chief mate (29)	12/41.4	17/58.6	16/55.2	13/44.8
Second officer (35)	22/62.9	13/37.1	22/62.9	13/37.1
Third officer (20)	8/40.0	12/60.0	11/55.0	9/45.0
Total (123)	61/49.6	62/50.4	68/55.3	55/44.7

The MoU annual reports on PSC show a significant number of shortcomings of the magnetic compasses. Respondents were asked about their experience with PSC inspections of a magnetic compass—90.2% of respondents answered that their ships had not been subject to PSC remarks regarding magnetic compass deficiencies.

The remaining 9.8% of respondents answered positively. Analysis of the responses revealed that the vast majority (9 out of 13) of deficiencies related to inadequate inspections, significant differences of the magnetic course, or entries in the compass observation book.

4. Discussion and Conclusions

Analysis of the magnetic compass compensation results shows that the first-class repair of a ship is the age limit for permanent magnetic field changes. We do not provide examples for the other age groups because the calculations confirm the above conclusion. In 92% of cases, especially for ships older than 20 years, the type of magnetic compass deviation is determined by the type of cargo carried. Relative stability is observed in tankers, reefers, and offshore supply ships. The cargoes carried by these types of ships are nonmetallic. In almost 100% of cases, the deviation has a semicircular character; that allows the use of an abbreviated procedure for its determination. The graph's expected shift can be along the abscissa in the direction of "+" or "−" within a maximum of $\pm 1^{\circ}$.

For the other types of ships studied, the experiments with Formulae (3) and (4) revealed significant differences in the deviation curves. In almost 60% of the cases, the type of deviation changed over time to quadrantal. This categorically rules out the possibility of using an abbreviated procedure to compensate and determine the residual deviation.

Figure 5 show the deviation curves of the general cargo vessel "Kalitihi Sea" for 2009 (blue line) and 2011 (red line). The ship was built in 1986. In this case, the tendency to change the type of deviation is evident.

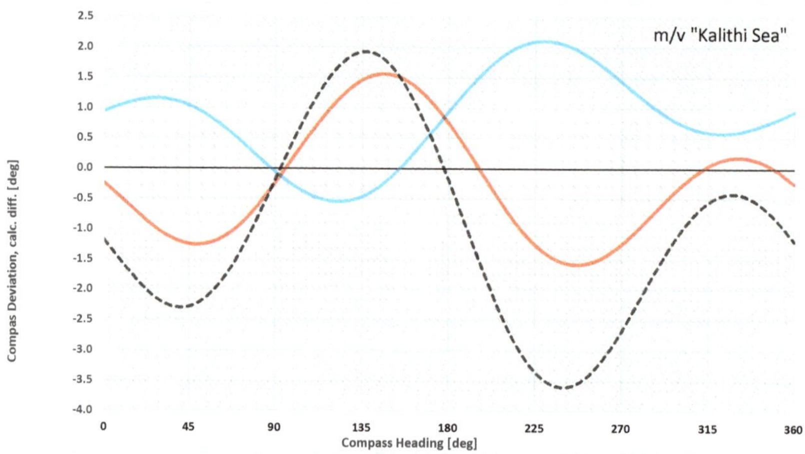

Figure 5. Graph of the deviation for 2009 (blue line) and for 2011 (red line), (m/v "Kalitihi Sea").

This and other similar cases suggest that compensation of deviation requires all parties' proper attitudes—ship owners, crew members, flag administrations, PSC inspections, and certified compass adjusters.

Analysis of the methods and techniques used to compensate for and determine the residual deviation of a ship's magnetic compass suggests the following conclusions:

- Ship's masters must instill in all bridge officers a responsible attitude toward magnetic compasses, standard and spare, and maintain them in good working order;
- The adjustment of the standard magnetic compass must be carried out under the requirements of the relevant IMO and flag state regulations;
- Before the commencement of deviation activities, the factors leading to the change in deviation should be analyzed;

- If the analysis shows that there are conditions for the change in the value and nature of the deviation, it is necessary to carry out its compensation by sailing in eight magnetic courses. The calculations should be made using Formula (3) instead of Formula (4) whenever the magnetic compass adjuster is unsure whether the ship's structural change during her life has had significant impact on the magnetic field of the vessel.
- It is important to emphasize that the Airy method is reliable enough and is applicable everywhere worldwide. Another advantage is that the method does not require any special equipment and extra complicated calculations, as required in other methods. The use and application of any innovative technological methods for determining the residual deviation are made based on the previously composed deviation table, in most of the cases calculated by the Airy method.

A regular check of deviation together with the maintenance and application of the deviation table (curve) are the main factors for the proper application of the magnetic compass in navigation. According to the survey results, 14.2% of the respondents said that they do not perform a regular deviation check. In comparing the answers to the question about the periodicities of a regular deviation check, we found that the percentage is significantly higher. Almost 50% of respondents do not perform this activity following the relevant STCW regulations. Analysis of the data on the importance and frequency of the deviation check confirms these findings. The fact that 45.1% of the respondents do not use the deviation chart or deviation table indicates that they do not use the magnetic compass in navigation correctly (if at all). This leads to remarks in PSC inspections where magnetic compass deficiencies are among the top five deficiencies per the analyzed annual reports of regional MoUs.

After analyzing the survey results, the following findings can be elaborated:

- About half of the respondents do not perform regular deviation checks of the magnetic compass, while about 45% of the respondents do not use the deviation table, including masters and chief mates, which is of particular concern;
- Comparing the results of this survey with the 2017 survey conducted by Pleskacz shows a 3.4% decrease in performing the deviation check, which shows how the proper use of a magnetic compass is decreasing;
- The reliability of modern compasses and navigation systems will further reduce the use of magnetic compasses;
- New generations of deck officers and masters will have less need for the use of magnetic compasses;
- These facts will lead to a further reduction in the level of knowledge of the proper use of magnetic compasses, especially in emergencies.

The proper use of a magnetic compass requires navigation officers' joint efforts in regular inspection and maintenance and compass adjusters in regular compensation.

The development of technology in the maritime sector has a significant impact on navigation. This development further reduces the need to use the magnetic compass in navigation. Nevertheless, the magnetic compass remains the only means of indicating course independent of any source of power. The navigation officers need to have the ability and readiness to switch to this "old fashioned" navigation using a reliable magnetic compass in case a gyrocompass malfunction is detected, which could endanger the ship's navigation, safety, and security. Therefore, regular compensation, an adequate level of knowledge, and the correct use of the magnetic compass remain a necessary condition for safe navigation, especially in emergencies.

Author Contributions: Conceptualization, A.A., B.B., and I.P.; methodology, A.A., B.B., and I.P.; data collection, B.B. and I.P.; validation, A.A. and M.P.; formal analysis, A.A., B.B., and I.P.; data curation, A.A., B.B., and I.P.; writing—original draft preparation, B.B. and I.P.; internal review, M.P. All authors have read and agreed to the published version of the manuscript.

Funding: The authors acknowledge the financial support of the Slovenian Research Agency (research core funding No. P2-0394, Modelling and Simulations in Traffic and Maritime Engineering).

Conflicts of Interest: The authors declare no conflict of interest.

Abbreviations

CoC	Certificate of Competency
CoG	Course over Ground
DGPS	Differential Global Positioning System
ECDIS	Electronic Chart Display and Information System
GNSS	Global Navigation Satellite System
GT	Gross Tonnage
IMO	International Maritime Organization
MoU	Memorandum of Understanding
OOW	Officer of the Watch
PSC	Port State Control
SOLAS	Safety Of Life At Sea
STCW	Standards of Training, Certification, and Watchkeeping

References

1. Lushnikov, E. Magnetic compass in modern maritime navigation. *TransNav* **2015**, *9*, 539–543. [CrossRef]
2. Perkovič, M.; Gucma, L.; Bilewski, M.; Muczynski, B.; Dimc, F.; Luin, B.; Vidmar, P.; Lorenčič, V.; Batista, M. Laser-based aid systems for berthing and docking. *J. Mar. Sci. Eng.* **2020**, *8*, 346. [CrossRef]
3. Marine Investigation Report M14C0106—Grounding Self-Discharging Bulk Carrier Atlantic Erie, Port Colborne, Ontario. 12 June 2014. Available online: https://www.tsb.gc.ca/eng/rapports-reports/marine/2014/M14C0106/m14c0106.pdf (accessed on 8 February 2021).
4. Lushnikov, E. The reliability of compass information at navigational safety. *Sci. J. Marit. Un. Szczecin* **2012**, *29*, 117–121.
5. Androjna, A.; Brcko, T.; Pavic, I.; Greidanus, H. Assessing cyber challenges of maritime navigation. *J. Mar. Sci. Eng.* **2020**, *8*, 776. [CrossRef]
6. Babineau, G.; Jones, R.; Horowitz, B. A system-aware cyber security method for shipboard control systems with a method described to evaluate cyber security solutions. In Proceedings of the 2012 IEEE Conference on Technologies for Homeland Security (HST), Waltham, MA, USA, 13–15 November 2012; pp. 99–104.
7. Masala, C.; Tsetsos, K.A. Cyber risks and threats: Demanding challenge for the maritime industry. In *Look Out 2016 Maritime Domain Cyber: Risks, Threats & Future Perspectives*; Lampe & Schwartze KG: Bremen, Germany, 2015; pp. 11–26. ISBN 9783000510144.
8. IMO. *The International Convention for Safety of Life at Sea*; Consolidated Edition; IMO Publishing: London, UK, 2020; ISBN 9789280116908.
9. IMO. Resolution A.382(X). 1977. Available online: https://www.register-iri.com/wp-content/uploads/A_X_Resolution_382.pdf (accessed on 15 April 2020).
10. IMO. *The International Convention on Standards of Training, Certification and Watchkeeping for Seafarers-78*; IMO Publishing: London, UK, 2017; ISBN 9789280115284.
11. Huey, C.; Taylor, C. Navigational compasses gyro versus magnetic. *J. Am. Soc. Naval Eng.* **1934**, *46*, 417–450. [CrossRef]
12. MGN 610 (M+F) SOLAS Chapter V: Guidance on the Merchant Shipping (Safety of Navigation) Regulations. 2020. Available online: https://www.gov.uk/government/publications/mgn-610-mf-solas-chapter-v-guidance-on-the-merchant-shipping-safety-of-navigation-regulations-2020 (accessed on 12 January 2021).
13. Port State Control in The Black Sea Region Annual Report. 2019. Available online: http://www.bsmou.org/downloads/annual-reports/BSMOU-AR-2019.pdf (accessed on 15 September 2020).
14. Indian Ocean MoU Annual Report. 2019. Available online: http://www.iomou.org/armain.htm (accessed on 14 October 2020).
15. Annual Report on Port State Control in the Asia-Pacific Region. Available online: http://www.tokyo-mou.org/doc/ANN19-f.pdf (accessed on 15 September 2020).
16. Paris MoU Annual Report. 2019. Available online: https://www.parismou.org/2019-paris-mou-annual-report-port-state-progression-detention-rate-down (accessed on 15 September 2020).
17. Paris MoU Inspection Result Deficiencies. 2019. Available online: https://www.parismou.org/inspection-search/inspection-results-deficiencies (accessed on 15 April 2020).
18. The Magnetic Compass. Available online: http://www.kompassjusterarna.com/dokument/NI_Magnetic_compass.pdf (accessed on 15 April 2020).
19. Airy, G. Account of experiments on iron-built ships, instituted for the purpose of discovering a correction for the deviation of the compass produced by the iron of the ships. *Phil. Transact. R. Soc. Lond.* **1839**, *129*, 167–213. [CrossRef]
20. Lushnikov, E.; Pleskacz, K. The ultimate solution to the deviation problem of magnetic compasses. *Sci. J. Marit. Un. Szczecin* **2018**, *53*, 74–80. [CrossRef]
21. Kozhukov, V.; Vorov, V.; Grigoriev, V. *Magnetic Compasses*; Publishing House Transport: Moscow, Russia, 1981. (In Russian)

22. Evans, F.J.; Smith, A. *Admiralty Manual for Ascertaining and Applying the Deviations of the Compass Caused by the Iron in a Ship*; Hydrographic Office of Admiralty: London, UK, 1863; ISBN 9781241702502.
23. Magnetic Deviation: Comprehension, Compensation and Computation. Available online: http://myreckonings.com/wordpress/2009/04/18/magnetic-deviation-comprehension-compensation-and-computation-part-ii/ (accessed on 19 March 2020).
24. Iribar, I.; Muñoz, J.; Labajos, C. Latitude error in compass deviation. Mathematical method to determine the latitude error in magnetic compass deviation. *PMR* **2014**, *3*, 25–31. [CrossRef]
25. Benković, F.; Piškorec, M.; Lako, L.; Čepelak, K.; Stajić, D. *Terestrička i Elektronska Navigacija*; Hidrografski Institut Ratne mornarice: Split, Croatia, 1996; ISBN 86-7033-001-6. (In Croatian)
26. Denne, W. *Magnetic Compass Deviation and Correction: A Manual of the Theory of the Deviations and Mechanical Correction of Magnetic Compasses in Ships*; Brown, Son & Ferguson Ltd.: Glasgow, UK, 1979; ISBN 978-0851743325.
27. Bowditch, N. *The American Practical Navigator, An Epitome of Navigation*; Defense Mapping Agency Hydrographic/Topographic Center: Bethesda, MD, USA, 2017.
28. Barber, G.; Arrott, A. History and magnetics of compass adjusting. *IEEE Trans. Magnet.* **1988**, *24*, 2883–2885. [CrossRef]
29. Moncunill Marimon, J.; Martínez-Lozares, A.; Martín Mallofré, A.; Francisco González la Flor, J.; Xavier Martínez de Osés, F. Compass adjustment by GPS and two leading lines. In Proceedings of the 8th International Conference on Maritime Transport: Technology, Innovation and Research, Barcelona, Spain, 17–18 September 2020; pp. 126–143, ISBN 978-84-9880-827-8.
30. Zorović, D.; Kos, S.; Vranić, D. *Brodski Magnetski Kompasi, Teorijske Osnove*; Pomorski fakultet u Rijeci: Rijeka, Croatia, 1998; ISBN 953-165-018-7. (In Croatian)
31. Kemp, J. Experiences with Compasses in the Mid-20th Century. *J. Navigat.* **2010**, *63*, 545–556. [CrossRef]
32. National Geospatial-Intelligence Agency. *Handbook of Magnetic compass Adjustment*; Defense Mapping Agency Hydrographic/Topographic Center: Bethesda, MD, USA, 2004; ISBN 1452860742.
33. Basterretxea-Iribar, I.; Sotés, I.; Uriarte, J. Towards an improvement of magnetic compass accuracy and adjustment. *J. Navig.* **2016**, *69*, 1325–1340. [CrossRef]
34. Pleskacz, K. Necessity for a change to the control procedures for merchant vessel course indicators. *Sci. J. Marit. Un. Szczecin* **2017**, *49*, 69–74. [CrossRef]
35. Kjerstad, N. *Electronic and Acoustic Navigation Systems*; Norvegian Institute of Science and Technology: Ålesund, Norway, 2016; ISBN 978-82-92186-57-2.
36. Harvey, G. Magnetic compass problems—A fresh approach. *J. Navigat.* **1949**, *2*, 230–242. [CrossRef]
37. IMO. *Model Course 7.01—Master and Chief Mate*; IMO Publishing: London, UK, 2014; ISBN 9789280115819.
38. Australian Maritime Safety Authority. Maintenance and Adjustment of Magnetic Compasses; Marine Notice 19/2016. Available online: https://www.irclass.org/media/2537/encl-1-amsa-marine-notice-no19-2016.pdf (accessed on 19 March 2020).
39. The Bahamas National Requirements. Available online: https://www.classnk.com/hp/pdf/activities/statutory/ism/flag/bahamas/BNR.pdf (accessed on 15 September 2020).
40. Magnetic Compasses—Deviation Table. Maritime Cook Island Circular 176/2018. Available online: https://www.maritimecookislands.com/wp-content/uploads/2015/11/Circular-1762018.pdf (accessed on 29 March 2020).
41. The Republic of Liberia, Bureau of Maritime Affairs—Guidelines on QUALSHIP 21 Program Qualification, Marine Operations Note: 03/2011. Available online: https://www.liscr.com/sites/default/files/online_library/MarineOperationsNote_3_2011_QUALSHIP_21.pdf (accessed on 15 September 2020).
42. US/UK World Magnetic Model—Epoch 2020.0—Main Field Declination (D). Available online: https://www.ngdc.noaa.gov/geomag/WMM/data/WMM2020/WMM2020_D_BoZ_MILL.pdf (accessed on 12 January 2021).
43. The World Magnetic Model—Accuracy, Limitations, Magnetic Poles and Error Model. Available online: https://www.ngdc.noaa.gov/geomag/WMM/limit.shtml (accessed on 12 January 2021).
44. Nguyen, V.S. Calculation of the deviation coefficients for marine magnetic compass. *J. Int. Marit. Saf. Environ. Affairs Shipp.* **2019**, *2*, 112–115. [CrossRef]
45. Felski, A. Application of the Least Squares Method for Determining Magnetic Compass Deviation. *J. Navig.* **1999**, *52*, 388–393. [CrossRef]
46. Lushnikov, E.; Pleskacz, K. Contemporary considerations of change regulations regarding use of magnetic compasses in the aspect of the technical progress. In Proceedings of the TRANSNAV 2017, Gdynia, Poland, 21–23 June 2017. [CrossRef]

Journal of
Marine Science and Engineering

MDPI

Article

Cost-Effective Design of Port Approaches Using Simulation Methods Based on the Example of a Modernized Port in the Ustka

Kinga Łazuga [1,*], Nguyễn Minh Quý [2] and Lucjan Gucma [1]

1 Faculty of Navigation, Maritime University of Szczecin, 70-500 Szczecin, Poland; l.gucma@am.szczecin.pl
2 Faculty of Hydraulic Engineering, National University of Civil Engineering (NUCE), Giai Phong 55, Hanoi 100000, Vietnam; quynm@nuce.edu.vn
* Correspondence: k.lazuga@am.szczecin.pl

Abstract: Port design and approaches are usually carried out using real-time computer simulation methods for ship manoeuvring. So-called ship real-time simulation methods are relatively expensive, especially in terms of survey time. Several real-time simulation scenarios carried out by masters and pilots are usually performed, with several simulation attempts in each scenario. Each such attempt can last up to one hour, which, with a large number of scenarios, prolongs the research and increases its cost. Particularly time-consuming is the repetition of many scenarios with alternative solutions for infrastructure development and in various hydrometeorological conditions. To reduce the time-consuming of the tests, a new two-stage method was used to design the target approach on the modernized Port of Ustka. In the first stage, the simulations were carried out with significantly reduced floating navigation marking, and in the second stage with the target marking. Moreover, the so-called "Soft-Bank" method was introduced, i.e., the effects of a collision with the seabed and infrastructure were excluded. Such a solution leads to significant time benefits in conducting research and at the same time does not reduce confidence in the results obtained.

Keywords: ship manoeuvring; design of ports; real-time ship manoeuvring simulations; navigation safety

Citation: Łazuga, K.; Quý, N.M.; Gucma, L. Cost-Effective Design of Port Approaches Using Simulation Methods Based on the Example of a Modernized Port in the Ustka. *J. Mar. Sci. Eng.* **2021**, *9*, 211. https://doi.org/10.3390/jmse9020211

Academic Editor: Jakub Montewka

Received: 18 January 2021
Accepted: 12 February 2021
Published: 18 February 2021

1. Introduction

The major aim of a case study [1] which is the illustration to presented Soft-Bank method was to design the new approach and breakwater solution for a general cargo ship of the following parameters: L = 133 m, B = 20 m, T = 7.9 m utilizing real-time ship manoeuvring simulation methods for the modernized Ustka Port located in the middle of the Polish coast. Additionally, the conditions of safe operation of port for the maximal vessel will be established. The main aim of the Ustka Port case study that was concerned with:

1. Determination of:
    - average ship dimensions and its characteristics like the power of the main engine, rudder type and its area or power of bow thruster,
    - safety waterway parameters needed for the safe operation of maximal ships,
    - turning place diameter concerning its shape.
2. Determination of safety conditions of port operation for:
    - admissible meteorological conditions for a given kind of ships and manoeuvres,
    - other navigational conditions and limitations like the presence of other ships on berths, use of position fixing systems on the approach, navigational markings, and vessel traffic service.
3. Determination of manoeuvring procedures during the entrance, berthing, un-berthing, exit port, and turning for different kinds of ships and propulsion systems.

4. Determining the conditions of ship mooring inside the port.

The above objectives are the objectives of the Ustka Port design case study that was the example of Soft-Bank method. The main objective of the paper is to present that method as the part of optimal port design strategy.

The major problem with designing the simulation experiment is the number of simulation trials (one trial is defined as one simulation run in given conditions), which have to be done within experiments. Each simulation trial is realized in real-time and it sometimes lasts more than half an hour in the presented case study.

The simulation trials are performed in groups (series) in the same controllable environmental condition but with different captains/navigators performing the simulation to achieve statistical variability of human (navigator) influence. The whole study must cover selected wind and current conditions. In single simulation series usually, 15 simulation trials are performed, which causes the design process is highly time-consuming. For example, in the Ustka case study, the 6 series are foreseen every 15 trials which make 90 simulation runs, each lasting more than 30 min in total which makes in total more than 60 simulation hours. When the infrastructure does not exist (like in the presented study) sometimes it needs to perform simulation in different infrastructure layouts, which again multiply the simulation time by the number of infrastructure solutions (proposed layouts).

In such cases, the simulation method is one of the most suitable to solve this research task. Several unsolved problems when ships outgrow the capacity of port infrastructure have been identified by Perkovic et al. [2]. The guidelines [3–6] also address the design issue. In parallel, the national and regional policy was identified [7]. Ports are also the subject of strategic risk assessments [8–11]. From the other side dedicated systems are used for met-ocean conditions monitoring near the ports and specialized quays [12]. The general rules of designing the ports and waterways for ships are presented in [13–15]. Methods applied in this study could be used also for other aspects of navigational safety like designing port regulations [16].

The methodology design of real-time simulations for waterway and port design purposes are presented in the latest PIANC guidelines [3,4]. The comprehensive study in this field, especially for the distribution of ship positions on the waterway, is presented by Irribaren [17]. Some general guidelines are presented in [18,19]. There are not many new types of research dealing with this topic mostly due to the process is highly related to the given case study. The design process also depends on the experience and knowledge of the simulation team. This is remarkable to the maritime simulation sector wherein compared to other branches of transportation like aviation or road engineering, there is still some freedom and art inside the process itself. The expert knowledge supported by pilots here is usually the key factor for conducting the simulations based on their previous design.

Benedict et al. [20] developed the computer-based support for the evaluation and assessment of ship handling simulator exercise results but dedicated mostly for training purposes of ship handling simulators.

Very useful in the navigator decision-making process are ship predictors, which are also quite common onboard usually achieved within ECDIS environment [21].

Zhang [22] presented a very comprehensive study on the assessment of the competence of seafarers trained on ship handling simulators in the scope of Bridge Team Management and implementing the International Convention on Standards of Training Certification and Watchkeeping for Seafarers (the STCW Convention).

Sarioz and Narli [23] presented the results of a real-time ship manoeuvring simulation study and its assessment intended to investigate the manoeuvring performance of large tankers in the Bosporus.

Inoue [24] developed a quantitative model for evaluating the difficulty of ship handling caused by a restricted manoeuvring area or by traffic congestion or by a combination of both. It includes acceptance criteria based on the mariner's perception of safety.

Lataire et al. [25] presented the systematic investigation of ship manoeuvring in restricted waters is performed by Flanders Hydraulics Research (FHR) and Ghent University (UGent) through five different simulations techniques including real-time human-controlled and fast time. Another Classification of ship maritime simulators is presented by Cross and Olafsson [26].

Donatini et al. [27] described the results of a survey performed by the authors to assess how hydrometeorological conditions are presently modelled in ship manoeuvring simulators. They found that while mathematical models for the manoeuvring behaviour of ships are well documented in the literature, an overview concerning hydrometeorological modelling does not exist yet. The results are based on a wide questionnaire of simulator end users.

Several types of research were made in the field of effects on the ship models implemented within ship handling simulators like wind [28], waves [29], ice, and current [30]. Delefortrie and Vantorre [31] presented the overview of research and practical applications of ship's behaviour and modelling in muddy areas.

Fast-time simulations (FTS) are the widely accepted preliminary study method in design of ports and its approaches [3–5]. The recent state of the art in this area is presented by the Benedict at all [32]. The drawbacks of FTS are the autopilot capabilities which is still far from human control and problems with automatic tug control which are the usual tool to enhance manoeuvres in ports which also was the case in the presented example.

2. The Ustka Port Case Study

Ustka (Figure 1) is a medium Polish port located on the Baltic Sea serving around 100 merchant ships per year with fishing and pleasure craft traffic. The maximal length of ships before the presented research was L = 60 m, B = 12 m, and T = 4.0 m with several operational and weather restrictions.

Figure 1. Ustka Port—the present layout.

The preliminary design of the breakwater layout is presented in Figure 2 (presented in yellow). After the extensive wave development inside port analysis, it was found that the wave was too high inside the port with NE winds and it was decided to investigate also the longer version of the breakwater (Figure 2 presented in red) together with the decreasing the entrance width and some works inside the port to reduce wave development inside the port.

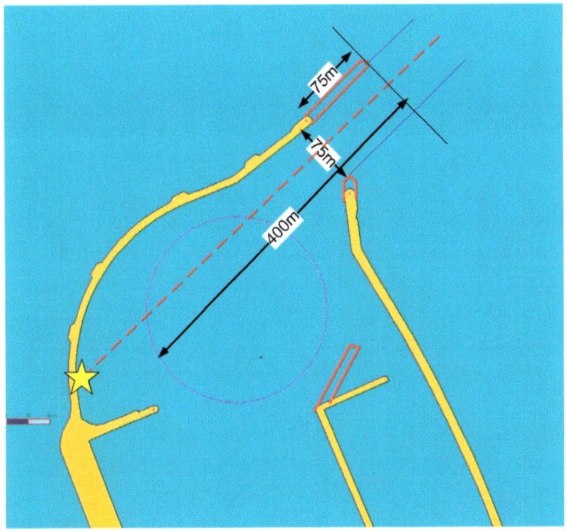

Figure 2. Ustka Port—the preliminary design (yellow) and proposed changes in the breakwater after wave conditions modelling inside the port (red).

2.1. Selected Elements of Ship Mathematical Model Creation

The values of hydrodynamic coefficients of particular forces and moments have been predetermined according to published literature data from hull model tests (surface and underwater part), propeller, and stern rudder, streamer rudder of similar dimensions and shapes as the "characteristic" ship. In the case of gross mismatch, appropriate extrapolation of test results to the technical and operational conditions of the model ship has been applied. This was all the more important because not all coefficients can then be optimized (tuned, identified) according to the measurements carried out during the nautical manoeuvre tests of the tested craft or similar. The models are usually identified based on sea trials of a modelled ship or similar ships according to the procedure presented by Artyszuk [33]. The model used for Ustka Port development was validated based on four following groups of tests:

1. Speed tests
2. Braking and acceleration tests
3. Circulation tests (Figure 3)
4. Zig-Zag tests (Figure 3).

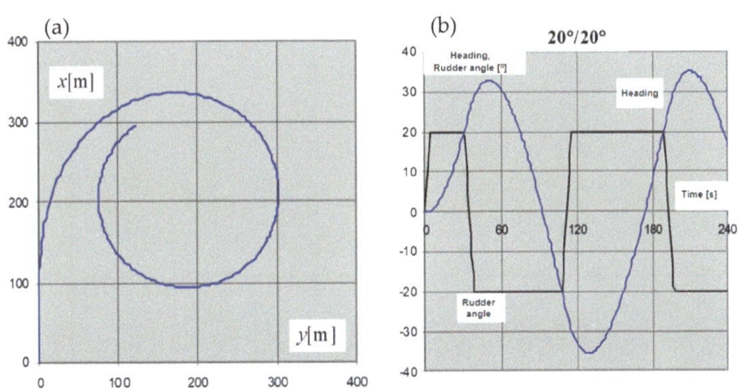

Figure 3. Selected results from model trials: (**a**) turning circle and (**b**) 20°-20° Zig-Zag manoeuvre.

Therefore, the model represents the average ship with the average parameters for typical Baltic Sea general cargo vessel of L = 133 m, B = 20 m and T = 7.9 m. The model was created on the base of several available sea trials of similar ships which results was averaged and used for fitting the hydrodynamic coefficients. The selected model sea trials i.e., turning circle and 20°-20° Zig-Zag manoeuvre are presented in Figure 3.

2.2. Selected Elements of Environment Modelling and Conditions Selection during Tests

The objective of the research: Research on the possibility of entrance and exit of a general cargo ship of L = 133 m, B = 20 m, T = 7.9 m in the modernized port of Ustka Modernization of external breakwaters and construction of an internal port.

Breakwater system: The corrected arrangement of breakwaters according to Figure 2.

Ship parameters:

Type: General cargo, coaster. Lc = 133 m; B = 20.0 m; T = 7.9 m.

Propulsion: Right-hand conventional propeller, bow thruster typical power for this vessel.

Indicative speeds: FA (Full Speed Ahead) = 12 kn, HA (Half Ahead) = 9 kn, SA (Slow Ahead) = 6 kn, DSA (Dead Slow Ahead) = 4 kn.

Towing operation: One tug with a 20 tf of pull power.

Hydrometeorological conditions: Manoeuvres were performed for conditions without wind and wave (as a reference and for a preliminary acquaintance of the ship and the area by captains) and with wind 11 m/s (lower value of 6°B), and 17 m/s (upper value of 7°B) blowing from the north-western direction, which is the most unfavourable direction for the entrance. For the exit and ships turning, the winds up to 11 m/s from the north-western and north-eastern directions were assumed to be the most influential for the ship during turning.

Wind cover by infrastructure was assumed (Figure 4). The steady wind (no gusts) was used for the analysis.

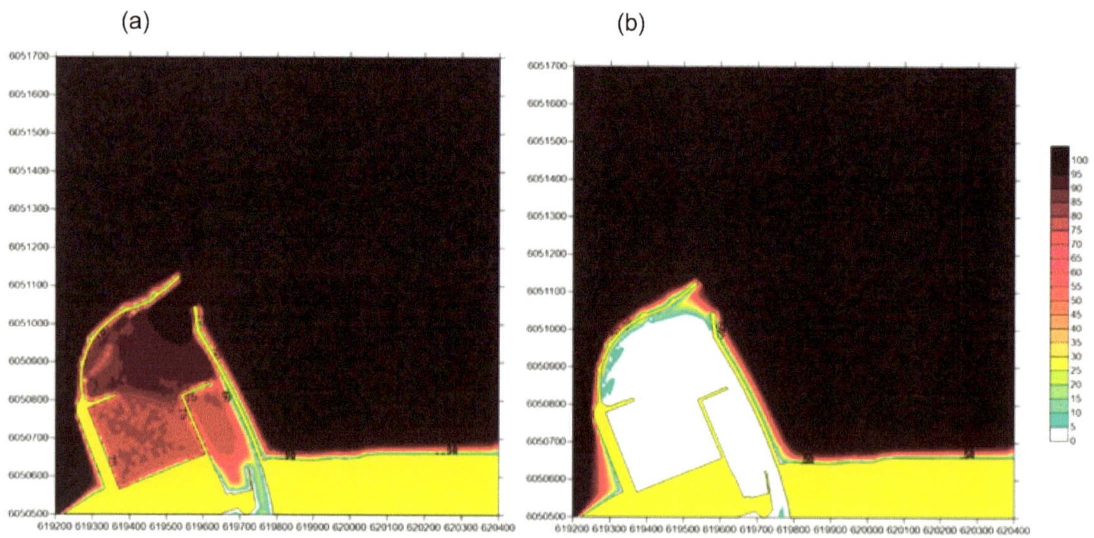

Figure 4. Selected elements of environmental modelling. (**a**) wind scale-up factors for NE winds, (**b**) wave scale-up factors for NE wave system.

The height of the wind-wave (the direction following the wind direction) was assumed equal: 0.9 m (wind 11 m/s) and 1.6 m (wind 17 m/s). The wave distribution is shown in Figure 4.

Good visibility and daytime visibility were assumed, which results from the type of simulator visualization.

One-way traffic of ships has been assumed.

The weather conditions, especially the wind and water level changes were analysed before planning the simulations. The influence of climate change is also considered as a mean water level change and changes in the severity of strong winds. There are hydro- and meteorological stations in the Ustka Port. The data were collected with the use of this station and then after statistical analysis applied to the project and final solution. The possibility of storm surges was included as a deterministic factor in the under-keel clearance analysis as an additional reserve for storm surges and water level changes appropriate for this region.

The selected wind conditions represent the one from most extreme to easiest in the scope of ship manoeuvring. It is done before simulations by very careful planning the experiment conditions. In this case, it was decided to consider the following conditions: the close to extreme operation conditions, maximum average conditions, and zero wind conditions. The selected wind directions are also considering infrastructure-related factors like breakwater layout. The 20 years wind rose for Ustka is presented in Figure 5.

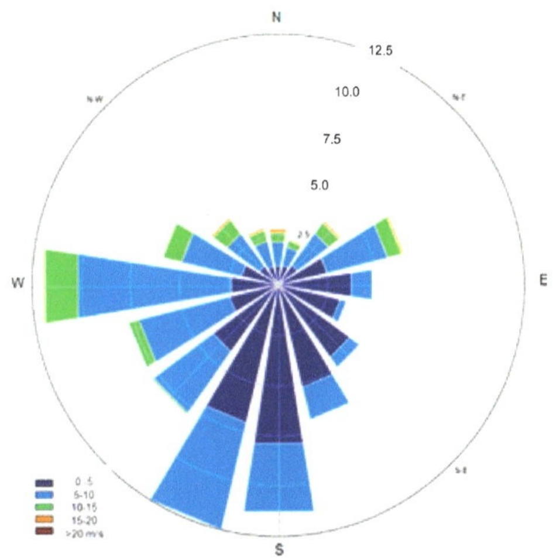

Figure 5. The wind rose for Ustka Port. (wind 1 h averaged, data from 2000–2020).

The prediction of weather is important for the operational stage of the process of ship entrance to the ports. Usually, ports use government regulated prediction platforms for their operation. As an appropriate tool for predictions, the Ensemble Prediction Systems (EPS) could be used. EPS are numerical weather prediction (NWP) systems that allow estimating also the uncertainty in a weather forecast as well as the most likely outcome [34]. This technique is based on running the NWP model several times with slightly different initial conditions [35]. Such a tool could be used also for hydrological problems like storm surges [36,37].

2.3. The Detailed Sea Trials Performed and Their Conditions

Five simulation series have been planned, each representing typical manoeuvres under different conditions, which have been selected to cause the greatest difficulty. The plan of research is presented in Table 1. The individual simulation series represent manoeuvring situations selected from the point of view of difficulty related to the operation of vessels in this area, i.e., successively:

1. series 1, 2, and 3—entry to the port of general cargo vessel L = 133 m without rotation and mooring. The purpose is to determine the parameters of the approach waterway and the safety of entrance and mooring energy.

2. series 4, 5 and 6—entry from the general cargo carrier port L = 133 m with rotation. The purpose is to define the parameters of the turning place, and waterways parameters during ship's departure.

Table 1. The plan of simulation research in Ustka.

No.	Name of File	Manoeuvre	Initial Speed (kn)	Wind Speed (m/s)	Wave on the Approach (m)	No. of Trials
1	1_L133_Wej_0	Entry into port and mooring on the starboard side	6	no	no	15
2	2_L133_Wej_NW11	as above	6	NW 11	0.9	15
3	3_L133_Wej_NW17	as above	6	NW 17	1.6	15
4	4_L133_Wyj_0	Unmooring, turning and leaving port	0	no	no	15
5	5_L133_Wyj_NW11	as above	0	NW 11	0.9	15
6	6_L133_Wyj_NE11	as above	0	NE 11	0.9	15

3. Methods

3.1. Proposed Method of Reducing Cost and Time of Simulation Analysis

The major change in the typical method used for designing the breakwater and port entrance in contrary to already existing methods [3–5,38] is the use of the so-called "Soft-Bank" method (Figure 6). This methodology is as follows:

1. Determine design water depth considering ships draft and under-keel clearance (H = 9.0 m in the presented study);
2. Design basic navigational aids with its minimum as possible number;
3. Design the navigable area without simulated embankments so no interaction between the ship and embankments is simulated;
4. Display the area layout on the electronic chart and inform the Captains performing the simulations that it is possible to passage the ship over the elements of infrastructure only in justified by environmental conditions cases;
5. Execute simulations and analyse the results.

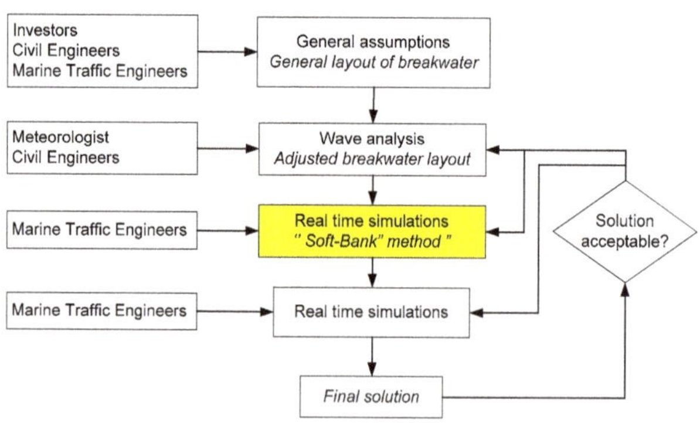

Figure 6. The flowchart of the new method of port approach area design applied in this study.

The procedure is presented in Figure 6 is alike the standard port design procedure [3] with the novel "Soft-Bank" module included.

It should be mentions that that in port and waterway design from ships manoeuvring perspective other preliminary methods could be applied like:

1. Analytical methods like PIANC, ROM or Japanese;

2. Statistical methods based on generalisation of simulation experiments;
3. Fast Time Simulations (FTS) method.

3.2. Real-Time Manoeuvring Simulation MethoD—Limited Task Simulator

The application of the real-time manoeuvring simulator with navigators (Captains and Pilots) as the control element in the loop, as applied in this study, is supposed to be the most reliable in port and waterways design studies [39]. There are several kinds of simulators with various applications, from the most advanced full mission simulators to limited task simulators. The latter was applied in the presented research. The simulator has 2D display and was made by the Maritime University of Szczecin research team and is described in [39,40]. The ship's hydrodynamic model applied in this simulator is based on detailed parameters of hulls, propulsors, and steering devices. External influences like current, wave, shallow water, and collisions are modelled. Usually, depending on the availability, the actual manoeuvring characteristics are applied for the model's validation. A special procedure for such validation is developed. The model of m/f Ustkamax used in the presented research is created with the modular methodology where all effects like hull hydrodynamic forces, propeller forces, and steering equipment forces together with given external influences are modelled as separate forces, and finally they are summed as perpendicular, parallel and rotational forces and later on movements [41].

The modular approach applied here for the ship manoeuvring simulator is presented in Figure 7. The graphical interface of the model is based on 2D display similar to the nautical chart (Figure 8). The interface includes also the data of ships basic parameters (position, course, speed, rotational speed, etc.), mooring pier and coastline location, navigational markings, soundings, external conditions information, tug steering interface, and line controller, and other control elements of the model. The model is implemented in Object Pascal with the use of Delphi™ environment and Visual C™ with the use of C++ language. Limiting to the usual 3DOFs (the horizontal planar motion) or in some application 4DOFs when the squat is included. The ship movement over the ground (thus the so-called dynamic effect of the water current is introduced) is given by Artyszuk [41]. The crucial element in modelling is the verification of the ship's hydrodynamic model [42]. Usually, the verification process is made until 10% or smaller error between model and real ship data in the selected trials like turning trial, zig-zag, and stopping—acceleration is achieved [39].

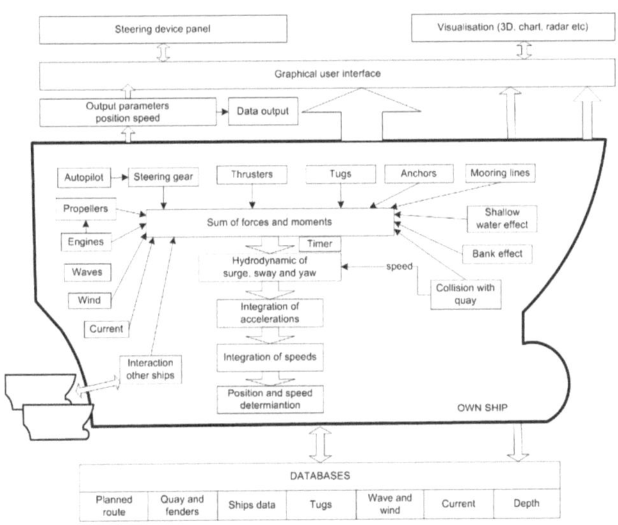

Figure 7. The modular model of ship manoeuvring for the port design that was applied in this study.

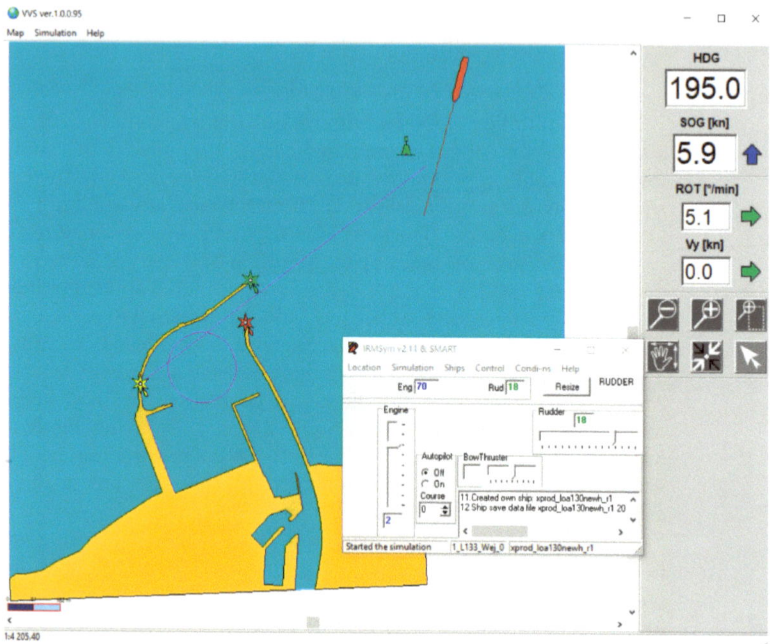

Figure 8. The 2D GUI (Graphical user interface) of the simulation model (with ships movement control panel activated).

3.3. *Statistical Methods of Data Processing*

Despite the real-time ship simulators are now very widely used and hydrodynamic models are becoming increasingly accurate, without efficient statistical data processing, it is usually not possible to draw reliable conclusions from the experiments. For the simulation data processing standard statistical models are proposed.

The Method of Simulation Result Data Processing

The most important safety factor is the horizontal area which is required for navigators to perform the safe manoeuvres [39,43]. Statistical processing of the simulation results allows the determination of the statistical parameters necessary to define a safe manoeuvring area (SMA). There are three specific values for the given waterway areas occupied by ships as the result of simulations. They are determined as (Figure 9):

1. Maximum waterway area needed for manoeuvring ships (extreme ships positions in all trials),
2. Average waterway area needed for manoeuvring ships (defined as mean SMA),
3. Waterway area on the given confidence level (defined as SMA on a given confidence level).

The analysis of simulation results is to determine the parameters of the ship's horizontal safe manoeuvring area. In simulation trials, these parameters are determined by the width of the ship's manoeuvring area, i.e., the area occupied by a single ship during a manoeuvre. The traffic lane (so-called PATH) is defined for a given ship and manoeuvre, while the safe manoeuvring area (SMA) is a term given to different ships and various manoeuvres (Figure 6). It can be seen from Figure 6 that the safe manoeuvring area (SMA) extends beyond the available water area (AWA) and encroaches on the navigational danger area (D), resulting in the need for some adjustments (e.g., dredging) to avoid potential accidents. SMA defined in such way is probabilistic 2D area which includes ships hydrodynamics with external effects (wind, wave, tugs, etc), position systems errors, and human-navigator performance during conducting the manoeuvre.

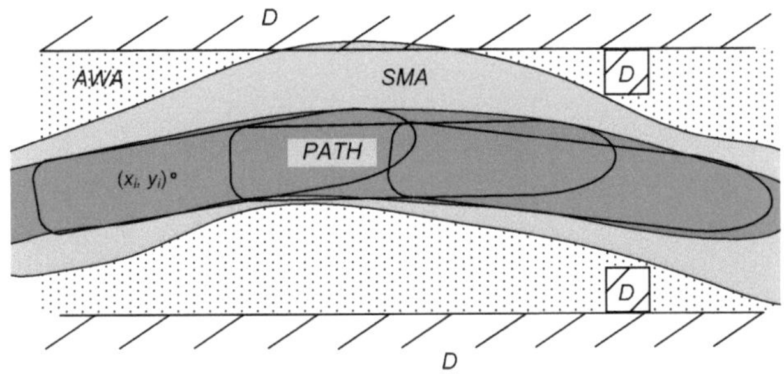

Figure 9. Definition of the ideas connected with horizontal areas taken by ships (PATH—lane of a single ship, AWA—available water area, SMA—safe manoeuvre area on the required confidence level, D—navigational danger).

A safe manoeuvring area is an area in which the probability of exit of the ship beyond the AWA is at a relatively low level. In the port design studies, usually, 95% is applied in typical ship operations and 99% in more critical operations such as in presence of passengers of dangerous cargo or the existence of hard bottom [40]. The basic safe navigation condition needs to satisfy the following dependency:

$$d_{i\alpha} \leq D_i \tag{1}$$

where:

Di—width i-th point of the waterway at the bottom for safe isobath,

$d_{i\alpha}$—width of safe manoeuvring area on the given confidence level (1-α).

It should be noticed that the general population here has the infinite number of variables of all possible simulation trials of the ship. The sample is defined as the series of simulation trials conducted with an adequate number at the same conditions. The width of the safe manoeuvring area of the ship is the range, which contains specified as a percentage part (fraction) of the general population. It can be defined as:

$$d_{i\alpha} = m_{di} + k_\alpha \sigma_{pi} + k_\alpha \sigma_{li} \tag{2}$$

where:

$$m_{di} = m_{pi} - m_{li} \tag{3}$$

or using the equivalent dependence in the form of:

$$d_{i\alpha} = d_{ip\alpha} - d_{il\alpha} \tag{4}$$

for:

$$d_{ip\alpha} = m_{pi} + k_\alpha \sigma_{pi} \tag{5}$$

where:

$d_{i\alpha}$—width of the safe manoeuvring area at i-th point of the waterway on the confidence level equals to (1-α);

m_{di}—average of the safe manoeuvring area;

k_α—factor dependent on the fraction of the general population, which should be covered by estimation (for SMA 95% k = 1.96);

mli, mpi—mean of the maximum distance of ship's extreme points on the port side and starboard side of the waterway;

σli, σpi—standard deviations of the maximum distance of ship's points to the port and starboard from i-th point of the waterway;

$d_{il}\alpha$, $d_{ip}\alpha$—width of the port and the starboard safe manoeuvre area at i-th point of the waterway at defined confidence level (1-α).

The presented approach leads to the probabilistic method of safe manoeuvring area establishing is presented in Figure 10.

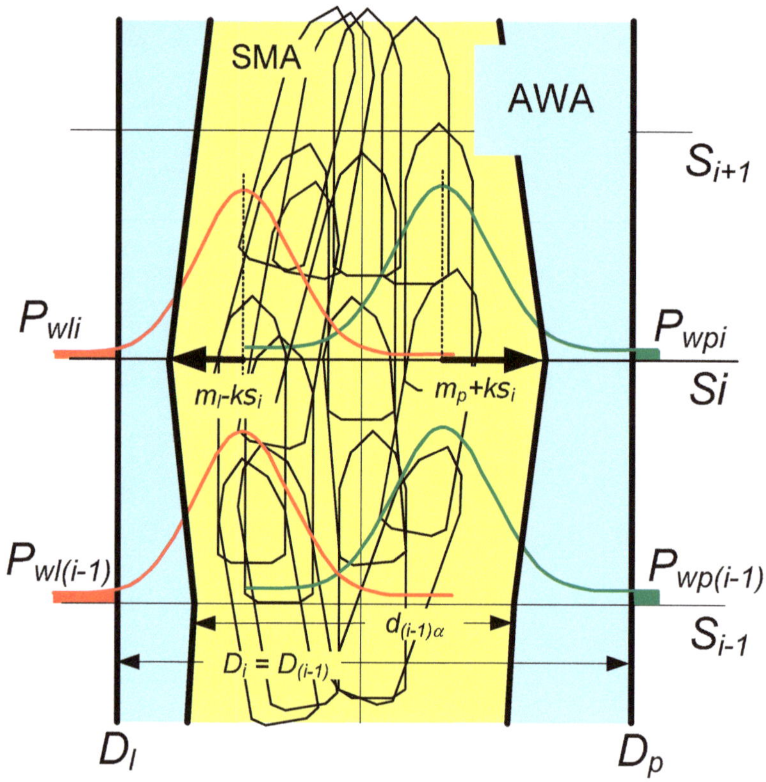

Figure 10. Definition of the probabilistic approach to determine horizontal areas from ships manoeuvring simulations (PATH—lane of a single ship, AWA—available water area, SMA—safe manoeuvre area on the given confidence level).

3.4. Conducting the Research

Several simulation series have been planned and then conducted as an illustrative to solve the port design problem:

1. Entrance and departure in no wind conditions (light conditions);
2. Entrance to the port with wind NW 11 m/s (moderate condition);
3. Entrance to the port with wind NW 17 m/s (severe condition);
4. Departure and turning manoeuvre in no wind;
5. Departure and turning manoeuvre with wind NW 11 m/s;
6. Departure and turning manoeuvre with wind NE 11 m/s.

The simulations according to the presented "Soft-Bank" methodology was made for the ship approach only (series 1–3). The shading effect of wind and wave were included in the simulation. Apart from the wind, the wave was modelled with a significant height of 0.9 m (moderate condition) for 11 m/s wind speed and 1.6 m (severe condition) for stronger winds. In total six experienced Captains were engaged to perform the simulations. The fifteen ship passages were conducted for each simulation series, which make 75 simulation runs in total. The single entrance manoeuvre of the ship for "Soft-Bank" method is presented in Figure 11.

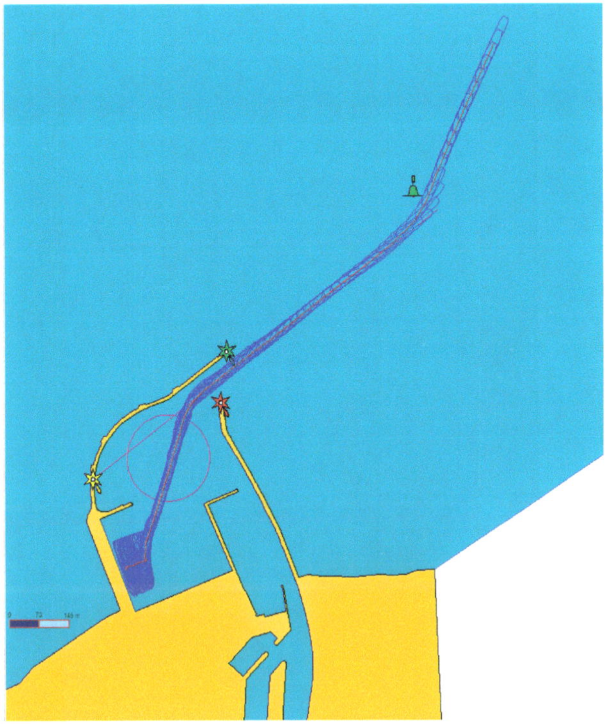

Figure 11. The single simulation passage—entrance and berthing in zero wind conditions.

4. Results of Simulation Research and Discussion

All simulations were conducted by qualified captains and pilots experienced in this type of vessel and manoeuvre. Simulation data were recorded and analysed. The analysis of the simulation results was carried out based on the horizontal safe manoeuvring area criterion at the 95% confidence level (SMA) typically used in marine operations [5,39] according to the previously presented method.

As it is presented in Figure 12. the navigational marking in "Soft Bank" method is reduced to a minimum—there is only one green buoy on approach to show the Captains the turning point but only in a very approximately manner. Figure 13 presents the results of a standard method (without "Soft-Bank") with final navigational markings and with modelled embankments and canal effects. The explanation for Figures 12 and 13:

- 95% is the Safe Manoeuvring Area (SMA) at a 95% level of confidence.
- Mean is the average waterway area.
- MAX is the maximal overbound area of all ships in series.

4.1. Comparison of Methods. Discussion

Table 2 shows channel widths obtained from different methods for two wind speeds. It can be noted that in the case of the simulation method, the channel widths under severe conditions may be smaller than under moderate conditions. This is because these manoeuvres are performed by experienced pilots and captains and not autopilot or artificial intelligence. Worse hydrometeorological conditions give fewer possibilities to manoeuvre freely (including the choice of speed and adjustments) and to manoeuvre correctly navigators have to do it very similar each time. In better hydrometeorological conditions, on the other hand, a man has more possibilities and selects the settings more freely. This is a common paradox when a man controls a ship. It should be noted that using analytical or empirical methods like PIANC, this phenomenon does not occur and the worse the conditions, the more manoeuvring space is necessary. Knowing this phenomenon, waterway designers use it by appropriately processing data from the simulation.

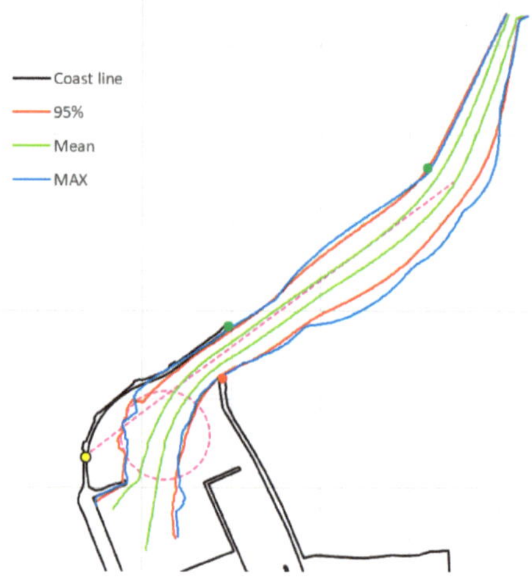

Figure 12. Manoeuvring areas of a general cargo ship of L = 133 m on approach to modernized Ustka Port by "Soft-Bank" with limited (preliminary) navigational marking.

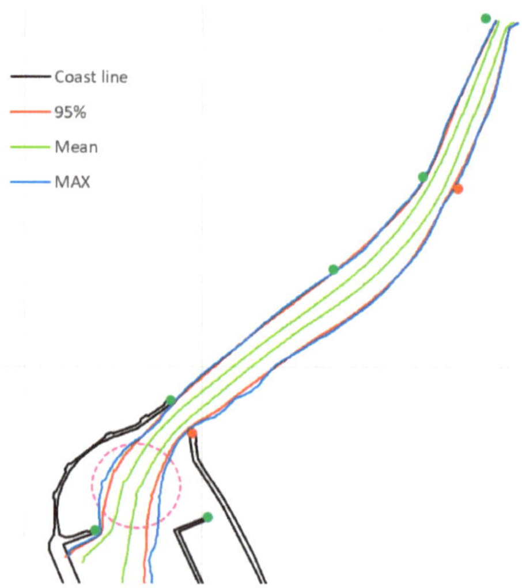

Figure 13. Manoeuvring areas of L = 133 m general cargo ship on approach to modernized Ustka Port using the standard method with final navigational marking.

The basic statistical parameters of the manoeuvring area namely: mean and standard deviation are presented in Figure 14 for the designed approach waterway. The number of sections is 250 and the section width is 5 m. It can be observed that the mean and standard deviation is significantly lower for the standard method in comparison to the "Soft-Bank" method. It is because of the design, after the first step, navigational marking limits the waterway. Moreover, some changes in the waterway layout have been made so the movement of the ship on the approach is more optimized. The green buoy on the approach in "Soft-Bank" method (Figure 12) has been removed and the final layout of navigational marking is proposed (Figure 13) that consists of the gate of red-green buoys and two green buoys marking the starboard side of approach waterway. Such a design of waterways together with the navigational marking need an experienced marine traffic engineer engaged in the process.

Table 2. Channel widths were obtained from different methods for two wind speeds.

No	Methods	Case	Channel Widths	
			Moderate	Severe
3	Simulations	Mean	30	33
		95% confident	83	80
		Max	79	75
1	PIANC		60 m (3.0 B)	70 m (3.5 B)
2	Japanese (OCDI 2009)		55.6 m (2.8 B)	NA

B: ship width.

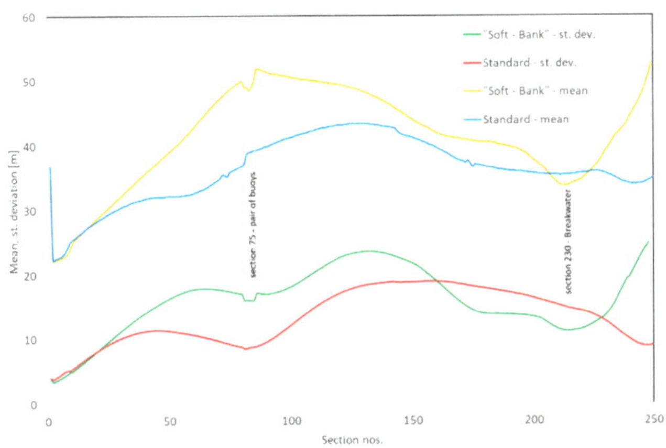

Figure 14. Comparison of two parameters of manoeuvring areas of the investigated ship on approach to modernized Ustka Port using the "Soft-Bank" method for preliminary design and standard method for the final design.

4.2. The Limitation of the "Soft-Bank" Method

The presented method has some limitations, mostly in the following form:

1. The bank effect cannot be taken into account;
2. Limited possibility of taking into account the settlement of the vessel;
3. Impossible to take into account an accident in the form of a stranding and collision with embankments or a moored vessel or another civil engineering technical object;
4. The psychological impact on the navigator due to a false sense of safety due to not considering the ship's collision possibility in simulation trials.

The first two limitations are not critical because there is the possibility to investigate them in the final stage of simulations if such need exists. In some research, however, depending on the investigated area effects presented in points 1 and 2 are not crucial and they could be neglected. Limitation number 3 is usually not critical since accidents are quite rare even in simulations. Limitation number 4 needs a deeper understanding of the Captains and Pilots. Usually, they are well debriefed before simulation (this was also the case in the Ustka Port study) and informed about such limitations. Clear instruction on how to behave shall be given because it could vast the overall effort of the simulation process. In the presented case, Captains and Pilots got the serious message in their task description namely: "The breakwaters and entrance waterway are modelled as "soft" and can be "passed over" by the ship—but this should be as much as possible avoided—this is done to determine possible changes in their layout. The quay is modelled as "hard" please impact with it to determine the energy of ship contact".

5. Conclusions

The presented study showed the applicability of real-time simulation methods for design the new ports, their approaches, and the breakwater layouts. The case study of Ustka Port was used as a working example. Moreover, the "Soft-Bank" method developed and presented could let to reduce almost 50% workload for conducting the simulation runs by reducing the number of possible port infrastructure solutions. The presented approach gives also the possibility of predesign and opening the discussion about the proposed solution in the middle of research work which gives always some opportunity to test more solutions and optimize the overall project. The method itself is useful because it comprises the most important effects related to ship's dynamics and enables suitable shaping of the layout of waterway and breakwaters without testing too many variants. The method has also great potential for designing the navigational marking like in this particular study the navigation marking was designed by the presented method, deciding on the possibility of mooring vessels in the vicinity of waterways (Some vessels could limit the existing waterways and decision could be done with presented method application. It should be noted that in comparison to the Fast Time Simulation (FTS) method, the Soft-Bank method seems to have both advantages and disadvantages. The Soft-Bank method is still more time consuming, especially when comparing different approach channel layouts and/or breakwater configurations. The Soft-Bank method enables to engage in the process of Captains and Pilots and tugs for manoeuvres which is usually problematic in the FTS method.

The case study was presented in [Gucma L., Łazuga K. & Perkovic M. 2019. Ship manoeuvres on existing turning places—when the size of the ship reaches the limits of port infrastructure on the example of Kołobrzeg Port. Proc. of European Navigation Conference (ENC) Warsaw.] to będzie 42), turning places design [11], and other types of port infrastructure dedicated for ships. The limitations of the presented method were selected and discussed. They generally do not negatively affect the overall process of waterway design since all its important drawbacks could be eliminated in the final step design like it was presented in the Ustka Port case study.

Moreover, it was observed that the real-time simulation method and the limited task simulator used here have proven their usefulness in port modernization work to increase the benefits of port operations without compromising navigational safety. The following precautions should be considered when planning and executing simulation experiments:

1. Simulators are widely used tools and proper verification, especially of the simulation hydrodynamic model and hydrometeorological conditions models, should be carried out at the outset to match simulations to reality as closely as possible.
2. The simulation method should be carried out by multiplying ship runs. The simulation studies based on a single or very small number of simulations without statistical data processing and experimental plan are questionable and do not present the proper value to port design needs.
3. In any real-time simulation project, a very good link between pilots with good local knowledge for validation and provision of domain expertise should be established.

Author Contributions: Conceptualization, L.G.; methodology, L.G.; software, L.G. and K.Ł.; validation, K.Ł., N.M.Q. and L.G.; formal analysis, K.Ł.; investigation, K.Ł. and N.M.Q.; resources, L.G. and N.M.Q.; data curation, K.Ł.; Writing—Original draft preparation, L.G.; Writing—Review and editing, K.Ł. and N.M.Q.; visualization, L.G.; supervision, K.Ł.; project administration, L.G.; funding acquisition, L.G. All authors have read and agreed to the published version of the manuscript.

Funding: This research received no external funding.

Institutional Review Board Statement: Not applicable.

Informed Consent Statement: Not applicable.

Data Availability Statement: Not applicable.

Conflicts of Interest: The authors declare no conflict of interest.

References

1. Maritime University of Szczecin. *Navigational Analysis of Ustka Modernization*; 2019; Unpublished. (In Polish)
2. Perkovic, M.; Brcko, T.; Luin, B.; Vidmar, P. Ship Handling Challenges When Vessels are Outgrowing Ports. In Proceedings of the IMLA-INSLC, Cape Town, South Africa, 16 September 2016.
3. Pianc MarCom WG 121: Harbour Approach Channels–Design Guidelines. Available online: https://www.pianc.org/publications/marcom/harbour-approach-channels-design-guidelines (accessed on 3 October 2020).
4. Pianc MarCom WG 158: Masterplans for the Development of Existing Ports. Available online: https://www.pianc.org/publications/marcom/masterplans-for-the-development-of-existing-ports (accessed on 3 October 2020).
5. ROM 3. *1-99: Recommendations for the Design of the Maritime Configuration of Ports, Approach Channels and Harbour Basins*; Puertos del Estado: Madrid, Spain, 2007; ISBN 978-84-88975-39-3.
6. IALA GUIDELINE 1058: Use of Simulation as a Tool for Waterway Design and Aids to Navigation Planning. Available online: https://www.iala-aism.org/product/use-of-simulation-as-a-tool-for-waterway-design-and-aids-to-navigation-planning-1058 (accessed on 3 October 2020).
7. Vuylsteke, A. *Scale Enlargement in the Flemish Port Area*; Vlaanderen, 2016.
8. Billington, C.J. Managing Risks in Ports. In *Managing Risks in Shipping: A Practical Guide*; The Nautical Institute: London, UK, 2001; pp. 57–69.
9. Port Marine Safety Code. Available online: https://www.gov.uk/government/publications/port-marine-safety-code (accessed on 3 October 2020).
10. Gucma, S.; Przywarty, M.; Dzwonkowski, J.; Bilewski, M. Dimensioning of Fairway Bends—Kinematic Method of Numerical Simulation. *J. Mar. Sci. Eng.* **2020**, *8*, 138. [CrossRef]
11. Vidmar, P.; Perkovič, M.; Gucma, L.; Łazuga, K. Risk Assessment of Moored and Passing Ships. *Appl. Sci.* **2020**, *10*, 6825. [CrossRef]
12. Pianc MarCom WG 117: Use of Hydro/Meteo Information for Port Access and Operations (2012). Available online: https://www.pianc.org/publications/marcom/use-of-hydro-meteo-information-for-port-access-and-operations (accessed on 3 October 2020).
13. McCartney, B.L. (Ed.) *Ship Channel Design and Operation*; ASCE manuals and reports on engineering practice; American Society of Civil Engineers: Reston, VA, USA, 2005; ISBN 978-0-7844-0770-7.
14. Ligterngen, H.; Velsink, H. *Port and Terminals*; VSSD: Kanpur, India, 2012.
15. Carl, A. *Thoresen Port Designers Handbook*, 3rd ed.; ICE Publishing: London, UK, 2014; ISBN 978-0-7277-6004-3.
16. Gucma, L.; Łazuga, K. The Support of Port Regulation Creation and Update by Real-Time Ship Manoeuvring Simulation Studies Exampled by Port of Kołobrzeg. *Trans Nav Int. J. Mar. Navig. Saf. Sea Transp.* **2019**, 13. [CrossRef]
17. Iribarren, J.R. *PIANC Bulletin No. 100*; PIANC: Bruxelles, Belgium, 1999.
18. William, C. *Webster Shiphandling Simulation: Application to Waterway Design*; National Academies Press: Washington, DC, USA, 1992; p. 2015, ISBN 978-0-309-04338-0.
19. Permanent International Association of Navigation Congresses; Permanent Technical Committee II; Working Group No. 20. *Capability of Ship Manoeuvring Simulation Models for Approach Channels and Fairways in Harbours: Report of Working Group No. 20 of Permanent Technical Committee II*; General Secretariat of PIANC: Brussels, Belgium, 1992; ISBN 978-2-87223-040-2.
20. Benedict, K.; Baldauf, M.; Felsenstein, C.; Kirchhoff, M. Computer-Based Support for the Evaluation of Ship Handling Exercise Results. *WMU J. Marit. Aff.* **2006**, *5*, 17–35. [CrossRef]
21. Benedict, K.; Kirchhoff, M.; Gluch, M.; Fischer, S.; Baldauf, M. Manoeuvring Simulation on the Bridge for Predicting Motion of Real Ships and as Training Tool in Ship Handling Simulators. *Int. J. Mar. Navig. Saf. Sea Transp.* **2009**, 3, 25–30.
22. Zhang, W. Assessing the Competency of Seafarers Using Simulators in Bridge Resource Management (BRM) Training. Dissertation Thesis, World Maritime University, Malmo, Sweden, 11 May 2017.
23. Sariöz, K.; Narli, E. Assessment of Manoeuvring Performance of Large Tankers in Restricted Waterways: A Real-Time Simulation Approach. *Ocean Eng.* **2003**, *30*, 1535–1551. [CrossRef]
24. Eloot, K.; Verwilligen, J.; Vantorre, M. A Methodology for Evaluating the Controllability of a Ship Navigating in a Restricted Channel. *Arch. Civil Mech. Eng.* **2007**, *7*, 91–104. [CrossRef]
25. Lataire, E.; Vantorre, M.; Delefortrie, G. The influence of the ship's speed and distance to an arbitrarily shaped bank on bank effects. *ASME* **2015**. [CrossRef]
26. Cross, S.J.; Olofsson, M. Classification of maritime simulators, the final attempt introducing dnv's new standard. In Proceedings of the International Conference on Marine Simulation and Ship Manoeuvrability 2000 (MARSIM 2000), Orlando, FL, USA, 8–12 May 2000; p. 7.

27. Donatini, L.; Vantorre, M.; Verwilligen, J.; Delefortrie, G. Description of Hydro/Meteo Data in Ship Manoeuvring Simulators: A Survey on the State of the Art. *Ocean Eng.* **2019**, *189*, 106344. [CrossRef]
28. Nam-Kyun, I.; Van-Luong, T. Ship's Maneuverability in Strong Wind. *J. Navig. Port Res. Int. Ed.* **2008**, *32*, 115–120.
29. Cieutat, J.M.; Gonzato, J.C.; Guitton, P. A New Efficient Wave Model for Maritime Training Simulator. In Proceedings of the Proceedings Spring Conference on Computer Graphics, Budmerice, Slovakia, 25–28 April 2001; pp. 202–209.
30. Artyszuk, J. A Uniform Current in Ship Manoeuvring Mathematical Model. *Annu. Navig.* **2004**, *8*, 67–76.
31. Vantorre, M.; Delefortrie, G.; Laforce, E.; Vlieger, H.D.; Claeys, S. Ship Manoeuvring at Very Small and Negative Under Keel Clearance. *IFAC Proc. Vol.* **2003**, *36*, 37–42. [CrossRef]
32. Benedict, K.; Kirchhoff, M.; Fischer, S.; Gluch, M.; Klaes, S.; Baldauf, M. Application of Fast Time Simulation Technologies for Enhanced Ship Manoeuvring Operation. *IFAC Proc. Vol.* **2010**, *43*, 79–84. [CrossRef]
33. Artyszuk, J. A Novel Method of Ship Manoeuvring Model Identification from Sea Trials. *Annu. Navig.* **2003**, *6*, 19–35.
34. World Meteorological Organization. *Guidelines on Ensemble Prediction Systems and Forecasting*; WMO: Geneva, Switzerland, 2012; ISBN 978-92-63-11091-6.
35. Molteni, F.; Buizza, R.; Palmer, T.N.; Petroliagis, T. The ECMWF Ensemble Prediction System: Methodology and Validation. *Q. J. Royal Met. Soc.* **1996**, *122*, 73–119. [CrossRef]
36. Mel, R.; Lionello, P. Storm Surge Ensemble Prediction for the City of Venice. *Weather Forecast.* **2014**, *29*, 1044–1057. [CrossRef]
37. Mel, R.; Viero, D.P.; Carniello, L.; Defina, A.; D'Alpaos, L. Simplified Methods for Real-Time Prediction of Storm Surge Uncertainty: The City of Venice Case Study. *Adv. Water Resour.* **2014**, *71*, 177–185. [CrossRef]
38. Technical Standards and Commentaries of Port and Harbour Facilities in Japan. *Japan Port and Harbour Association 1999 Translated by the Overseas Coastal Area Development Institute of Japan in 2002. Jpn. Port Harb. Assoc.* **2002**, *1*, 455–473.
39. Perkovic, M.; Gucma, L.; Przywarty, M.; Gucma, M.; Petelin, S.; Vidmar, P. Nautical Risk Assessment for LNG Operations at the Port of Koper. *SV-JME* **2012**, *58*, 607–613. [CrossRef]
40. Gucma, L. *Zarządzanie Ryzykiem w Rejonie Mostów Usytuowanych Nad Drogami Wodnymi w Aspekcie Zderzenia z Jednostkami Pływającymi. (Risk Management in the Area of Bridges Situated on Waterways in Respect to Ships Collisions)*; Scientific Publications of Maritime University of Szczecin: Szczecin, Poland, 2012; ISBN 978-838-990-70-5.
41. Artyszuk, J. Towards a Scaled Manoeuvring Mathematical Model for a Ship of Arbitrary Size. *Zesz. Nauk. Akad. Morska Szczec.* **2005**, *78*, 21–37.
42. Kobyliński, L. Capabilities of Ship Handling Simulators to Simulate Shallow Water, Bank and Canal Effects. *TransNav Int. J. Mar. Navig. Saf. Sea Transp.* **2011**, *5*, 247–252.
43. Iribarren, J.R. Determining the Horizontal Dimensions of Ship Manoeuvring Areas. *Bull. Perm. Int. Assoc. Navig. Congr.* **1999**, *100*, 5–26.

Journal of
Marine Science and Engineering

MDPI

Communication

The Simulation of Sloped Bank Effect Influence on Container Ship Trajectory

Mate Baric [1,*], Robert Mohovic [2], Djani Mohovic [2] and Vinko Pavic [1]

1 Maritime Department, University of Zadar, 23000 Zadar, Croatia; vpavic1@unizd.hr
2 Faculty of Maritime Studies, University of Rijeka, 51000 Rijeka, Croatia; mohovic@pfri.hr (R.M.); dmohovic@pfri.hr (D.M.)
* Correspondence: mbaric@unizd.hr

Abstract: The latest container vessel grounding in the Suez Canal, which occurred on 23 March 2021 (the Ever Given), raised many questions regarding the safety of navigation. The sudden concern about safety is due to fears that traffic flow through the Suez Canal could be blocked for longer periods of time. Besides external forces imposed by wind, in this case bank effect had a significant influence on the ship's grounding. Bank effect occurs due to restricted water flow caused by narrow waterways. Many fairway design standards consider sloped banks such as those of the Suez Canal as unsubstantial in bank-effect forces. This paper analyses the impact of sloped banks on container ship trajectory and proposes minimal distances that may decrease bank-effect forces in order to reduce the risk of vessel grounding and increase the safety of navigation. However, this type of accident has happened before and may occur again due to a small sailing distance from the bank in cases where vessel speed is increased.

Keywords: sloped bank effect; grounding; fairway width; full mission ship handling simulation

Citation: Baric, M.; Mohovic, R.; Mohovic, D.; Pavic, V. The Simulation of Sloped Bank Effect Influence on Container Ship Trajectory. *J. Mar. Sci. Eng.* **2021**, *9*, 1283. https://doi.org/10.3390/jmse9111283

Academic Editor: Marko Perkovic

Received: 22 October 2021
Accepted: 15 November 2021
Published: 18 November 2021

1. Introduction

The grounding of container vessel Ever Given raised concerns of possible Suez Canal blockage. The Suez Canal is one of the main fairways between the Indian Ocean and the Mediterranean Sea/Atlantic Ocean. According to Suez Canal Authorities [1], 18,880 vessels transited the Suez Canal carrying over 1,207,087,000 tons of cargo in 2019. The daily average was 51.7 ships per day and 3,307,000 tons of cargo. Container vessels comprise 28% of all vessels in transit. According to reports [2], the Suez Canal has a good safety record and in 2020 had 75 incidents reported, with grounding being the most common type of accident. In the past 10 years, every third accident was related to grounding.

Detailed information about accidents is scarce and most available information is in the form of media reports where the initial causes are not stated. One accident that was investigated was the grounding of container vessel APL Danube [3]. The vessel (length 299.95 m) grounded on 19 April 2019 at southbound transit due to bank effect according to an investigation. The vessel veered off from the middle of the canal close to the eastern bank, where the bow was then pushed away from the bank. The rudder was turned towards the bank to counter bow movement, but this was ineffective due to low water flow. Besides bank effect, investigation determined that crew members did not detect that the vessel was closing toward the bank.

The latest grounding of Ever Given received the most attention due to a period of blockage of the Suez Canal. An investigation will be carried out, with the only existing information about the accident being AIS (automatic identification system) track data. The vessel track published by Maritime Casualty Specialists [4] was one of the first videos that showed the vessel's track and speed from AIS data. The track shows that the vessel was pushed by a southerly wind onto the western channel bank. In order to obtain better responsivity, the vessel speed was increased to about 13.5 knots in the final moments. That high speed, in combination with the proximity to the bank, caused bank effect, and the vessel consequently grounded. However, this is yet to be confirmed by official accident investigation, which will provide more detailed information.

While previous groundings were caused by the same effect, this accident showed that vessel navigational safety may be in danger in some cases. Sloped banks are considered less influential than vertical embankments, however these real situations and research data show that sloped bank effect may have a significant external influence on vessel trajectory.

2. Bank Effect and Channel Design Standards

Bank effect occurs due to asymmetric flow around the vessel hull when sailing near a bank, which results in a force acting on the vessel and a yawing moment pushing the bow away from the bank. The magnitude of this effect depends on bank shape, water depth, vessel distance from the bank, vessel properties and speed [5]. A decreased distance between vessel and bank influences manoeuvring performance considerably and makes it difficult to steer [6]. Besides distance, bank effect is pronounced with increased vessel speed [7]. The influence of sloped banks was researched in a towing tank using a vessel model, which showed that sailing above sloped banks generated larger forces than vertical banks [8].

Fairway design guidelines recognize and consider bank effect in horizontal fairway dimensions. Harbour Approach Channels Design Guidelines Report No. 121-2014 [9] by PIANC (Permanent International Association of Navigation Congresses) defines horizontal fairway dimensions as the sum of the basic manoeuvre lane, additional width due to external forces and additional bank clearance on both fairway sides. The additional fairway width is determined by the fairway location (inner protected or outer unprotected channel), ship speed and bank type. The clearance is greater if the vessel speed is higher and the bank is steeper. Recommendation for Maritime Works ROM 3.1-99 [10] by Puertos Del Estado includes bank effect in fairway width as additional banks' safety distance and bank safety margin. The bank safety margin is a method for the calculation of error margin, and is added to the additional distance to compensate for bank effect. The sum of these two elements provides total additional fairway width to account for bank effect. The necessary fairway width is determined based on slope type and vessel speed. In the previous method, vertical banks have a larger recommended additional width than sloped banks. Technical Standards and Commentaries for Port and Harbour Facilities in Japan, by the Japanese Ministry of Land, Infrastructure, Transport and Tourism (MLIT) [11], only provide the additional width due to bank effect for vertical banks. Data are given by vessel type, which is determined in towing tanks and represents the distance from the bank where a small rudder action (up to 5°) can overcome bank forces.

These methods all state that vertical banks have far greater vessel bank distance than sloped banks. Design standards recommendations are shown in Table 1, and the recommended additional fairway width due to bank effect for vertical banks is greater than for sloped banks. This comparison may lead to the conclusion that current recommendations are not appropriate and do not ensure an adequate level of safety in navigation, because the effect of the sloped banks may be also significant.

Table 1. Comparison of recommended fairway design standards for vessel–bank distances.

Vessel Speed (kn)	Bank Slope	PIANC	ROM 3.1	MLIT
5	Vertical bank	0.5 B [1]	-	1.1 B
7.5		1.0 B		
10		1.3 B		
5	1:2 slope	0.3 B	0.6 B	-
7.5		0.5 B	1.0 B	

Table 1. *Cont.*

Vessel Speed (kn)	Bank Slope	PIANC	ROM 3.1	MLIT
5		0.0 B	0.3 B	
7.5	1:4 slope	0.1 B	0.5 B	-
10		0.2 B	0.7 B	

[1] B represents ship breadth.

3. The Bank Effect of Sloped Banks

As the latest event of grounding showed, sloped banks may have a significant effect on vessel trajectory. Figure 1 indicates that the Suez Canal has slopes with 1:4 gradient (the slope length is four times larger than the slope height) and an approximate surface width of 313 m. However, at a depth of 24 m the width is approximately 121 m. A container ship of 400 m length and 59 m width at static draft of 15.7 has approximately 180 m of canal width. This gives around 60.5 m of free space on each side of the vessel before grounding. That distance can be even smaller due to higher dynamic draft caused by the squat effect. However, this horizontal distance is not only used for avoiding bank effect, but also must allow additional distance for other external forces (such as drift due to wind), steering drift and other factors that cause vessels to veer off from their trajectory. While the dimensions of the Suez Canal have increased over time, the breadth of container vessels also increased significantly with increases in their capacity (Figure 1).

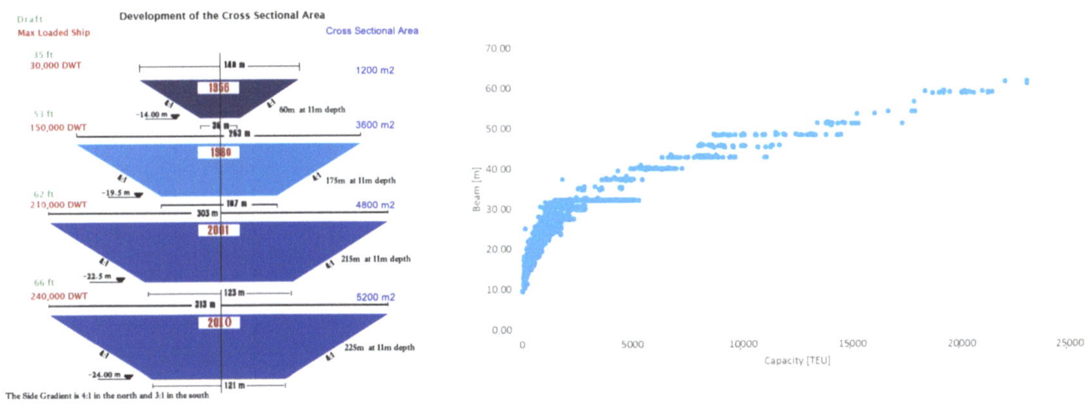

Figure 1. Development of the Suez Canal cross section (**a**) [12] and growth of container vessels size in TEU capacity and beam (**b**) [13].

The research analysing the effect of sloped banks suggests that the generated forces are significant. A study [5] that included container vessel models sailing near different bank types in towing tanks showed greater forces at sloped banks compared with vertical banks. The vessel model represents a 350 m long and 42.9 m wide container vessel. The model was towed at different speeds, at different drafts and at different distances from the banks. Figure 2 shows the sway force of a 1:3 slope ratio bank and vertical banks at different speeds and different distances from the bank. It can be seen that sway force is greater with higher vessel speed and bank slope. Additionally, the maximum sway force occurs at a shorter distance from sloped banks than vertical banks. This indicates that when the vessel is over the bank, the under-keel clearance decreases in comparison with vertical banks, where that phenomena does not occur. This conclusion is in accordance with research [14] suggesting yawing moment increases substantially with decreasing under-keel clearance.

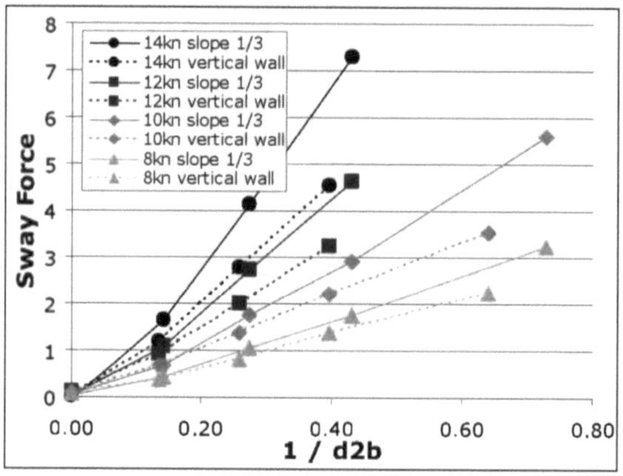

Figure 2. Sway force (scaled to model) induced by different bank types at different vessel speeds and distances from the bank [5].

Similar results were obtained in [15] using a real-time navigational simulator for container vessels in different types of banks, vessel speeds and distances from the bank. The results suggest the same pattern of sway force and proposed minimum distances from the bank to avoid vessel grounding.

4. Simulation of Sloped Bank Effect on Large Container Vessels

The analysed recommendations suggest that sloped banks may have less influence on vessel trajectory than vertical banks. However, real events prove that significant bank effects may occur when sailing near sloped banks. To confirm the sloped bank effect, simulation tests were conducted using the Transas NaviSailor NTPro 5000 v5.35 real-time simulator. Previously, the ship-handling simulator has proved to be a suitable tool for basic and advanced ship-handling operations and for advanced manoeuvres required in search and rescue operations [16], and for the analysis of accidents at sea [17] or incidents (near misses) [18], including the impact of AIS spoofing on nautical safety [19]. More recently, the simulator has increasingly been used for seafarer competency research [20,21]; port, waterway and canal planning [22] and pilot training [23]. This simulator is certified by a DNV certificate [24] which confirms that simulations, vessel models and generated data are tested and certified as reliable. During simulation, the additional forces caused by banks are calculated as hydrodynamic components of vector forces and moments affecting the vessel. The principal parameters in vessel/channel interaction are vessel breadth; draft and length ratios; distance between vessel and channel boundaries; vessel speed; angle between vessel centreline and channel line of the wall; vessel rate of turn; and channel width, depth and slope angle. The effect of these parameters on vessel hydrodynamic forces is calculated based on an analysis of various model-testing results [25]. However, to validate the results, simulated data will be compared with towing tank results [5] (Figure 2).

During simulations, the following vessel models and conditions were used. The model represented a container vessel 347 m long and 43 m wide with 14 m of draft. The summer deadweight of the vessel was 104,696 tons with a block coefficient of 0.65. The fairway was sloped in a ratio of 1:4 to simulate the Suez Canal banks and the canal depth was adjusted to 1.2 and 1.5 vessel drafts. The fairway was simulated on one side because the vessel only approached one side of the channel, with the other side too far away to produce a significant effect. Weather and sea conditions, along with sea current, were set to zero.

Simulations were repeated several times for the same settings to obtain stabilised data. The distance from the bank (d) (distance between ship and closest horizontal bank point) was set from 50 m to 10 m, at a frequency of approach of 10 m. Three sailing speeds were tested: 10, 7.5 and 5 knots. The steering was set to track control interconnected auto-pilot on

"Precise" mode in order to simulate the return of the ship to its original route, as in-course control setting ensures that a vessel can continue to sail in the same course but on a parallel route. This eliminates human error in steering and enables uniformity of simulations. To confirm that the ship deviated from its route, simulations with rudder in midship were also conducted.

Figure 3 shows the vessel trajectory at a speed of 10 knots, using distances from the bank of 50 m (Figure 3a), 30 m (Figure 3b) and 10 m (Figure 3c). It can be seen that if there is no rudder action the bank effect will push the vessel bow away from the bank regardless of the distance from the bank.

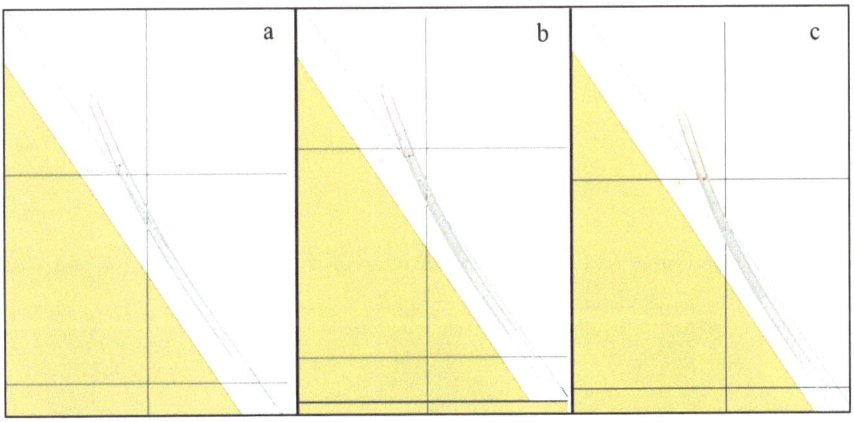

Figure 3. Ship trajectory under 1:4 ratio sloped bank effect and speed 10 knots when rudder is set to midship: (**a**) 50 m distance from slope; (**b**) 30 m distance from slope; (**c**) 10 m distance from slope.

More than 100 simulations were conducted and the lateral force and cross-track error were measured. The data provide information regarding the influence of the sloped bank effect on vessel trajectory and at what distance a vessel should sail in order to avoid grounding due to bank-effect-induced trajectory oscillation.

Figure 4 displays simulated sloped bank force data. The full line represents the force for a depth draft ratio of 1.2 and the dotted line represents the force for a depth draft ratio of 1.5. It can be seen that increments of vessel speed increased the acting force, resulting in larger oscillations of vessel trajectory. Simulated force was then compared with bank force in Figure 2 [5] gathered from the towing tank test that validated and confirmed the simulated data were reliable and accurate.

The difference between two depth-draft ratios is occurred due to vessel movement over a sloped bank (Figure 5). Larger depth-draft ratio allows larger fairway width and the vessel may move more over the bank.

Figure 6 represents the maximum cross-track error caused by the sloped bank effect and rudder action. As expected, the greater cross-track error was obtained at the highest sailing speed. However, the data show that for lateral force the greatest offset and force will occur at a 1.5 depth draft ratio. This occurs due to increased vessel movement inside the bank. With a higher depth–draft ratio (Figure 5a), the vessel can move over the sloped part of the bank more than it can with a lower depth–draft ratio (Figure 5b), and thus the bank force and effect on trajectory are increased.

The simulated data can provide recommendations for ship bank distance in order to avoid significant bank effects. For the analysed bank type and vessel, the recommended distances from the sloped bank are shown in Table 2.

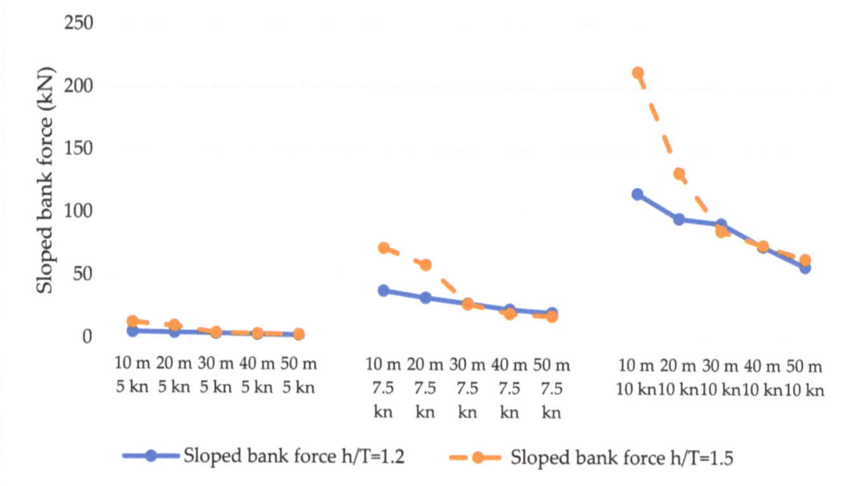

Figure 4. Sloped bank force acting on vessel when sailing at different speeds and distances from a sloped bank (h-depth; T-draft).

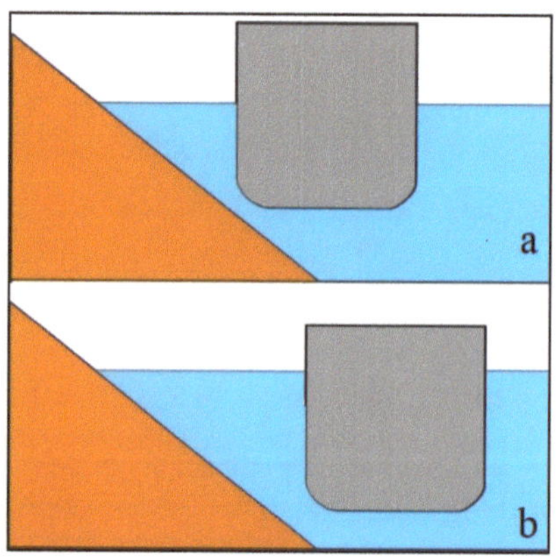

Figure 5. Vessel movement over a sloped bank: (**a**) vessel depth–draft ratio 1.5; (**b**) vessel depth–draft ratio 1.2.

The distances in Table 2 are relative to vessel breadth and may be used for larger similar vessel types. If we compare these data to the situation in the Suez Canal, which has banks with a slope ratio of 1:4, the following is obtained. A container vessel that is 59 m wide and with a draft of 15.7 m has 180 m of underwater fairway width. Based on the canal depth and vessel draft, the ratio is 1.5. At an approximate distance of 50 m from the bank (when the vessel is sailing approximately at the centre of the canal) and sailing at 7.5 knots, the vessel needs 0.2 B (breadth) distance from the bank, which in this instance is approximately 12 m. Additionally, if the vessel comes closer to the bank at a distance of 20 m, then the required vessel bank distance is equal to the minimum vessel bank distance, and this may lead to bank-effect-induced grounding (depending on the action taken). In this instance, the minimum required distance to avoid significant influence of bank effect at speed 7.5 knots and 1:4 slope ratio bank is 20 m. The vessel, when sailing at the centre of the canal, has only 43.5 m at either side to correct its trajectory oscillation and consequently avoid grounding. This implies that when other external forces act (e.g., wind force), the vessel has only 43.5 m in width to correct its trajectory and with greater cross-track error the bank force is even stronger, which makes the vessel harder to control.

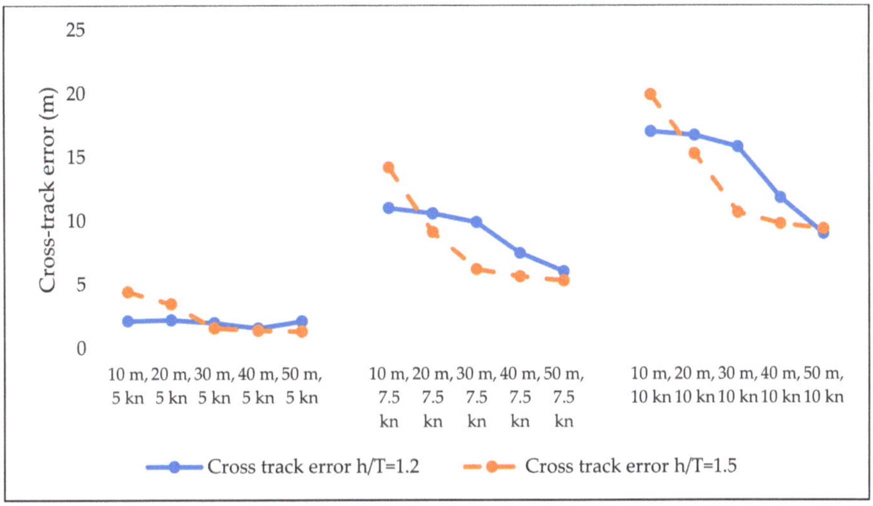

Figure 6. Maximum recorded cross-track error of a vessel when sailing at different speeds and distances from a sloped bank.

Table 2. Suggested additional fairway distances to accommodate bank effect.

Vessel Speed (kn)	Bank Slope	h/T = 1.2T Ship Bank Sailing Distance					h/T = 1.5T Ship Bank Sailing Distance				
		10 m	20 m	30 m	40 m	50 m	10 m	20 m	30 m	40 m	50 m
5	1:4 bank	0.1 B	0.1 B	0.1 B	0.1 B	0.1 B	0.1 B	0.1 B	0.0 B	0.0 B	0.0 B
7.5		0.3 B	0.3 B	0.3 B	0.3 B	0.3 B	0.4 B	0.3 B	0.3 B	0.2 B	0.2 B
10		0.4 B	0.4 B	0.3 B	0.3 B	0.3 B	0.4 B	0.3 B	0.3 B	0.2 B	0.2 B

5. Discussion

This research was conducted using real-time navigational simulation. This simulator was originally used as a teaching aid. However, as the models were upgraded, it has increasingly been used as a research tool. The simulator's models have been developed by real ship data or ship models in towing tanks. In most cases, the mathematical models are based on model tests carried out in towing tanks [26]. Since the models used in simulation are based on model test data, many research and fairway tests are conducted using navigational simulators in order to reduce costs and time. Navigational simulation is also the only tool to test new fairways before they are built. The continuous development and improvement of mathematical-physical models, together with increasing computing power and photo-realistic displays, make simulations paramount to any other approach, as long as the flaws, drawbacks and imperfections of this virtual reality are known to the involved parties and are considered in the evaluation phase of the results [27,28]. The results of similar studies have proved that the simulator is a suitable tool, capable of supporting designs and verifying designs in sea-traffic engineering [29]. This is proof that the real-time simulation method is highly applicable when the effectiveness of safety improvement planning of port and harbour facilities must be assessed [30].

In the present study, the simulation was conducted using only one bank since the main goal was to investigate the influence of sloped-bank-induced yawing and cross-track error. This effect occurs when a vessel veers off the middle of a canal and comes too close to one bank. In such instances, only one bank induces a force on the vessel hull since the other side of the bank is too far away to induce significant counter-reacting forces. When these forces are present, the vessel trajectory will start to oscillate between both banks and in extreme cases may result in grounding.

The results obtained from simulation showed that a sloped bank generates sufficient force to induce significant yawing moment and route oscillation. Similar research showed that 1:3 and 1:5 sloped banks and larger under-keel clearance generate similar yaw moments as vertical banks [31]. One study [26] showed that the magnitude of a lateral force at the forward perpendicular, either an attraction or a repulsion force, will increase with the forward speed of the ship, when the ship sails closer to the bank, or in a more confined cross section. Similar results were obtained in research using a fixed water depth, where larger bank angles led to greater transverse force and yaw moment [32].

6. Conclusions

Bank effect is generally considered to be a force that significantly occurs when sailing near vertical banks, and sloped banks were previously considered non-influential in terms of bank effect. However, sloped banks are superior to vertical banks and provide more sea area for vessels. The latest research in towing tanks and navigational simulators suggests that sloped banks also generate significant bank effect. This is even more important because sloped banks are often found in narrow channels and canals and, in combination with low depth and larger speed, could lead to grounding. Modern vessel size is increasing and fairways are lagging behind, which may lead to greater pressure to achieve economic results. The present research shows that in one of the most important fairways, a large 20,000 TEU container vessel has approximately only 50 m of manoeuvring space at each side of the fairway before bank effect significantly influences its trajectory and the vessel grounds. Grounding due to bank effect has happened before and the latest occurrence due to high speed and hard grounding shows that the safety of navigation is in danger and relies only on the diligence of crew and pilot in keeping the vessel in the centre of the fairway. Grounding can be prevented by upgrading the education of vessel bridge teams and pilots in terms of improving training methods and familiarity with the bank effect forces that influence vessel trajectory. Consideration should also be given to upgrading vessel wheelhouse posters in order to provide more information, even in electronic form, as well as Pilot/Master information exchange cards.

Author Contributions: Conceptualization, M.B. and V.P.; methodology, M.B.; validation, M.B., R.M. and D.M.; data curation, M.B.; writing—original draft preparation, V.P.; writing—review and editing, V.P., R.M. and D.M. All authors have read and agreed to the published version of the manuscript.

Funding: This research received no external funding.

Institutional Review Board Statement: Not applicable.

Informed Consent Statement: Not applicable.

Data Availability Statement: Not applicable.

Conflicts of Interest: There are no conflict of interest.

References

1. Suez Canal Authority. *Annual Report 2019*; Planning & Research Department, Information Center, Suez Canal Traffic Statistics: Ismailia, Egypt, 2019.
2. Allianz Global Corporate & Speciality. *Safety and Shipping Review 2020*; Alliant Global Corporate and Speciality: Berchem, Belgium, 2020.
3. Transport Malta. *Marine Safety Investigation Report, APL Danube, Report No. 08/2020*; Marine Safety Investigation Unit: Furjana, Malta, 2020.
4. Marine Casualty Specialists. Available online: http://www.maricaspe.com/ (accessed on 10 May 2021).
5. Lataire, E.; Vantorre, M.; Laforce, E.; Eloot, K.; Delefortrie, G. Navigation in Confined Waters: Influence of Bank Characteristics on Ship-Bank Interaction. In Proceedings of the International Conference on Marine Research and Transportation, ICMRT, Naples, Italy, 28–30 June 2007.
6. Serban, P.S.; Panaitescu, V.N. Simulation of ship to shore interaction in shallow and narrow waters. *Sci. Bull. Mircea Cel Batran Nav. Acad.* **2015**, *18*, 112.
7. Sian, A.Y.; Maimun, A.; Priyanto, A.; Ahmed, Y.M. Assessment of Ship-Bank Interactions on LNG Tanker in Shallow Water. *J. Technol.* **2014**, *66*, 141–144. [CrossRef]

8. Eloot, K.; Vantorre, M. Ship behaviour in shallow and confined water: An overview of hydrodynamic effects through EFD. In Proceedings of the AVT-Specialists' Meeting on Assessment of Stability and Control Prediction Methods for Air and Sea Vehicles, Portsdown, UK, 11 October 2011.
9. Permanent International Association of Navigation Congresses (PIANC). *Harbour Approach Channels Design Guidelines*; Report No. 121–2014; PIANC MarCom Working Group 30: Brussels, Belgium, 2014.
10. Del Estado, P. Recommendations for Maritime Works (Spain) ROM 3.1-99: Designing Maritime Configuration of Ports, Approach Channels and Floatation Areas. Spain CEDEX. 1999. Available online: https://www.puertos.es/es-es/BibliotecaV2/ROM%203.1-99%20(EN).pdf (accessed on 14 November 2021).
11. Ministry of Land, Infrastructure, Transport and Tourism (MLIT). *Technical Standards and Commentaries for Port and Harbour Facilities in Japan*; OCDI: Tokyo, Japan, 2009. Available online: https://ocdi.or.jp/tec_st/tec_pdf/tech_00H.pdf (accessed on 14 November 2021).
12. Suez Canal Authority. Available online: https://www.suezcanal.gov.eg/English/About/SuezCanal/Pages/CanalCharacteristics.aspx (accessed on 10 May 2021).
13. Container-Ship Size: What Dimensions Can We Expect to See? Available online: https://piernext.portdebarcelona.cat/en/mobility/container-ship-size/ (accessed on 10 August 2021).
14. Vantorre, M.; Delefortrie, G.; Eloot, K.; Laforce, E. Experimental investigation of ship-bank interaction forces. In Proceedings of the International Conference MARSIM, Kanazawa, Japan, 25–28 August 2003.
15. Baric, M.; Mohovic, R.; Mohovic, D. Determining Restricted Fairway Additional Width due to Bank Effect for Fine Form Vessels. *J. Navig.* **2019**, *72*, 1435–1448. [CrossRef]
16. Kim, I.; Chae, C.; Lee, S. Simulation Study of the IAMSAR Standard Recovery Maneuvers for the Improvement of Serviceability. *J. Mar. Sci. Eng.* **2020**, *8*, 445. [CrossRef]
17. Hörteborn, A.; Ringsberg, J.W. A method for risk analysis of ship collisions with stationary infrastructure using AIS data and a ship manoeuvring simulator. *Ocean Eng.* **2021**, *235*, 109396. [CrossRef]
18. Szlapczynski, R.; Szlapczynska, J. A ship domain-based model of collision risk for near-miss detection and Collision Alert Systems. *Reliab. Eng. Syst. Saf.* **2021**, *214*, 107766. [CrossRef]
19. Androjna, A.; Perkovič, M.; Pavic, I.; Mišković, J. AIS Data Vulnerability Indicated by a Spoofing Case-Study. *Appl. Sci.* **2021**, *11*, 5015. [CrossRef]
20. Park, D.J.; Yim, J.B.; Yang, H.S.; Lee, C.K. Navigators' Errors in a Ship Collision via Simulation Experiment in South Korea. *Symmetry* **2020**, *12*, 529. [CrossRef]
21. Yim, J.-B.; Park, D.-J.; Youn, I.-H. Development of navigator behavior models for the evaluation of collision avoidance behavior in the collision-prone navigation environment. *Appl. Sci.* **2019**, *9*, 3114. [CrossRef]
22. Łazuga, K.; Minh, Q.N.; Gucma, L. Cost-Effective Design of Port Approaches Using Simulation Methods Based on the Example of a Modernized Port in the Ustka. *J. Mar. Sci. Eng.* **2021**, *9*, 211. [CrossRef]
23. Žagar, D.; Svetina, M.; Košir, A.; Dimc, F. Human factor in navigation: Overview of cognitive load measurement during simulated navigational tasks. *J. Mar. Sci. Eng.* **2020**, *8*, 775. [CrossRef]
24. DNV Statement of Compliance. Bridge Operation Simulator, Navi-Trainer NTPro 5000, Class A Standard for Certification of Maritime Simulators No. 2. Sandefjord. 2012. Available online: https://www.transas.com.ua/certificates-sim/SIM%20-%20002%20121218%20DNV.pdf (accessed on 14 November 2021).
25. Transas MIP Ltd. *Description of Transas Ship Motion Mathematical Model*; Version 2.11; Transas MIP Ltd.: Cork, Ireland, 2012; pp. 62–63.
26. Lataire, E. Experiment Based Mathematical Modelling of Ship-Bank Interaction. Ph.D. Thesis, Ghent University, Ghent, Belgium, 2014.
27. Böttner, C.-U. Ship handling simulation in approach channel and harbour design. In Proceedings of the 34th PIANC World Congress: Connecting Maritime Hubs Globally, Panama City, FL, USA, 7–11 May 2018.
28. Senčila, V.; Zažeckis, R.; Jankauskas, A.; Eitutis, R. The Use of a full mission bridge simulator ensuring navigational safety during the Klaipeda Seaport development. *TransNav Int. J. Mar. Navig. Saf. Sea Transp.* **2020**, *14*, 417–424. [CrossRef]
29. Czaplewski, K.; Zwolan, P. Using the software of Navi-Trainer PRO 5000 simulator for assessment of the designed navigational infrastructure. *Pol. Nav. Acad. Annu. Navig.* **2011**, *18*, 55–68.
30. Perkovic, M.; Brcko, T.; Luin, B.; Vidmar, P. Ship handling challenges when vessels are outgrowing ports. In Proceedings of the 19TH INTERNATIONAL NAVIGATION SIMULATOR LECTURERS' CONFERENCE, Western Cape, South Africa, 5 September 2016.
31. Duffy, J.T. Modelling of Ship-Bank Interaction and Ship Squat for Ship-Handling Simulation. Ph.D. Thesis, University of Tasmania, Hobar, Australia, 2008; p. 103.
32. Luo, W.; Yang, B.; Sun, Y. Hydrodynamic Analysis of KVLCC2 Ship Sailing near Inclined Banks. *Math. Probl. Eng.* **2021**, *1*, 6655971. [CrossRef]

Journal of
Marine Science and Engineering

MDPI

Article

Reconstructing Maritime Incidents and Accidents Using Causal Models for Safety Improvement: Based on a Case Study

Lucjan Gucma [1], Andrej Androjna [2], Kinga Łazuga [1], Peter Vidmar [2] and Marko Perkovič [2,*]

[1] Faculty of Navigation, Maritime University of Szczecin, 70-500 Szczecin, Poland; l.gucma@am.szczecin.pl (L.G.); k.lazuga@am.szczecin.pl (K.Ł.)
[2] Faculty of Maritime Studies and Transport, University of Ljubljana, 6320 Portorož, Slovenia; andrej.androjna@fpp.uni-lj.si (A.A.); peter.vidmar@fpp.uni-lj.si (P.V.)
* Correspondence: marko.perkovic@fpp.uni-lj.si

Abstract: No advance in navigation has yet to prevent the occurrence of accidents (incidents are always implied when we discuss accidents) at sea. At the same time, advances in accident models are possible, and may provide the basis for investigations and analyses to help prevent future adverse events and improve the safety of marine transport systems. In such complex socio-technical systems models that treat accidents as the result of a chain or sequence of events are used most commonly. Such models are well suited to damage caused by failure of physical components in relatively simple systems. Although these often include methods for modeling human error, they do not cover broader aspects related to the management of the organization using the means of transport itself (shipowners) nor errors that may occur in the design phase. In particular, they do not cover changes in the systems over time. The paper presents accident investigation approaches and uses a modified causal model to analyze an incident that occurred in January 2019 on the city ferry in Świnoujście. The results of the analysis were used to provide guidelines for increasing safety at the crossing and to evaluate the accident analysis model used. Additionally, incidentally, through the study of this case we uncovered a problem in communication among stakeholders that unnecessarily complicates the models for the models for the improvement of safety.

Keywords: accident investigation model; ferry incident; accident model

Citation: Gucma, L.; Androjna, A.; Łazuga, K.; Vidmar, P.; Perkovič, M. Reconstructing Maritime Incidents and Accidents Using Causal Models for Safety Improvement: Based on a Case Study. *J. Mar. Sci. Eng.* **2021**, *9*, 1414. https://doi.org/10.3390/jmse9121414

Academic Editor: José-Santos López-Gutiérrez

Received: 23 November 2021
Accepted: 10 December 2021
Published: 11 December 2021

1. Introduction

Maritime transport is a complex socio-technical global system and many of the risks particular to sea transport are greater than those of the other three main modes of transport. Aside from basic environmental threats such as oil pollution, cruise ships, which carry up to 8000 people, represent the threat of extraordinary loss of life in large numbers at one time in one place; if this seems alarmist, one can find innumerable cases of passenger ferries capsizing with the result of great loss of life. The International Maritime Organization (IMO) has addressed the issue of accidents investigation and reporting in its regulations in the Casualty Investigation Code [1]. This Code, in contrast to the earlier, outdated Resolution A.884(21) (Appendix 1) [2], requires that an accident investigation be carried out for any "very serious marine casualty," which is defined as an accident at sea resulting in the total loss of the ship or death or serious environmental damage. The European Maritime Safety Agency (EMSA) is also obliged by the European Union (EU) Directive 2009/18/EC [3] Article 8 to establish the principles for accident investigation and each EU Member State should establish an independent and permanent accident investigation body. In Poland, the State Marine Accident Investigation Commission (PKBWM) is responsible for accident investigation and the present case study was investigated within the framework of the PKBWM [4].

By definition, a maritime accident means an occurrence that involves a ship and in which a person is seriously harmed or the ship sustains damage or structural failure, while the marine incident means any occurrence, other than an accident, that is associated with the operation of a ship and affects or could affect the safety of operation [5]. In other words, when a mishap causes damage it is considered an accident, and when it merely could have, it is considered an incident. Although there is a legal framework for marine casualty investigation, there is no mention of the models to be used. For the purpose of our research the following classification of event which changes operational state of ship into failure state are: near-miss, incident, accident. According to Stoop [6], several questions are asked in the accident (or incident) investigation. What happened? How did it happen? Why did the accident happen? What can be done to prevent the accident from happening again? What can be done to minimize the consequences of such accidents? According to Awal and Hasegawa [7], several studies, including Laflamme [8], Quereshi [9], and Khanzode et al. [10], have attempted to classify an accident investigation model that should readily provide these answers as a first step, and marine casualty models, which can be classified into several groups from different points of view, attempt to answer these questions by various means [11]:

1. Traditional linear models (Sequence models [12], Heinrich's [13] domino model of accident causation, Bowtie models [14], Reason's [15] Swiss Cheese model);
2. Nonlinear models (Epidemiological accident models [16,17], Complex socio-technical systems [18]),
3. Systems Theory Approach like Functional Resonance Accident Method (FRAM), Systems Theory Approach (STAMP), SHEL (Software–Hardware–Environment–Liveware) [12,19–21].

This does not reflect a static or inflexible approach. For instance, Awal and Hasegawa [22] use different categories of accident reconstruction and suggest:

1. statistical analysis and trends,
2. risk analysis,
3. domino theory,
4. epidemiological theory,
5. control and system theory models.

It should come as no surprise—because accidents and incidents continue to occur—that new models are constantly being developed [11,22,23]. Along these lines, it may be worth considering larger issues, such as the economic model that appears to require constant growth at a time when environmental concerns, no matter how serious they sound, are never serious enough to affect the workings of large businesses. The fact is, the changes forced on stakeholders in the maritime industry themselves create a danger, as the best safety measures available are generally outmoded in too short period of time. An element of this principle emerges in this case study, though primarily we are concerned with a particular type of incident that occurs all too often among ferry services. Taking into account the variety of models used for accident investigation, we propose some modifications based on the work of Rasmussen [24,25]. The slightly amended proposed model is based on a risk management framework that includes a socio-technical and legal environment. The practical results of these studies will make it possible to evaluate the causes of such unanticipated abnormalities as the problematic motion of the ferry "Bielik IV" (the subject of our case study) after the failure of the main propulsion and to develop recommendations for improving the safety of navigation of ferries of this type.

2. Material: Contextual, Technical, and Met-Ocean Facts

The ferry crossing in Swinoujscie is one of two ferries connecting the island of Usedom with the mainland of Poland. The intensity of ferry traffic is 65 per day in one direction. The ferry crossing is usually operated by two ferries of the Bielik type running simultaneously. The ferry owner Żegluga Świnoujska has 4 ferries of the Bielik type. These ferries are built to sustain all but the most catastrophic conditions to be reasonably expected in these waters, such that service is literally never interrupted. They can load 33 passenger cars and take onboard up to 700 passengers at one time. Clearly a ferry accident can have catastrophic consequences.

The incident involving the Świnoujście–Warszow city ferry near the Kosa peninsula occurred on 9 January 2019 (Figure 1). The west station is marked with green lights and the east with red lights. On the east side, a pier is used for the ferries to separate them and allow safe berthing. The ferry crossing has a depth of up to 14 m and even 16 m near the emergency anchorage. The ferry anchored in an emergency situation about 50 m from the Kosa peninsula, but the anchor did not hold and the ferry was adrift, an incident that puts lives in danger. The sequence of incidental events in local time was as follows:

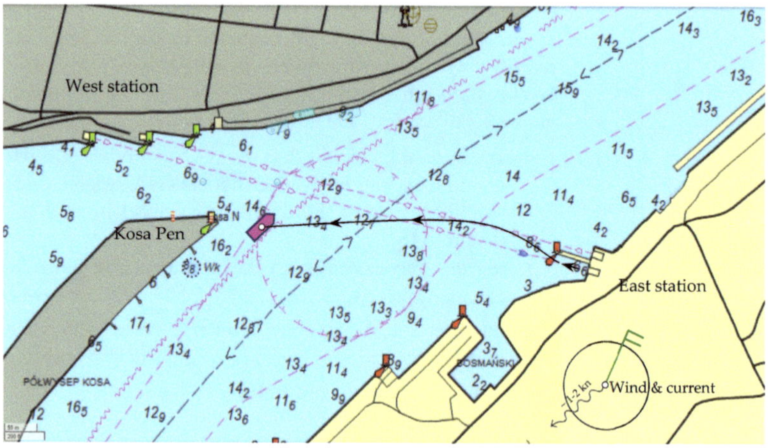

Figure 1. Navigation map with bathymetry of the area and ferry routes and real tracks, with wind and currents (electronic map from MarineTraffic).

18:30—The ferry is loaded and unmooring from East ferry station. 18:45—Main engine failure. 18:46—Changeover to an emergency engine. The "Uran" tugboat was called for emergency via the Shipowner. 18:48—Notification of the incident to VTS Świnoujście. 18:48—Both anchors were dropped almost simultaneously. The chain length was 1.5 shackle (approx. 40 m). Both anchors dragged slightly. The emergency engine was operated forward throughout emergency anchoring. 18:53—Report to VTS Świnoujście, information about the stop at anchors. From 19:00 to 19:25—Ship moves due to the action of wind and current, stern anchor probably drags and the ship lines up with its bow to the wind and current. 19:25—The "Uran" tugboat approaches from the port side and passing tow lines fore and aft. 20:15—Bielik IV is safely moored at station no. 4 on the west side of Świnoujście Municipal Ferry Terminal.

Based on the port regulations [4], it can be stated that there are no conditions restricting the traffic of city ferries in Świnoujście. Anchoring during extreme weather should not in itself present any difficulty. The following particulars of the Bielik ferry (shown in Figure 2) that was designed for inland conditions with winds up to 5°B, were determined for purposes of investigation:

- Lenght overall L = 49.90 m.
- Width B = 15.60 m.
- Constructional draught T = 2.25 m.
- Main propulsion power 750 kW.
- Auxiliary (emergency) power 170 kW.
- Max. number of passengers 695.
- Anchor type (number of anchors): patented Hall (2).
- Single anchor weight 450 kg.
- One chain length 40 m (1.5 shackles).

Figure 2. The general plan of the ferry *Bielik III* (sister ship to the one surveyed). View from starboard and stern.

3. Methods and Calculation

Based on video signal recordings of the Vessel Traffic Service (VTS) Szczecin–Świnoujście, the event was reconstructed and displayed on a map of the basin. The reconstruction was done by analyzing the position and course of the ferry at half-minute intervals. The video images of the ferry's unfiltered radar echo were used for this purpose (it is a radar set up on the Kosa peninsula in the vicinity of the event).

The use and reading of the position determined by the VTS system recording (i.e., the position automatically estimated from the radar echo by the system manufacturer's filtering algorithms) is impossible due to the errors inherent in the system. The filtered vessel position has significant jumps in both the position and course of the vessel and rarely (about 20% of the time) shows a correct echo. The reason for this is, as the operators claim, the difficult radio propagation situation in the area where the event took place. It should be noted that the very data used by the VTS operators is unreliable to a degree that raises the question of whether this jeopardizes the safety of traffic control on the route.

However, manually reconstructed data from the radar video images was used to obtain the ship position and heading that is shown in Figure 3.

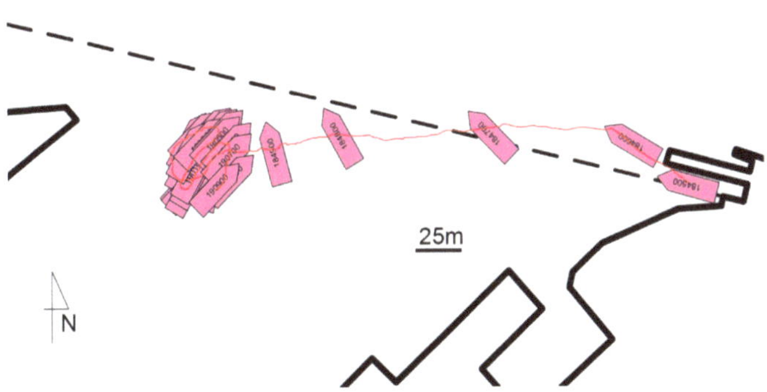

Figure 3. The sequence of ferry positions during the event (with a 1 min time step).

The sampling frequency is 1 min, beginning at the ship's berth—i.e., 18:45. The sequence covers 24 min—until 19:09. It can be seen that the ferry does not anchor steadily but moves in a circle of about 40 m in diameter. This is confirmed by the records shown in the local Cartesian coordinate system, as shown in Figure 4.

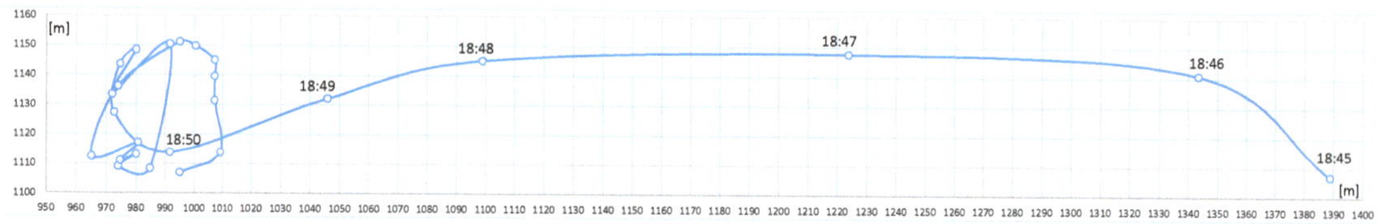

Figure 4. Repetition of the unit position during the event (local coordinates in meters, every 1 min).

The met-ocean forces on the ship are shown in the Figure 5, including emergency propulsion thrust direction and possible anchor circumstances.

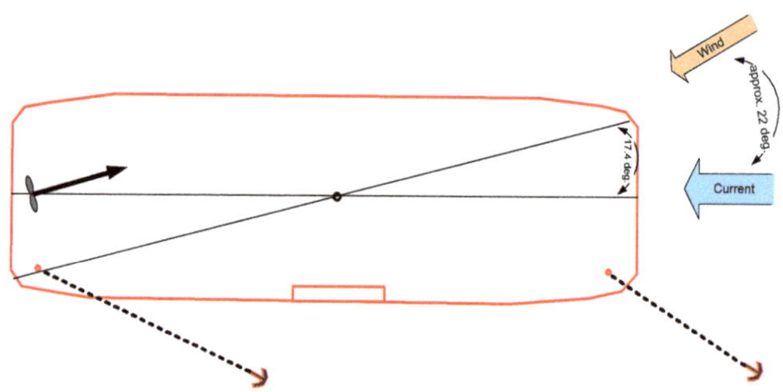

Figure 5. Possible force directions on the ferry during emergency anchoring.

During anchoring, the ferry was oscillating as the master struggled to remain in one place, the speed varying, at times up to 1 knot (0.5 m/s). The average heading of the ferry from 18:54 to 19:09 (i.e., after relative stabilization) was 046 degrees (NE). The angle to the wind of the shuttle, which according to VTS records came from the NNE direction, was therefore 22.5 degrees and the wind was blowing from the port side. The ferry stabilized its position and heading using the incoming current.

Unfortunately, there are no records of what the anchors looked like during the event and how the stern thruster worked, and it is difficult to accurately recreate the forces acting on the ship.

3.1. Estimation of Wind and Current Forces

The next step was to determine the wind and current forces using the commonly known dependencies in hydrodynamics and ship maneuvering.

Resultant wind forces:

$$F_{nw} = \sqrt{{F_{nx}}^2 + {F_{ny}}^2} \tag{1}$$

where the component of the wind resultant force in vessel longitudinal and transverse directions is:

$$F_{nx} = 0.5 \cdot c_x \cdot \rho_p \cdot {V_{wx}}^2 \cdot P_{nx} \tag{2}$$

$$F_{ny} = 0.5 \cdot c_y \cdot \rho_p \cdot {V_{wy}}^2 \cdot P_{ny} \tag{3}$$

Cx, cy—aerodynamic factors,
ρp—air density,
Vwx, Vwy—longitudinal and transverse wind speed,
Pnx, Pny—longitudinal and transverse wind exposed area.

Resultant horizontal force of the currents:

$$F_p = \sqrt{{F_{px}}^2 + {F_{py}}^2} \tag{4}$$

where the component of the resultant force in vessel longitudinal and transverse directions is:

$$F_{px} = 0.5 \cdot c_{pp} \cdot C_{px} \cdot \rho_w \cdot V_{px}{}^2 \cdot P_{px} \quad (5)$$

$$F_{py} = 0.5 \cdot c_{pp} \cdot C_{py} \cdot \rho_w \cdot V_{py}{}^2 \cdot P_{py} \quad (6)$$

where:

C_{pp}—shallow water factor,
C_{px}, C_{py}—current pressure factor,
ρ_w—water density,
V_{px}, V_{py}—longitudinal and transverse current speed,
P_{px}, P_{py}—longitudinal and transverse current exposed area.

Diagrams of the coefficients Cx, Cy, Cpx, and Cpy were constructed, which are shown in Figure 6. Since the unit is almost symmetrical, they were determined for one quarter only. It can be seen that the coefficient of transverse aerodynamic resistance (Cx), which is the most important for the anchoring calculations, reaches its highest values at a wind angle of about 15 degrees.

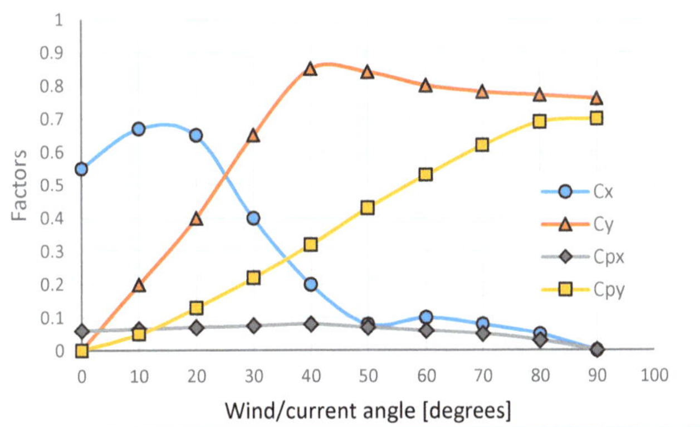

Figure 6. Estimated aerodynamic factors of transverse (Cx) and longitudinal (Cy) air pressure and transverse (Cpx) and longitudinal (Cpy) current pressure coefficients.

It should be noted that the actual wind angle of about 22.5 degrees determined from the video recording is very close to the theoretical wind angle of 17.4 degrees determined using the ship's diagonal (see Figure 5), and that this is also the angle at which the maximum wind force occurs (see Figure 6), which is about 15 degrees.

Figure 7 shows the coordinate system used for the calculations, taking into account the wind direction and current. The calculations did not take into account the resistance of the bow propeller nacelle.

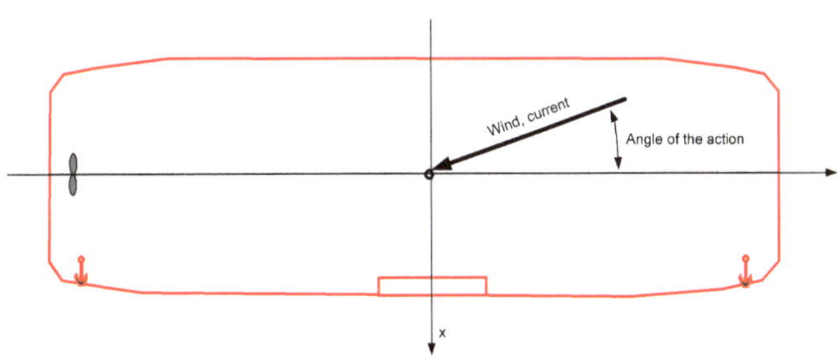

Figure 7. Definition of the angle of attack of wind and current.

In addition, the increase in the hydrodynamic resistance coefficient against the current was determined for the shallow-water vessel and presented in Figure 8. It shows how the resistance to the current increases as a result of the forces related to the proximity of the bottom. However, in the case studied this is not relevant as the ferry has a shallow draught and the h/T (h—depth of area for the mean water level; T—the draft of ferry) ratio is 6.6, which is considered deep water, therefore the Cypp value = 1.

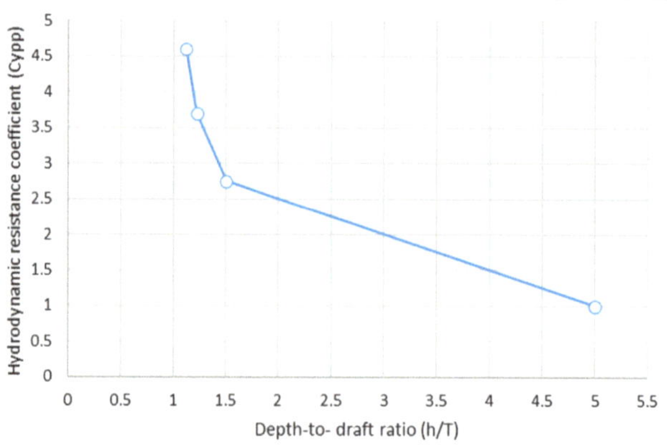

Figure 8. The additional coefficient for a shallow-running vessel [26].

Furthermore, the calculation of the forces resulting from the action of the current has been verified with good results using the methods recommended by Oil Companies International Marine Forum (OCIMF) [27]. For the wind, an angle of 17.5 degrees determined by the ship's diagonal was used, which is close to the actual angle the ship was at during the event. The angle of the current is assumed to be zero, which also corresponds to the actual situation.

The calculations of all forces were intentionally made in metric units of tonne force, marked T (usual) or tf (new, recommended). The reason for this is the intuitive understanding of these units in the shipping industry and their use, e.g., for towing. When converting to the units SI, please note that 1T(tf) = 9.81 kN.

The analysis of the results of the wind and current impact assessment shows that:

1. The wind load is predominant. It is over 5 tf at 10.5 m/s wind in the crossing area.
2. The load from the current is minor and does not exceed half a tonne of force, as the ferry has a small chassis area and an excellent hydrodynamic shape in forward and reverse, for which it was designed.

3.2. Determination of the Shuttle Emergency Propulsion Thrust

Following Gerr [28], the thrust of the propeller operating at zero or low speed of the vessel, called pulling force on the hull, was determined. Of course, the flow can slightly influence the calculation of the efficiency and the forces generated by the propulsion. The calculations are shown in Table 1.

It should be noted that the thrust of typical 200 kW jet rudders is about 3 tf. In total, 250 HP units generate about 2.5 tf of pull, which confirms the calculations presented. It can therefore be assumed that the Bielik ferry's propulsion units can generate just over 3 tf of thrust with an emergency engine, as shown in Table 1.

Table 1. Determination of the emergency thrust of the drive unit.

No	Symbol	Relation	Value	Unit
1	Fs	diameter of propeller	1.57	m
2	Ss	propeller pitch	1.635	m
3	Pa	power of emergency engine	170	kW
4	SHP	shaft horse power $SHP \approx 0.96 \times Pa$	163.2	hp
4	Ta	thrust of emergency engine [28] $Ta = 0.005384141 \times [SHP \times Fs(inch)]^{0.67}$	3.2	tf

3.3. The Forces and Effectiveness of the Emergency Stopping of the Ferry at Anchor

The next step was to determine the forces on the anchors and evaluate the effectiveness of the anchors in stopping the ship [28–33].

Anchor holding force:

$$F_k = w_\alpha \cdot K \cdot w_k + \mu \cdot w_{cb} \tag{7}$$

where:

w_α—coefficient of reduction of the holding force of the anchor due to the lifting of its shank by angle α,
K—anchor holding force coefficient,
μ—coefficient of friction,
wcb—the weight of the chain lying on the ground.

At the beginning, the so-called anchor holding force coefficient (K) was determined, which is usually the factor by which the anchor weight (Wk) is multiplied. The anchor holding force (Fk) is expressed as: Fk = K·Wk. Some authors [28–34] propose a slightly modified relationship in the form $Fk = K\cdot(Wk^{0.92})$, where the power factor indicates an increase in force. However, this is a minor difference that is not commonly used in research, so the classical dependence was used. The values of the holding coefficients and the chain friction against the bottom of the basin are given in Table 2, which is an example of AC-14 and Hall type, which we have given only for comparison and indeed the anchor type for Bielik was Hall type and we assumed a coefficient for anchor hold force equal to 3.5. It should be noted that the assumption about the type of the bottom we are dealing with, and therefore the holding coefficient of the anchor, is crucial and introduces the greatest uncertainty. In this study it is assumed that the bottom is muddy with a layer of sedimentary river sand underneath. The anchor's holding factor is conservatively chosen as K = 3.25, which means that the anchor will hold more than three times its mass.

Table 2. Anchor holding force coefficient and friction coefficient chain-bottom.

Type of Bottom	AC-14	JIS/Hall	Coefficient of Friction
Mud	10	3	1
Sand	7	3.5	0.75
Gravel	8	3.5	
Rock/mud	2.4	1.8	
Clay	10	3.5	

Figure 9 shows the symbols used for further calculations regarding anchoring. The relationships that allow a calculation based on the load force of the chain by the ship, the weight of the chain and its length, the so-called chain curve (catenary), including the point of contact of the chain with the bottom of the basin (xh) and the length of the chain resting on the bottom (Sd), were used.

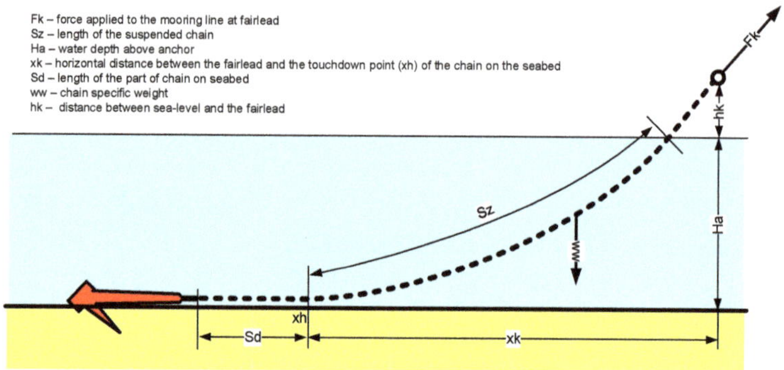

Figure 9. Chain curve and adopted markings.

For the Bielik type ferries, the length of the anchor chain is only 40 m. Usually commercial vessels use chains of up to 4–6 times the depth of the basin to effectively moor. The reason is not the weight of the chain itself and the extra resistance it creates, but the fact that the angle of the lift of the anchor shank that occurs when the chain is too short has an extremely negative impact on the anchor's holding power. Based on the literature review in this article [35–38] and OCIMF [27], coefficients for Hall or similar (stockless) anchors were determined for the patented anchor, their failures to sufficiently grip resulting from the upward lifting of the anchor shank that occurs when the chain is too short and light. The values of the coefficients are shown in Figure 10.

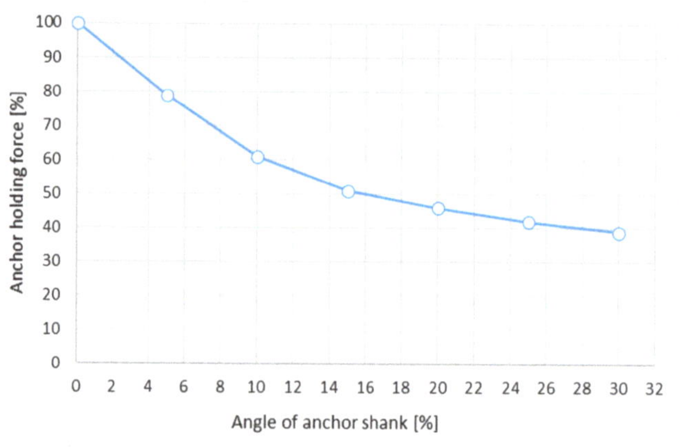

Figure 10. Coefficient of reduction of the anchor holding force due to the lifting of the anchor shank.

A probable anchoring pattern of the Bielik unit with two anchors ejected to the maximum chain length is shown in Figure 11, made on scale. It can be seen that the angle formed by the anchor shank with the chain is considerable (it can theoretically be up to 25 degrees, but in practice, it is smaller due to the weight of the chain). This is due to the short and light chain and the deep water. In addition, the chain may not be in contact with the bottom of the water, which obviously has a negative effect on the anchor's holding force.

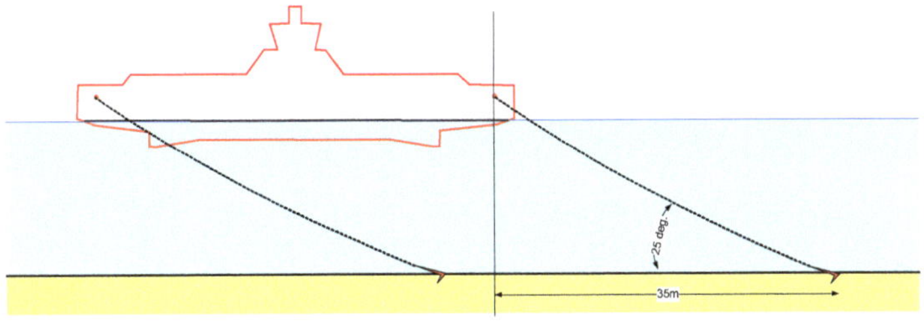

Figure 11. Probable angles of the chain and anchors at the Bielik emergency anchorage radius.

The problem of the fading of the chain curve due to increasing wind or current is illustrated in Figure 12.

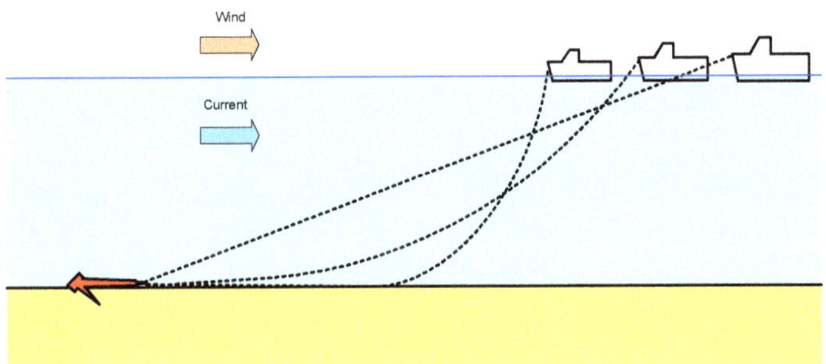

Figure 12. The problem of the disappearance of the chain curve and the lifting of the anchor shank when wind/current or the size of the unit increases.

In a further step, the possibility of lengthening the anchor chain by a certain value until the anchor curve forms and the chain reaches the bottom with its part to minimize the lifting angle of the anchor shank was considered as a natural solution. Table 3 shows the calculated forces acting on the ship when the chain is lengthened and the wind speed in the crossing area is increased and may serve as a recommendation to choose the optimal chain length in such an area. For this purpose, formulae 1., 2., 3., 5. and 6 were used to determine the aerodynamic force from the wind and the hydrodynamic force from current (assumed to be constants) for winds in the range 10–13 m/s. Two anchor holding forces were calculated using formula 7. assuming coefficients for the reduction of the holding force and for different chain lengths from 3 to 6 shackles. Thrust of the emergency engine was calculated as shown in Table 1.

Table 3. Total forces per radius with increasing chain length and wind speed in the crossing area.

Wind at the Crossing (m/s)	3 Shackles	4 Shackles	5 Shackles	6 Shackles
10	1.6	2.0	2.3	2.7
11	0.4	0.7	1.1	1.5
12	−2.5	−0.5	−0.2	0.2
13	−3.7	−1.9	−1.5	−1.2

The value of the chain length at which this effect occurs under the considered conditions is 3 shackles (82.5 m). In the case considered, the chain rests on the bottom at a distance of about 10 m from the anchor shank without any lifting of the chain. A further extension of the chain is of course even more effective in terms of holding force, but it is not justified due to the water restriction and the proximity to the shore. The additional holding force of the anchor is about 1 T.

4. Discussion

One key finding from this study is the lack of knowledge about the differences between design and operating conditions in extreme situations and the fact that the ship was designed for shallower depths than those prevailing in the Ferry Crossing. The management (shipowner) and the classification (Classification Society) and control bodies (Port State Control) were, one hopes, not aware of this. The detailed information flow between these actors is shown in Figure 13. The missing link in the information flow and the unknown links are also shown. The lessons learned after the incident relates not only to the simple redesign of the ferry (lengthening the anchor chains) but also to the establishment of a permanent flow of information between the decision-makers during the planning process.

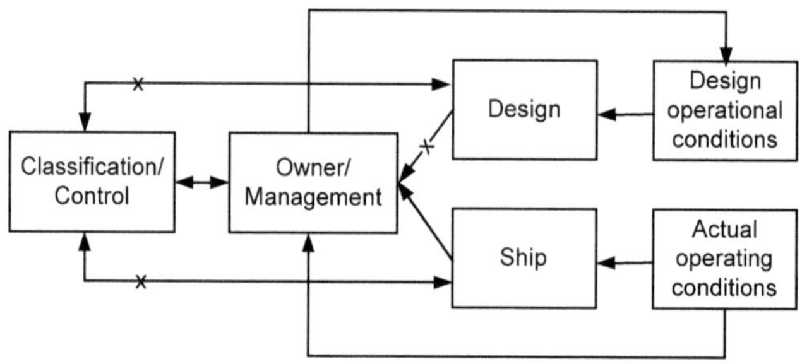

Figure 13. The information flow between Management and Classification and Control Bodies together with connections that are unknown or missing.

From the theoretical perspective the study showed that there is no link between actual operating conditions and the classification society, or that at the very least in this particular case the link failed to connect and with potentially dangerous results. A similar problem was observed in deep sea Polish ferry operators serving in the ferry link between Poland and Sweden. Several secondhand Ro-Pax ferries originally designed for the Mediterranean Sea were introduced into service in the Baltic Sea (for example, the m/f Gryf and m/f Galileusz ferries). Apparently, the power of the main engines and thrusters was not sufficient and caused problems with delays and safety of operation in more harsh wind conditions.

From the practical perspective we would like to highlight that a number of models are available, but unfortunately changing conditions and the evident lack of coordination among stakeholders unnecessarily complicates the accident/incident investigation and modeling process. The design of the hull was mentioned, but all aspects of the vessel must be designed for the purpose for which it was used. The ferry is not designed for emergency anchoring since anchors are only on one side and for shallow water (which is usually the case with this type of cross-service ferry). The missing link in the information flow between the operating body (owner) and the classification society, which focused on the condition of the ferry, which has a direct impact on safety, rather than on the operating conditions, which are different from the design, could present a relatively simple error that could have cost dozens of lives.

5. Conclusions

The main finding of this paper is that models based on socio-technical theory and systems theory can be useful for reconstructing accidents (incidents, too, of course), but that the application of such models does not take into account that external systemic issues will be predicted, such as that a vessel built for one particular maritime environment will be used in another where adverse issues may arise. This confirms the fact that even if these models are mostly proactive in nature, the maritime business environment may not be. The reconstruction carried out as part of this work led to several findings related to regulations, lack of cooperation between management and control bodies, and lack of knowledge on the part of the shipowners about the design conditions used in the construction of the ships. The ferry is designed to be anchored on one side only. This could also contribute to the postulates for a change in the design. However, such a change would mean a significant intervention in the structure of the ferry and was not proposed in the accident/incident report. Changing the anchor position on the opposite sides of the ferry would increase the ferry's anchoring capacity. The main conclusion on accident reconstruction models is that they are reactive given the instability caused by the maritime industry's leading decision makers.

Given this seemingly perpetual situation, our study shows that the accident investigation model of Rassmussen [24] should be improved because there is no correlation between the design of the ship and the given conditions on the ship. The accident investigation model should be dynamic—if the conditions on the ship change, the criteria must be considered a second time in order to include a dynamic aspect in risk assessment, i.e., a model to dynamically adapt the risk model to the current operating conditions.

On the basis of the calculations and analyses carried out, it was established:

1. that the Bielik ferry performed significant movements with an amplitude of up to 40 m during the emergency mooring on 9 January 2019.
2. the wind and current forces acting on the ferry were in excess of 5 T and could not be compensated by the auxiliary drive, which was estimated at 3 T.
3. the anchors in the area were only able to operate sub-optimally, due to their short chain and demonstrated a lifting of their shanks, which did not provide the necessary forces to keep the vessel in place.
4. the missing forces to keep the ferry in place could be about 1 T under the given conditions.

In such a situation, the best solution would be to extend the anchor chains to at least twice as long as 3 shackles (about 80 m) without changing the chain diameter and anchor size. This results in the chain resting on the bottom of the basin about 10 m from the anchor shackle and minimizes the angle of lifting of the shank itself.

It should be assumed that, based on the models and calculations, the wind speeds up to and including 11 m/s, i.e., up to and including 5°B (constant without taking gusts into account) in the area of the crossing itself is maximum for the operation of the ferry in case of main propulsion failure and emergency anchoring.

The analytical tools in the form of records of the Szczecin–Świnoujście VTS system have major limitations and do not allow accurate tracking of objects in restricted areas, i.e., almost in the entire area of the Świnoujście–Szczecin waterway.

Further studies should be of an investigatory nature to determine to what degree problems of oversight undermine the science of risk management. At a different level, interdisciplinary studies should be made regarding the instability and expense, not to mention dangers, caused by the economic model applied by financiers of the maritime industry; which is to state quite frankly, that the driving force behind the industry should not be the absolute maximization of profit. Under such conditions, which are the very reason for the constant pressure to build larger, more 'economical' vessels, ports are under pressure to expand beyond reasonable capacities [39,40], and aside from enriching a few stakeholders have an adverse effect on small economies, place the environment in danger at all times (for solutions are always temporary), and, not to attempt to be comprehensive, in the case of marine science experts, engage the best maritime minds of our generation in the study of short-term problems and take innumerable hours away from the studies of more important long-term issues regarding the environment.

This paper is practical and identifies an aspect of ship operations that is sometimes neglected in operations and therefore potentially increases the overall safety of maritime and awareness of stakeholders and the maritime community. Of course, this would require an interest on the part of the stakeholders and a delivery mechanism that communicates our concerns; we urge that such should be done, and thoroughly.

As for limitations, this paper is not comprehensive in any particular way; models have been simplified, but the main points are irrefutable.

Author Contributions: Conceptualization, L.G. and M.P.; methodology, L.G., K.Ł., A.A., P.V. and M.P.; data collection, K. Ł. and L.G.; formal analysis, L.G., M.P. and K.Ł.; investigation, L.G., M.P., P.V. and K.Ł.; data curation, L.G. and A.A.; visualization, A.A., P.V. and L.G.; supervision, L.G. and M.P.; writing—original draft preparation, L.G., K.Ł., A.A., P.V. and M.P.; internal review, A.A., M.P. and L.G. All authors have read and agreed to the published version of the manuscript.

Funding: The research was based on a case study prepared by the authors for the Polish State Marine Accident Investigation Commission (PKBWM). The publication of the paper is partially financed by the research project (L7-1847; Developing a sustainable model for the growth of the "green port") and the research group (P2-0394; Modelling and simulations in traffic and maritime engineering) at the Faculty of Maritime Studies and Transport, financed by the Slovenian National Research Agency.

Institutional Review Board Statement: Not applicable.

Informed Consent Statement: Not applicable.

Data Availability Statement: The data for this analysis came from Maritime Office in Poland and the owners of Bielik IV—Żegluga Świnoujska.

Acknowledgments: Not applicable.

Conflicts of Interest: The authors declare no conflict of interest.

Abbreviations

EMSA	European Maritime Safety Agency
EU	European Union
FRAM	Functional Resonance Accident Method
IMO	International Maritime Organization
OCIMF	Oil Companies International Marine Forum
PKBWM	Poland State Maritime Accident Investigation Commission
SHEL	Software–Hardware–Environment–Liveware
STAMP	Systems Theory Approach
VTS	Vessel Traffic Service

References

1. IMO. Code of the International Standards and Recommended Practices for a Safety Investigation into a Marine Casualty or Marine Incident (Casualty Investigation Code)—MSC-MEPC.3/Circ.2. 2008. Available online: http://emsa.europa.eu/retro/Docs/marine_casualties/msc-mepc_3-circ_2.pdf (accessed on 12 September 2021).
2. Amendments to the Code for the Investigation of Marine Casualties and Incidents, Resolution A.884(21), Appendix 1 the IMO/ILO Process for Investigating Human Factors, adopted on 4 February 2000. Available online: https://puc.overheid.nl/doc/PUC_3003_14/1/#23099 (accessed on 3 December 2021).
3. Directive 2009/18/EC of the European Parliament and of the Council of 23 April 2009 Establishing the Fundamental Principles Governing the Investigation of Accidents in the Maritime Transport Sector and Amending Council Directive 1999/35/EC and Directive 2002/59/EC of the European Parliament and of the Council. Available online: https://eur-lex.europa.eu/legal-content/EN/TXT/PDF/?uri=CELEX:32009L0018&from=EN (accessed on 10 September 2021).
4. Przepisy Portowe 2019. Zarządzanie nr.3 Dyrektora Urzędu Morskiego w Szczecinie z 26.06.2013. Consolidated Text as at 4 March 2019. Available online: https://www.ums.gov.pl/po/2013/Zarz_3_2013_1.pdf (accessed on 15 September 2021).
5. Cassama, F. A Study on Marine Accident Causation Models Employed by Marine Casualty Investigators. World Maritime University, Dissertations. Malmö: The Maritime Commons: Digital Repository of the World Maritime University. 2015. Available online: https://core.ac.uk/download/pdf/217237205.pdf (accessed on 27 October 2021).
6. Stoop, J.A. Maritime accident investigation methodologies. *Inj. Control Saf. Promot.* **2003**, *10*, 237–242. [CrossRef] [PubMed]
7. Awal, Z.I.; Hasegawa, K. A study on accident theories and application to maritime accidents. *Procedia Eng.* **2017**, *194*, 298–306. [CrossRef]
8. Laflamme, L. A better understanding of occupational accident genesis to improve safety in the workplace. *J. Occup. Accid.* **1990**, *12*, 155–165. [CrossRef]

9. Qureshi, Z.H. A Review of Accident Modeling Approaches for Complex Critical Socio-Technical Systems. In Proceeding of the Twelfth Australian Workshop on Safety Critical Systems and Software and Safety-related Programmable Systems; 2007; Volume 86, pp. 47–59. Available online: http://citeseerx.ist.psu.edu/viewdoc/download?doi=10.1.1.294.4125&rep=rep1&type=pdf (accessed on 3 December 2021).
10. Khanzode, V.; Maiti, J.; Ray, P.K. Occupational injury and accident research: A comprehensive review. *Saf. Sci.* **2012**, *50*, 1355–1367. [CrossRef]
11. Mullai, A.; Paulsson, U. A grounded theory model for analysis of marine accidents. *Accid. Anal. Prev.* **2011**, *43*, 1590–1603. [CrossRef] [PubMed]
12. Leveson, N. A new accident model for engineering safer systems. *Saf. Sci.* **2004**, *42*, 237–270. [CrossRef]
13. Heinrich, H.W. *Industrial Accident Prevention: A Scientific Approach*, 2nd ed.; McGraw-Hill Book Company: New York, NY, USA, 1941; ASIN: B0007E73LS.
14. Hollnagel, E. Risk + barriers = safety? *Saf. Sci.* **2008**, *46*, 221–229. [CrossRef]
15. Reason, J. *Managing the Risks of Organizational Accidents*, 1st ed.; Ashgate Publishing Ltd.: Aldershot, UK, 1997; ISBN 1840141050.
16. Hollnagel, E.; Woods, D.D.; Leveson, N. *Resilience Engineering: Concepts and Precepts*, 1st ed.; Ashgate Publishing Ltd.: Aldershot, UK, 2006; ISBN 9780754649045.
17. Hollnagel, E. Understanding accidents-from root causes to performance variability. In Proceedings of the IEEE 7th Conference on Human Factors and Power Plants, Scottsdale, AZ, USA, 19 September 2002; IEEE: Scottsdale, AZ, USA, 2002; p. 1. [CrossRef]
18. Righi, A.W.; Saurin, T.A. Complex socio-technical systems: Characterization and management guidelines. *Appl. Ergon.* **2015**, *50*, 19–30. [CrossRef] [PubMed]
19. Hollnagel, E.; Goteman, O. The Functional Resonance Accident Model. *Proceedings of Cognitive System Engineering in Process Plant*. 2004, pp. 155–161. Available online: https://citeseerx.ist.psu.edu/viewdoc/download?doi=10.1.1.579.1930&rep=rep1&type=pdf (accessed on 12 September 2021).
20. Laracy, J.R. A systems theoretic accident model applied to biodefense. *Def. Secur. Anal.* **2006**, *22*, 301–310. [CrossRef]
21. Larsson, P.; Dekker, S.W.; Tingvall, C. The need for a systems theory approach to road safety. *Saf. Sci.* **2010**, *48*, 1167–1174. [CrossRef]
22. Awal, Z.I.; Hasegawa, K. Accident analysis by logic programming technique. In *Paper Submitted for the Proceedings of the European Safety and Reliability Conference (ESREL)*; Taylor & Francis Group: London, UK, 2015; pp. 13–21. ISBN 978-1-138-02879-1.
23. Wang, J.; Yan, M. Application of an Improved Model for Accident Analysis: A Case Study. *Int. J. Environ. Res. Public Health* **2019**, *16*, 2756. [CrossRef] [PubMed]
24. Rasmussen, J. Risk management in a dynamic society: A modelling problem. *Saf. Sci.* **1997**, *27*, 183–213. [CrossRef]
25. Puisa, R.; Lin, L.; Bolbot, V.; Vassalos, D. Unravelling causal factors of maritime incidents and accidents. *Saf. Sci.* **2018**, *110*, 124–141. [CrossRef]
26. Hensen, H. *Tug Use in Port: A Practical Guide*, 1st ed.; The Nautical Institute: London, UK, 2003; ISBN 9781870077392.
27. Oil Companies International Marine Forum (OCIMF). *Prediction of Wind and Current Loads on VLCCs*; Witherby and Company: London, UK, 1994; ISBN 9781856090421.
28. Gerr, D. *Propeller Handbook: The Complete Reference for Choosing, Installing and Understanding Boat Propellers*; International Marine Ragged Mountain Press: Camden, NJ, USA, 2001; ISBN 0071381767.
29. Vryhof Manual—The Guide to Anchoring. Vryhof Anchors B.V. 2015. Available online: https://www.plaisance-pratique.com/IMG/pdf/Vryhof_Anchor_Manual2015.pdf (accessed on 15 September 2021).
30. Shin, H.; Seo, B.; Lee, J. Experimental study of embedding motion and holding power of drag embedment type anchor on hard and soft seafloor. *Int. J. Nav. Archit. Ocean. Eng.* **2011**, *48*, 183–187. [CrossRef]
31. Hancox, M. *Oilfield Seamanship Series-Volume 3: Anchor Handling*; Oilfield Publications: Ledbury, UK, 1994; ISBN 9781870945493.
32. Hinz, E. *The Complete Book of Anchoring and Mooring*, 2nd ed.; Schiffer Publishing Ltd.: Atglen, PA, USA, 2009; ISBN 0870335391.
33. Ren, Y.X.; Lei, Z.M.; Sun, L.Q.; Yan, S.W. Model tests of dragging hall anchors in sand. *J. Mar. Sci. Technol.* **2016**, *24*, 4. [CrossRef]
34. Brix, J. *Maneuvering Technical Manual*; Seehafen Verlag: Hamburg, Germany, 1987; Volume 36, ISSN 0036-603X.
35. Edwards, R.Y. Hydrodynamic Forces on Vessels Stationed in a Current. In Proceedings of the 17th Annual Offshore Technology Conference, OTC 1985, Houston, TX, USA, 6–9 May 1985; OTC: Houston, TX, USA, 1985; pp. 99–105. [CrossRef]
36. Gunnu, G.; Wu, X.; Moan, T. Anchor handling vessel behavior in horizontal plane in a uniform current field during operation. In Proceedings of the 2nd Marine Operations Speciality Symposium, Singapore, 6 August–8 August 2012; Research Publishing Services: Singapore, 2012. [CrossRef]
37. Remery, G.; Oortmerssen, G. The Mean Wave, Wind and Current Forces on Offshore Structures and their Role in the Design of Mooring Systems. In Proceedings of the Offshore Technology Conference, Houston, TX, USA, 29 April–2 May 1973; OTC 1741. [CrossRef]
38. Batista, M.; Perkovič, M. Computation of mooring chain with the touchdown on an inclined seabed. *J. Mar. Eng. Technol.* **2019**, *29*, 1–14. [CrossRef]
39. Gucma, L.; Perkovič, M.; Przywarty, M. Assessment of influence of traffic intensity increase on collision probability in the Gulf of Trieste. *Annu. Navig.* **2009**, *15*, 41–48. Available online: http://yadda.icm.edu.pl/baztech/element/bwmeta1.element.baztech-article-BATA-0011-0020 (accessed on 21 September 2021).
40. Vidmar, P.; Perkovič, M.; Gucma, L.; Łazuga, K. Risk Assessment of Moored and Passing Ships. *Appl. Sci.* **2020**, *10*, 6825. [CrossRef]

Journal of
Marine Science and Engineering

MDPI

Article

Berthing Assistant System Using Reference Points

Jan Mentjes *, Hilko Wiards * and Sebastian Feuerstack *

German Aerospace Center (DLR), Institute of Systems Engineering for Future Mobility, Escherweg 2, 26121 Oldenburg, Germany
* Correspondence: jan.mentjes@dlr.de (J.M.); hilko.wiards@dlr.de (H.W.); sebastian.feuerstack@dlr.de (S.F.)

Abstract: With more goods to be transported oversea, traffic and vessels' dimensions increase while berthing areas merely remain constant and thus challenge ship masters and pilots to maneuver in small basins with dense traffic even in bad weather situations. Too fast approaches or steep angles of attack result in damages to fenders, quay walls, or even impact the hull structure. We propose a shore-based, vessel-independent berthing assistant system to support sailors by Reference Points that are aligned to a quay's meter markings and identify the precise berthing location by measuring distance and approach speed. For this purpose, we define the concept of a Berthing Support Area (BSA), which specifies an area in which, subject to constraints, safe berthing is provided. Within this area there are Reference Points, perpendicular distance measurements at arbitrary positions, which are implemented with a set of LiDAR sensors that have been integrated into the quay wall. In a test campaign with a vessel equipped with DGPS sensors, we sailed seven different maneuvers and evaluated the precision and the accuracy of the Reference Points for speed and distance measurements.

Keywords: berthing aid system; laser scanner; precise positioning; docking; scenario-based testing

Citation: Mentjes, J.; Wiards, H.; Feuerstack, S. Berthing Assistant System Using Reference Points. *J. Mar. Sci. Eng.* **2022**, *10*, 385. https://doi.org/10.3390/jmse10030385

Academic Editor: Marko Perkovic

Received: 10 February 2022
Accepted: 4 March 2022
Published: 8 March 2022

1. Introduction

An increasing amount of goods to be transported around the globe resulted in continuously increasing ship dimensions [1]. Since 1996, container vessels' size has increased by 90% [2]. Maneuvering in such dense traffic and in ports areas becomes more challenging the bigger a vessel is. Ports are limited in their growth and cannot be expanded at will [3]. Space in harbors is often limited and areas are difficult to overlook and to access. Sometimes even full ship rotations need to be performed in narrow port basins. Stringent time slots and high workload of ship masters and pilots to coordinate supporting tugboat actions increases the likelihood of accidents resulting in damage to ships and port infrastructure. Maintenance of damages to port infrastructure might hinder port access for long periods of time [4]. Because of this, berthing maneuvers are considered to be a highly complex task [5].

To support captains and pilots in challenging berthing maneuvers, Berthing Aid Systems (BAS) are being developed to enhance their situational awareness in high-workload situations. In general, two different approaches for BAS can be distinguished [6]: ship-based systems and shore-based assistance systems. Ship-based systems enhance vessels with sensor technology, such as port radars and Portable Pilot Units (PPU) that connect to on-board systems to create situational pictures based on GPS and automatic identification system (AIS) information to support pilots during their assistance. AIS is an automated vessel tracking system that communicates a vessel's unique identification, position, course, and speed in certain time slots [7].

Shore-based systems depend on sensors installed at the shore. They integrate sensors, such as LiDAR, into quay walls to measure the distance to approaching ships. LiDAR technology typically offers centimeter precise distance accuracy by emitting light pulses, which are reflected by the targeted objects, while also archiving a high angular resolution in contrast to radar solutions, such as automotive mmWave radar, where the state of the art seems to reach a resolution of 1° [8]. The distance is determined according to the Time of

Flight (TOF) and the speed of light by measuring the time between sending and receiving the reflected pulse [9]. In the case of 1D LiDAR sensors, light pulses are directed to a single target point destination so that the distance then is measured. The 2D and 3D LiDAR sensors extend this approach by additionally considering horizontal (2D) and vertical (3D) distance measurements. These measurements are displayed on huge screens installed in the sight of the pilot close to the berthing location or are digitally transmitted to a pilot PPU to support berthing maneuvers [10,11]. Shore-based systems have the advantage that they do not need vessels to be equipped with additional systems.

Assistance systems are often considered as safety-critical systems if system failures result in substantial damage to people, property, or the environment (i.e., pollution) [4]. System verification and validation is therefore an essential part of the system engineering process to proof system dependability properties, such as availability, reliability, and safety. However, if non-deterministic approaches or black-box methods are used, the functional safety can hardly be assured [12]. In the former case, the algorithms generate a different result for the same input, making them difficult to test. In black box methods, the internal implementation is not accessible for inspection, so it is not possible to understand how the system generates the output [13]. In interviews with experienced pilots, we learned that specifically; system reliability, thus, is the ability of a system to deliver its services as specified, and is therefore of major importance. In the automotive domain, the term Operational Design Domain (ODD) is used to describe conditions under which a system is designed to function. Among other things, it contains restrictions for environmental influences or geographical areas [14]. The conditions for the functionality of the system and the range in which it must operate must therefore be constrained to ensure high reliability within the system specifications.

In this paper, we therefore propose a deterministic, ship-independent, and shore-based BAS using LiDAR sensors, which implements Reference Points placed in line with the positioning marks of a quay wall to measure and communicate the distance, speed, and acceleration of an object in relation to a quay wall to a pilot's PPU. At the same time, we combine the algorithm with the determination of an area in which functional safety can be ensured. The goal is to secure the system by defining the ODD based on vessel and environmental characteristics, expressed as a well-defined polygon. In the following section, we summarize interviews that were performed with pilots (harbor and port access pilots) as part of two workshops and also consider the relevant guidelines to derive the requirements for a shore-based BAS (Section 2). Thereafter, in Section 3, we give an overview about the current state of the art equipment, focusing on shore-based BAS. In Section 4, we present our concept based on deterministic Reference Point measurements for such a system followed by the derivation of a Berthing Support Area. A description of the implementation and installation in the port of Wilhelmshaven in Germany is described in Section 5. Finally, we present the system evaluation based on a field test (Section 6) and close with a discussion of the results and state further work (Sections 7 and 8).

2. General Requirements for Berthing Assistance Systems

For the requirement analysis, we performed a task analysis based on information gained by interviews with experienced pilots (Section 2.1) and by deriving further requirements based on regulations, guidelines, and common practice (Section 2.2).

2.1. Task Analysis with Pilots, Common Practices, and Local Port Regulations

We had two workshops with different groups of maritime pilots: four river pilots that support navigation in dense traffic regions to two oversea ports in Germany, and five harbor pilots that are responsible for supporting navigation in the port of Hamburg. Figure 1 (taken from [15]) summarizes the main tasks of pilots during berthing. Conducting a berthing maneuver involves three basic subsequent tasks: initial preparations (T1—e.g., requesting pilot support, positioning crew for observation, configuring the bridge system, and connecting the pilot plot), performing the berthing maneuver (T2), and finally mooring the ship (T3). We focused on identifying relevant information for the berthing maneuver (T2).

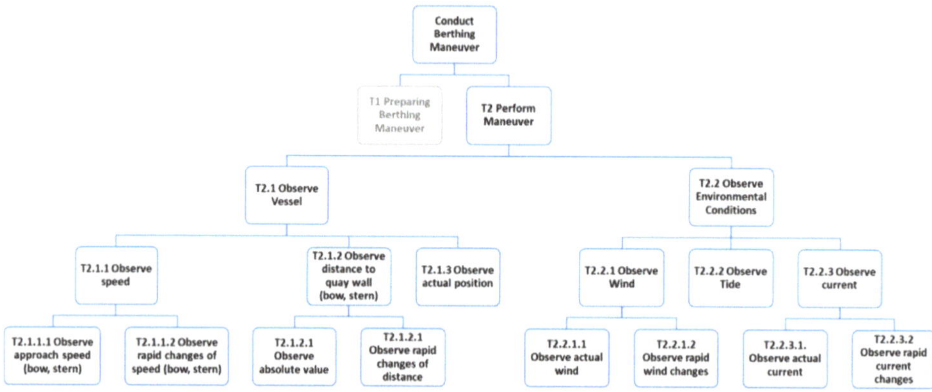

Figure 1. Analysis based on interviews with harbor and river pilots taken and adapted from [15].

Pilots reported the berthing speed for stern and bow (T2.1.1.1) and the corresponding distances to the quay (T2.1.2) combined with information about changes in velocity (T2.1.1.2) and the actual absolute position of the ego ship (T2.1.3) as the most relevant information during berthing. Further on, wind speed and direction (T2.2.1.1) and significant changes of those (T2.2.1.2—e.g., changes can occur during bridge passing or at locations with heavy gusts of wind), as well as currents (T2.2.3.1) and the tidal system (T2.2.4.2) were also mentioned to be required to be carefully observed by the pilots. For this contribution, we specifically focus on BAS that support pilots in observing the ego vessel (T2.1). Berthing areas in highly frequented ports are identified by meter markings that appear every 10–15 m and there are strict rules that require a vessel to stop exactly at meter mark zero ("red flag") with a targeted discrepancy of less than 2 m. In the interviews, the pilots also explicitly stated system integrity as the most important acceptance factor for a berthing support assistant (the pilots agreed: "The BAS should only communicate information if it is 100% sure. In situations with less confidence, it should simply communicate no information at all").

For some of the information that the pilots communicated as the most important and challenging, (c.f. Figure 1), guidelines and common practices have also been reported. The PIANC Guidelines for the Design of Fender Systems [16] considers berthing velocities from 0.08 m/s (over 50,000 DWT under favorable conditions) up to a maximum of 0.6 m/s for vessels under 10,000 DWT under unfavorable conditions). Berthing angles are assumed to be below 6 degrees for vessels larger than 50,000 DWT and for smaller vessels (mainly those without tug boat assistance) between 10 and 15 degrees based on measurements in Japan [16]. Hein [17] observed a total of 1082 berthing activities of vessels between 200 m and 400 m in Bremerhaven, Germany, and reported average berthing velocities (i.e., perpendicular approach speed to quay) between 0.051 m/s (200 m) and 0.057 m/s (400 m) with outliers up to 0.2 m/s. Comparable values were also reported by the local port pilots that stated an "absolute maximum" perpendicular approach speed of 0.3 m/s and a maximum pressure the fenders caused by an approach speed of 0.15 m/s. The maximum speed over ground (SOG) in port areas depends on port regulations. For our local testbed in Wilhelmshaven, there is a speed restriction of 6 kn for all port areas [18].

Average berthing angles were observed between 0.34 degree (200 m) and 0.18 deg (400 m) with outliers up to 1.25 degree (<300 m) and up to 1 degree (>300 m), respectively. Specifically, for larger vessels, a steep berthing angle up to 5 degrees would result in overhanging, curved hull sections of the ship, and would add additional risk for damage of constructions and crane systems located at the pier area.

We also asked the pilots for the required distance that they favor to receive distance measurements to support the berthing activities. It transpired that a sensing distance of 100–120 m (four times the width of a vessel) seems to be the preferred distance in that the pilots started to observe the approach in relation to the quay.

2.2. IMO Regulations

Integrity has been defined as "The ability to provide users with warnings within a specified time when the system should not be used for navigation" in the IMO Resolution A.915 (22) [19]. For port navigation, the resolution defines an alert limit (AL) of 2.5 m, a time to alarm (TTA) of 10 s, and an integrity risk of 10^{-5} per 3 h as the main system integrity parameters. The integrity risk is defined as "The probability that a user will experience a position error larger than AL being raised within the specified TTA at any instant of time at any location in the coverage area." Besides integrity, the IMO states an absolute horizontal accuracy of 1 m as a minimum maritime user requirement for general navigation in ports together with service level parameters for an availability of 99.8% (per 30 days), a continuity of 99.97% per 3 h, and a position fix interval of 1 s for port navigation.

3. Related Work

Many shore-based assistant systems are using LiDAR technology to detect approaching ships. One example is the SmartDock developed by Trelleborg [10], for which multiple 1D LiDAR sensors have been installed at quay walls to measure the distance to approaching ships. By calculating the change in distance, the speed and acceleration of a vessel in relation to the quay wall is determined. These measurements are then shown on displays installed at quay walls. However, often the sensors cannot be placed at arbitrary locations on the quay wall. Limited availability of cable niches [6], the dangers of mooring lines, high tides or flooding situations, and sight blocking fenders limit installation locations. In addition, the fixed range measurement locations limit BAS support to vessels of a certain predefined size to obtain appropriate bow and stern distance measurements of a vessel. With respect to the use of LiDAR for BAS, cost, short detection ranges, and additional problems with dark-hulled vessels are reported by [6].

DockAssist is a similar system proposed by Metratek [11], which consists of four parts: a Laser Berthing Aid System, an Advanced Detection and Automation System (ADAS), an Environmental Monitoring System (EMS), and an Audio and CCTV Surveillance System. For the BAS part, LiDAR sensors are installed on the quay wall and measure the distance, speed, and heading of the vessel. We could not ascertain whether the 1D or 2D LiDAR sensors are used. The ADAS uses the Automatic Identification System (AIS) to detect incoming vessels at an early stage to provide distance information beyond the range of the LiDAR sensors. Environmental information is collected by the EMS through the integration of wind, wave, tide, and current information. Moreover, audio and video data are recorded by the AVS to enable monitoring of the berthing location. We were not able to ascertain details about the applied algorithms (i.e., if DockAssist implements a deterministic approach). For visualization, the collected data and measurements are then sent to a mobile device that can be used on the ship. Thus, no display is installed on the quay wall itself, but a portable unit that can be used by the operators is.

Perkovič et al. report on a comprehensive BAS for detecting approaching vessels, determining the stern and bow of the vessels, and measuring the distance to the quay walls using 3D LiDAR sensors [20]. Additionally, wind and current sensors are used to provide context information for pilots. A roll-on, roll-off bridge is considered by using two 3D LiDAR sensors. One of them is used to detect the side (port or starboard) of approaching objects and the other to detect the bow or stern. They apply the Random Sample Consensus (RANSAC) algorithm to detect the side of a ship. This algorithm is able to identify geometries in a point cloud based on a reference model (e.g., line and plane). They evaluated their system under real conditions and can achieve more accurate results than those obtained by AIS. RANSAC uses a random set of points to determine a vessels' side and therefore implements a non-deterministic approach.

In [21], a shore-based assistance system based on cameras is proposed that is also capable of detecting partly obscured vessels in multi-ship situations (e.g., tug-supported berthing). Therefore, the distance of objects to the quay wall is determined by an artificial intelligence vision-based monitoring system (AVMS) consisting of a camera, a Differential Global Positioning System (DGPS), and an Inertial Measurement Unit (IMU) sensor. The image data from the camera are first processed by a neural network using the Ski-ENet model to detect ship contours. Then, the position and orientation received by the DGPS and IMU sensor of the AVMS are used to determine the relative position of the vessel to the quay wall. The approach is evaluated using data sets from a field test in Korea and compared with a conventional LiDAR BAS using a 16-channel LiDAR sensor. Measurements of the conventional LiDAR-based BAS under good weather conditions (no rain and daytime, etc.) were considered reference values. The authors report that in contrast to the LiDAR system, the camera system shows, even under bad weather conditions, more stable results. Because this approach uses a neural network to detect a vessel, the actual detection mechanism is a black box for the evaluation, which makes it difficult to test the reliability criteria of the system.

In [22], a method for berthing information extraction is presented using 3D-LiDAR sensors. It features bow and stern recognition and measures distance, velocity, and approach angles in relation to the quay walls. First, they projected LiDAR measurements into a berthing coordinate system using the offset between the sensors' location and filtered fixed infrastructure (e.g., cranes). Then, a statistical outlier removal algorithm is applied to the remaining points, removing points based on the distance distribution in the point cloud. Using principal component analysis, eigenvectors and eigenvalues of a point cloud are extracted and the direction of the longest vector determines the ship's course. After this, region growing is applied to the point cloud to segment this into sides (bow, stern, and hull). By combining bow and stern feature points, as well as the result of region growing, all sides of a ship can be determined. In the last step, the authors apply a visibility analysis to differentiate between six ship attitudes, showing which sides of a ship are possibly visible by a LiDAR sensor. Using this approach, a bow and stern point of a ship can clearly be identified and further used for the berthing parameter calculation (distance, speed, and angle). Using field tests and simulation approaches, the authors show that their method provides stable and accurate measurements. However, the use of a single LiDAR could compromise the robustness of the system in case of failure.

In [23], a comparison of mooring systems is conducted. In most cases, ropes and windlasses are used for mooring. However, this creates an increased risk due to equipment failures or safety procedure errors and new technologies have emerged that can therefore improve the mooring process. Both magnet-based and vacuum-based systems are presented as possible alternatives for securing the vessel to the quay wall. The authors conclude that vacuum-based systems offer the most advantages as they are safer, faster, and more environmentally friendly. Compared to our approach, these systems are suitable for the final step of mooring, while we are focusing on assisting the vessel's approach.

In [24], an approach to solve the berth allocation and quay crane assignment problem is presented. For efficient port operations, berths and cranes for unloading goods must be assigned to arriving vessels. The authors propose to partition the berthing space and assign berths. They investigated the division of the berth in 10 m, 20 m (distance between bollards), and dynamic (based on vessel length) distances. The results suggest that a finer partitioning is very efficient. This approach has been extended in [25] by a weekly planning of resources in combination with a reactive planning to handle if ships arrive later or earlier than planned. This helps to make port operations more efficient and to save energy. In [26], it was determined that energy can be saved by optimizing port processes. The potential for savings can be achieved primarily through improved planning, but also by shortening processes. Berthing assistance systems can support this by making processes safer and more time-efficient. Therefore, they can support the greening of ports.

To the best of our knowledge, current commercially available BAS focus on support for specific berthing situations. For instance, in the European TWIN-PORT 3 project, different vacuum-based auto-mooring systems from Cavotec [27] and Trelleborg [28] provide quick and fast mooring to secure passenger and cargo ramp access. For huge RoRo and cruise vessels, saving berthing and mooring time significantly reduces harmful emissions [29]. Recent research evaluates the combined sensor systems towards vessel-independent BAS and the potential of AI to also handle specific complex situations such as partially hidden vessels. Based on the interviews with the pilots, the trustworthiness of a BAS is a currently underestimated factor. Therefore, we propose in the following section a deterministic approach for a BAS, making use of a deterministic approach to ensure the functional safety of the assistant system.

4. Reference Points

Based on the interviews with the pilots, the meter markings and individual spots (such as specific fenders and constructions on the quay) transpired to be the main source of orientation with which the pilots visually estimate the vessels' positioning and orientation in relation to a quay. We therefore propose Reference Points that are aligned to the meter markings to ease and improve the situation awareness of the pilots with respect to the perpendicular distance and berthing approach velocity in relation to a quay wall. Current PPU systems calculate this information based on the AIS data and the topographic information encoded into an electronic chart. Because the AIS data sending frequency is connected to the actual SOG of a vessel (which, in general, means that less SOG results in less AIS updates) [30], the distance and approach speed calculation measurements are considered as very unreliable by the pilots as their readings significantly jump, become impacted by deteriorating GPS information caused by signal disturbances close to port constructions, and become less frequent with minor speed the closer the vessel is to approaching a quay. Current maritime radar systems, which are installed on vessels or at the port, do not offer an appropriate minimal detection range and resolution to support berthing.

The technical concept of the Reference Points originates from the way 1D LiDAR sensors work. These emit a light pulse that is reflected by the targeted object. The distance is then measured using the time deviation between emitting and receiving the light pulse, using the speed of light. In some cases, LiDAR sensors can receive multiple echoes of an emitted light pulse. Based on the opening beam of the laser, an object can thus reflect several light beams, so that several distances are measured [31]. Finally, the sensors' software decides which distance to use. Thus, the first echo corresponds to the shortest distance, the last to the maximum distance measured.

Similar to this approach, a Reference Point also considers a set of data points. However, instead of the individual LiDAR spot reflection echoes (which are handled in the sensors firmware), it calculates the perpendicular distance and approach speed from the quay to a vessel's hull based on a set of LiDAR spots. Figure 2 shows the structure of a Reference Point, which consists of an origin point p (left side) and a filter box defined by its length l and the width w. In Figure 2 the point cloud is shown as black dots, where five of six points are in the filter. If points are detected inside the filter, the distance to each point is calculated. As it can be seen in the figure, we choose the perpendicular distance, which determines the distance at a right angle to the Reference Point. The point with the smallest perpendicular distance d to p sets the distance that is then reflected by the Reference Point. Therefore, based on the working principle of LiDAR, the first echo is used as the output distance. Changes in the distance are used to calculate a vessels' velocity and acceleration relative to the quay.

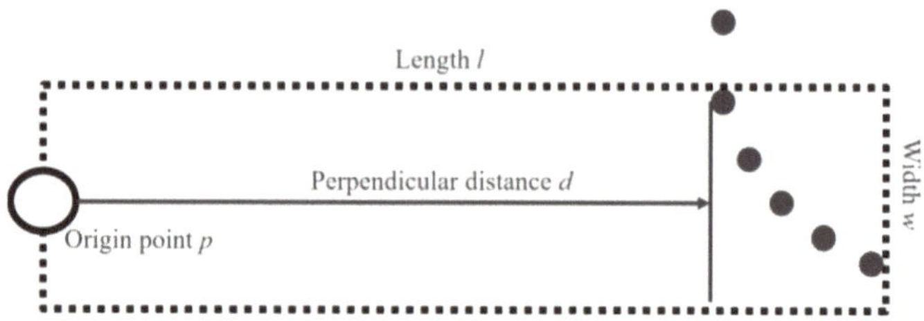

Figure 2. Concept of a Reference Point. Origin p is the point to which the distance of a vessel is calculated. A filter box is marked by dotted lines, which is defined by a length l and a width w. Starting from the point, the perpendicular distance to a ship is calculated, which is marked by a point cloud (black dots).

Reference Points are defined for positions for that detailed distance and relative speed calculations are relevant. For instance, for port entries and locks, they can support identification of approach angles and check for appropriate speeds. Coupling them with landmarks, such as cranes, quay meter marks, or fenders eases sailing by sight during the very last meters of an approach. Depending on the minimum ship length that the BAS is to support, at least three Reference Points per ship length should be placed evenly distributed along the quay. This ensures that a vessels' bow and stern (and therefore also the ROT) can be tracked during the entire berthing process even with steeper and uncommon berthing angles. Figure 3 shows the general setup of the proposed BAS.

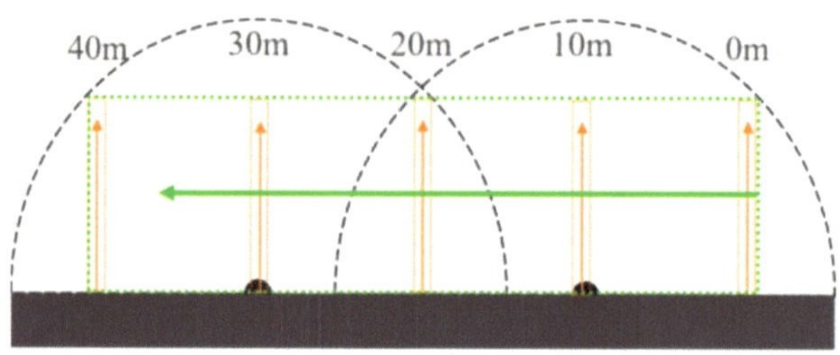

Figure 3. Berthing Assistant System Concept. LiDAR sensor positions shown as dots and their detection range in semi circles. Vertical Reference Points measure the distance in relation to quay wall. Horizontal Reference Points shown in green measure the distance to meter mark zero.

Figure 3 depicts two LiDAR sensors, five vertical and one horizontal Reference Point placed along the quay wall. The LiDAR Sensor detection range is shown by semicircles. Vertical arrows and boxes mark the position of vertical Reference Points, which measure distance in relation to a quay wall. Also, a horizontal Reference Point is shown, which measures the forward SOG of a ship. This is also used to measure the distance to a stopping point (i.e., end of a berthing location). The dashed rectangle identifies the area for that the BAS offers support. The characteristics of the LiDAR Sensors (i.e., opening angle, resolution, and supported distance), hull forms, and coatings as well as the environmental conditions (e.g., rain and snow) determine the size of this rectangle. The width of the Reference Point boxes determines the amount of LiDAR beam measurements to be considered for.

In this example, Reference Points have been placed in 10 m intervals on the quay wall to support vessels larger than 30 m. The position of these Reference Points also corresponds to meter marks along the quay. This physical mapping eases the pilots' orientation with respect to the electronically communicated values to the pilots PPU. Because berths in the harbor usually have fixed dimensions, a Reference Point at meter mark 0 m indicates where a ship has to stop. Due to a ship's mass and the resulting relatively long breaking distance, pilots and ship masters need to be informed early on how fast the ship is moving towards the end of the quay. Therefore, the definition of horizontal Reference Points sometimes also make sense, e.g., to ease precise RoRo ramp berthing. Figure 3 illustrates such a horizontal Reference Point at meter mark 0 m.

Besides the ship length and its mass, the overall harbor layout, it's corresponding berthing areas, and also application-specific requirements are further aspects that determine the amount of the required Reference Points. For instance, RoRo ramp berthing benefits from horizontal Reference Points, and crooked port areas and areas with strong currents or winds might require a higher density of Reference Points. Finally, the amount of LiDAR sensors and also their opening angle and resolution limits the amount of Reference Points.

Berthing Support Area Derivation and Operation Process

A Berthing Support Area (BSA) defines the precise geospatial area on a sea chart for which the BAS system offers support. Following the concept of the Operational Design Domain (ODD) from the automotive domain, this Support Area defines an area and a set of operating conditions in which the assistance system is specified to function [14]. It can generally be understood as a well-defined polygon A_{poly} covering the berthing area and its immediate surroundings and a set of constraints C on parameters. The parameters considered here consist of the sets $P_{control}$, $P_{construction}$, and $P_{environmental}$. The set $P_{control}$ contains all parameters that define the control of a vessel (e.g., the speed over ground, the heading, or the distance to the quay wall). The parameters that are defined by the general construction of the ship, for example the hull size and coating, are defined in the set $P_{construction}$. Moreover, the set $P_{environmental}$ contains all parameters that are defined by the environment. These parameters are for example the current visibility, tide, winds, or currents. These sets of parameters are not fixed and must be adapted to the local circumstances and intended use of the system.

The total set of relevant parameters P_{constr} for possible constraints results from the union of these sets. A constraint $c_p \in C$ on $p \in P_{constr}$ is the restriction of such a parameter. For numeric parameters, these are validity intervals, and for corresponding categorical parameters, sets of valid values. With these the validity of the BSA comes down to:

$$\text{BSA is valid} \iff \forall c_p \in C : p.\text{value} \in c_p \vee \text{ship.hull} \subset A_{poly}$$

To derive a specific BSA and thus the polygon A_{poly} and all constraints on the set of parameters P_{constr} to finally gain a running BAS, we propose a BSA derivation and operation process, which is depicted by Figure 4.

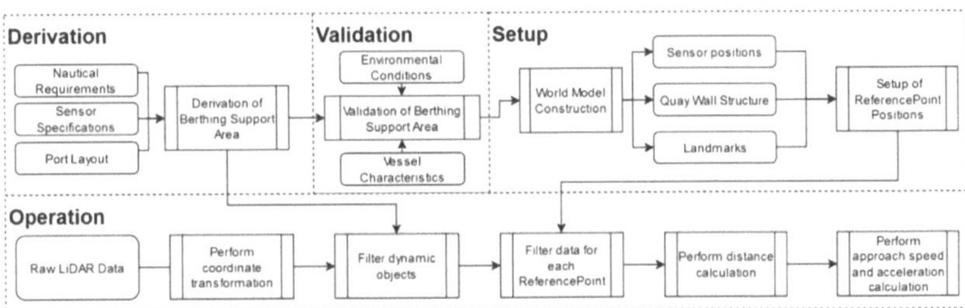

Figure 4. The BSA derivation and operation process. The former one includes set setup and validation to gain Reference Points, while the latter summarizes the process for the Reference Points to be used to determine the distance and speed measurements.

The port layout (e.g., quay size and layout, and accessibility) together with the nautical requirements (port specific speed limits, and pilots and shipmasters' demands for support area e.g., based on relevant situation awareness criteria) are used to derive an initial polygon.

Based on the ideal polygon requirements, suitable sensors (e.g., LiDAR or short-range-radar) are selected and the corresponding sensor specifications (e.g., opening angle and measurement distance and precision) might confine the size of the ideal polygon and also constrain the $P_{construction}$ and $P_{control}$ parameters (e.g., opening angles and sensor position) based on realizations of $P_{environment}$ (e.g., visibility for LiDAR). A basic mathematical model (c.f. for details sec 5.1) is used to determine that the sensors are able to sense the vessel within the BAS and within pre-defined $P_{control}$ constraints (e.g., common berthing angles). For the resulting BSA, a world model of the berthing location (i.e., quay) is created that stores the sensor positions, relevant quay wall structures and landmarks (that are used by the ship masters and pilots for orientation) in a global coordinate system.

By a simulation, the BSA is validated with respect to relevant $P_{construction}$ variants and corresponding constraints are derived. Finally, Reference Points are manually set in the world model based on pilots' and shipmasters' demands (e.g., Reference Points attached to meter marks, fenders, or other relevant landmarks used for orientation during the berthing process).

Regarding the operation phase, the first step is the acquisition of the raw sensor data. These measurements are converted to X, Y, and Z points using a coordinate transformation and a global coordinate system is established. After this step, a filter process is applied to the resulting point clouds. Because a Berthing Support Area was derived in the creation process, the LiDAR points can be filtered with respect to that. As a result, only points located within the Berthing Support Area are further processed. The next step is the filtering of LiDAR points for each Reference Point. Based on the structure of these, the dimensions of a filter area were defined. This is used to filter data points for each Reference Point. For each filtered data point, the perpendicular distance to the Reference Point is calculated. Then the point with the minimal distance to the Reference Point is chosen. Changes in distance are used to calculate approach speed and acceleration.

5. Use Case: SmartKai, a LIDAR- and Shore-Based BAS

We implemented the process as part of the SmartKai project [32]. In this, a shore-based laser system for the detection and support of berthing maneuvers of seagoing vessels is being developed. It focuses on a shore-based infrastructure to support pilots and nautical personnel on a ship's bridge. Based on laser sensors, a situational picture of the berthing process should prevent accidents and damage to port infrastructure. Besides LiDAR sensors, SmartKai considers further sensors to observe a docking area such as AIS, video cameras, and sensors to collect environment-related data such as weather and visibility. During berthing, these data are transferred via a mobile network to the pilot's PPU, which visualizes a sea chart with a vessel's hull form (i.e., the pre-filtered LiDAR points) together with the perpendicular distance, approach speed, and acceleration for each Reference Point.

The implementation of the proposed system was performed in front of the Hannoverkai in Wilhelmshaven, Germany. We chose this berthing location because it lies in an enclosed area with no tidal influences and currents to minimize influencing factors. The experiment setup of our prototype is shown in Figure 5.

The bold dots identify the position of the two 2D LiDAR sensors. The sensors are positioned in a straight line along the quay wall and have been positioned at the same level of height with a distance of 80 m between them. The resulting LiDAR point cloud can be seen in Figure 5b. Points on the left and bottom belong to the quay wall itself, and on the right a ship can be seen.

For the experimental setup, two 2D LiDAR sensors from SICK are used (LD-LRS 3611). Table 1 summarizes the relevant information of these sensors. The LD-LRS 3611 provides a maximum detection distance of 250 m at 90% remission with an opening angle of 360°. For the experiment, the sensors were configured to operate at 5 Hz, with an angle resolution of 0.125° and an opening angle of 300°. Because the ships are moving relatively slowly, we trade scanning frequency for a higher angular resolution to improve the recognition of smaller vessels. Measurements of these sensors are collected using two sensor processing units as illustrated by the two rectangular boxes close to the sensors (Figure 5a) and shown by Figure 6.

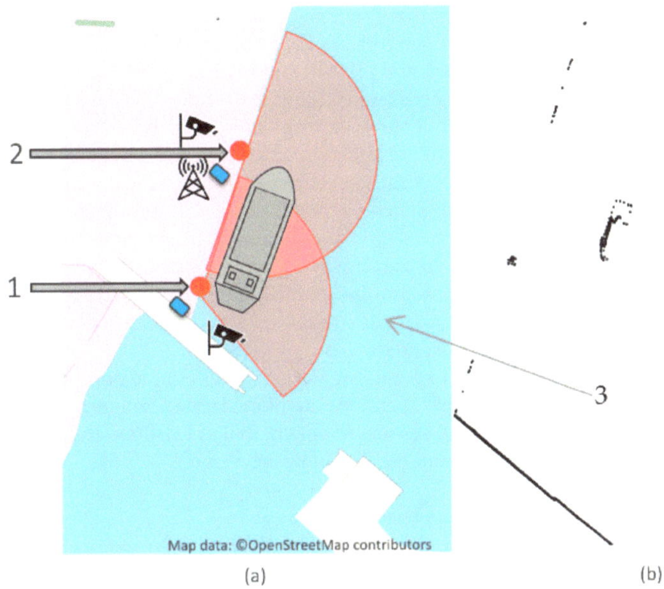

Figure 5. (**a**) Prototype implementation of SmartKai in the port of Wilhelmshaven, Germany. Red dots mark the position of two 2D LiDAR sensors. (**b**) Resulting LiDAR data from 2D LiDARs installed at the quay wall. Quay wall outline and a ship on the right image can clearly be seen.

Table 1. LiDAR specifications SICK LD-LRS 3611.

Model	SICK LD-LRS 3611
Light source	Infrared (905 nm)
Scanning frequency	5 Hz–15 Hz
Angular Resolution	0.0625° (interlaced), 0.125°, 0.1875°, 0.25°, 0.375°, 0.5°, 0.75°, 1°
Working Range	5 m–250 m
Opening Angle	360°
Scanning Range at 10% Remission	120 m
Systematic Error	±38 mm
Statistical Error	30 mm

Figure 6. Sensor processing unit with battery backup that records the data from the LiDAR sensors together with various other sensors, such as AIS, camera, and weather-related information, and also communicates processed information via the mobile network.

These sensor processing units are equipped with an Industrial PC (IPC), backup battery, and a network switch. Their task is to aggregate data from all sensors (AIS, camera, environmental, and LiDAR) and to process these in a distributed setup. To keep the latency between a sensor and IPC low, everything is connected via ethernet cables. Each sensor's processing unit runs our implementation based on a real-time multi-sensor framework. Using this, we are able to record and replay data from multiple sensors, and save them with highly accurate, synchronized timestamps. Thus, for SmartKai, this software is used to record time synchronized data from all sensors in Wilhelmshaven.

One of these processing units is also equipped with an LTE router, enabling information transfer to the ship using a mobile network. Data transfer from processing units to pilots on the ship is realized using the text-based Message Queuing Telemetry Transport (MQTT) protocol. This allows us to use web sockets in addition to supporting a high bandwidth of end devices.

For the experiment, we had access to the port operation ship Argus with a length of 16 m (c.f. Table 2) and we therefore defined Reference Points equally distributed every 5 m between both sensors to ensure that at least two Reference Points can capture the ship at any time within the BAS support area.

Table 2. Test campaign ship Argus.

Attributes	Argus (NPorts)
MMSI	211327610
Ship Type	Other
Length	16.08 m
Width	4.8 m
Draught	1.1 m

Figure 7 shows an image of the ARGUS during the test campaign in Wilhelmshaven. The vessel has a white cabin on top. The hull is black, with the bow higher than the side.

Figure 7. NPorts Port operation ship ARGUS.

5.1. Derivation of Berthing Support Area (BSA)

In order to create a reliable BAS, we defined an area in which the BAS can provide support. In this area, we can ensure the functional safety of our approach. The size of this area is significantly influenced by how well a vessel is detected by the LiDAR sensors, measured by the point density. If only a few points are available for a ship, these can also be considered as point outliers. In the worst case, an object is not detected at all. Therefore, in order to define the area, influencing factors for the sensors must first be defined. These can be subdivided into hardware limitations and environmental influences. The former is defined by the installed components (i.e., laser) and is reflected in the sensor specifications (i.e., range at 10% remission). Regarding environmental influences, precipitation and visibility (e.g., fog) are named in most cases, which limit the maximum range of the sensors [33,34]. Furthermore, the point density for the detection of an object is influenced by the angle between the object and the LiDAR sensor, due to the angular resolution of the device. If the hull of the object is at a right angle to the sensor beam, many points are reflected while the density decreases as the angle gets smaller [35].

In our experiment, we focus on the specification of the sensors and the angle between the sensor and the ship to define the BSA. In this we will consider the possible illumination of vessels based on the sensor specification and setup. The influence of environmental conditions on LiDAR sensors is difficult to estimate and depends on the sensor model [34]. Therefore, the influence of weather is difficult to measure, and we have too little information to consider this for the BSA.

In the following, the geometric model on which the BSA is based is described. Starting from a sensor at position $p_{sens} = (x_{sens}, y_{sens})$ and a ship side surface at position $p_{ship} = \left(x_{ship}, y_{ship}\right)$ with an angle of attack α.

The equation for the approximated hull is thus defined as:

$$\vec{hull} = \begin{pmatrix} x_{ship} \\ y_{ship} \end{pmatrix} + \lambda * \begin{pmatrix} \cos(\alpha) \\ \sin(\alpha) \end{pmatrix}$$

The two points $hull_1$ and $hull_2$ around p_{ship} are now defined on this linear equation by choosing the following values for λ.

$$\lambda_1 = \frac{n}{2 * \sin(\alpha)} \text{ and } \lambda_2 = -\frac{n}{2 * \sin(\alpha)}$$

By choosing the λ values, the two points now have the property of a fixed distance of n meters in the y-dimension. This is chosen because the Reference Points have a width of n meters and are defined for this model along the y-axis. This relationship is visible in Figure 8a.

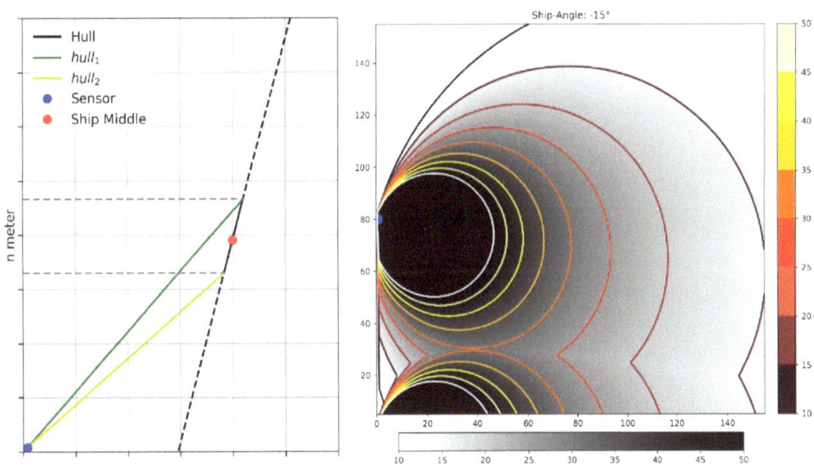

(a) (b)

Figure 8. (**a**) Relationship between sensor position, ship-angle, and section width. (**b**) Contour-map of the number of rays that hit a section surrounding each position on the map.

As a criterion for safety, a minimum number of M points is therefore required, which fall within the range of a Reference Point. That is why it is important to determine how many sensor beams can actually hit the n-meter wide Reference Point optically. For the number of sensor beams, the angular resolution ω of the sensor and the angle opened between the two hull points and the sensor are relevant.

Starting from the sensor, the direction vectors to the two hull points are therefore as follows and the angle δ results from this to:

$$\vec{l_1} = \left(\begin{pmatrix} x_{\lambda_1} \\ y_{\lambda_1} \end{pmatrix} - \begin{pmatrix} x_{sens} \\ y_{sens} \end{pmatrix} \right) \text{ and } \vec{l_2} = \left(\begin{pmatrix} x_{\lambda_2} \\ y_{\lambda_2} \end{pmatrix} - \begin{pmatrix} x_{sens} \\ y_{sens} \end{pmatrix} \right)$$

$$\delta = \cos^{-1}\left(\frac{\vec{l_1} * \vec{l_2}}{\left|\vec{l_1}\right| * \left|\vec{l_2}\right|} \right)$$

From $num\_rays = \delta / \omega$ we gain the number of rays hitting the ship's hull at position p_{ship} in an n meter section along the y-axis at an angle of α degrees to the quay wall. By performing this calculation over a grid of positions, a map with the corresponding number of possible rays can be created for each position. For a mooring angle of $-15°$ relative to the quay wall, this is shown in Figure 8b with a minimum of 10 rays for a 5-m section.

As can be seen in the figure, this approach allows it to create a contour map for an area depending on application angles, sensor positions, and section widths. Depending on the desired safety level, the contour can now be exported for the required minimum number of beams and used as a polygon in the further process.

With the described model, the nautical requirements, and the sensor range, a BSA could be defined. As a result of the discussions with the pilots, a range of 100–120 m is required to ensure a safe approach and docking. Because the sensors also have a 10% range of 120 m, the BSA was defined accordingly to a range of 120 m × 120 m.

5.2. World Model Generation

Regarding the data processing pipeline, we first extract a model of the quay wall from the raw LiDAR data (Figure 5b) and set the Reference Points. For this, a box filter is used to extract the points which define the quay wall. These are then processed by a concave hull algorithm to compute a polygon of the quay wall. Thus, only the outer hull of these points is used to define the quay wall geometry. This model is then used to set Reference Points along the quay wall. In Figure 9, the extracted quay wall model is shown.

Figure 9. Quay wall model extracted from a point cloud. Black dots show the position of the LiDAR sensors. Polygon on the left shows the extracted quay wall shape. Orange boxes show the location and box filter of the defined vertical Reference Point filters. Additional horizontal Reference Point is marked as green rectangle.

On the left side, the determined quay wall can be seen as a black polygon. In addition, the positions of the LiDAR sensors are indicated by black circles. The vertical and horizontal Reference Points are shown as orange and green rectangles. The horizontal Reference Point (green) measures the distance and forward speed of the ship in relation to the end of the quay wall (meter mark 0 m). The vertical Reference Points measure the transverse distance and docking speed. According to the concept, vertical Reference Points can be placed at the corresponding meter marks, which, however, are not available at the Hannoverkai in Wilhelmshaven. In relation to our concept, we defined reference points along the quay wall every 5 m to reflect the size of the Argus ship with a total of 24 reference points. An additional one was defined at the southeastern part of the quay wall, since this is where the ship is supposed to come to a stop. The horizontal Reference Point was also placed at this point.

5.3. Processing Pipeline Implementation

In this chapter, the data processing pipeline for live LiDAR data is implemented. The first step is to read the data from all the sensors. These send an array of distances, which are transformed to points to create point clouds. After this, using coordinate transformation, all point clouds are aligned relative to each other to create a global coordinate system. The result of this step is shown in Figure 10.

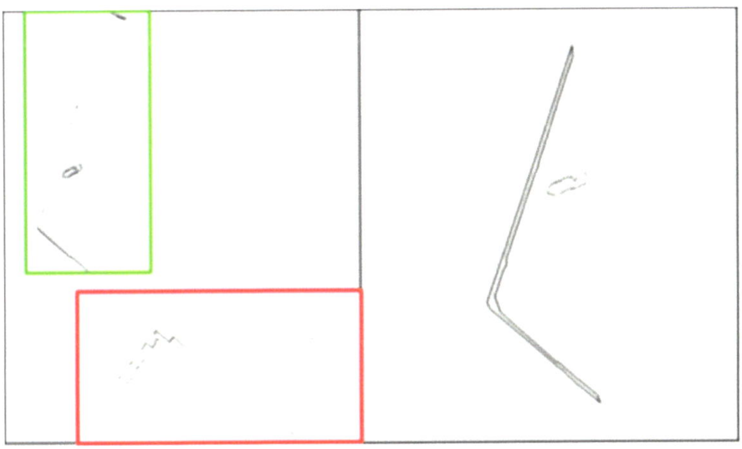

(a) (b)

Figure 10. (**a**) Raw LiDAR data transformed into a global coordinate system. In green: identified quay wall, red: static infrastructure removed from further processing. (**b**) resulting data after point cloud preprocessing. Quay wall points were replaced by the quay wall shape resulting from the world model.

As can be seen in the left image, a quay wall, ships, and fixed infrastructure can clearly be identified. Fixed infrastructure is marked by red boxes, while the quay wall is outlined using green boxes. After the filtering process, only points of ships are left (b). After this step, the LiDAR measurements are synchronized on a temporal level. Because multiple sensors are used for the BAS, the measurement must be time synchronized. In our case, the sensors operate in 5 Hz intervals. For time synchronization, we define a time window of 200 ms, so that older measurements are discarded if the difference to the newest measurement is higher. Point clouds are then further processed by Reference Points. The length of the box filter was set at 120 m. This value is based on the specifications of our BSA calculation. The width of our Reference Point is set to 5 m, as the ship under consideration has a length of 16 m, and we defined that at least three Reference Points need to be defined per ship length. This value compensates for gaps in the LiDAR data, as shown in Figure 5b. A higher width can thus compensate for larger gaps in point clouds. For the horizontal Reference Point, 120 m in length was also defined, so that these correspond to the vertical Reference Points. The width extends over the entire quay and thus corresponds to 120 m.

6. Evaluation

This section presents the evaluation of our BAS. Firstly, we describe the conducted test campaign and the defined scenarios. Secondly, we evaluate the precision, robustness, and stability of our system by using parallel sailing scenarios. Then we show how our system behaves in berthing scenarios in the port of Wilhelmshaven and compare our measurements with a DGPS sensor as a reference measurement. We will close with a discussion of the results and further improvements for our system.

6.1. Test Campaign Design

For the evaluation, we conducted a test campaign in Wilhelmshaven using the SmartKai prototype implementation and evaluated two types of maneuvers: parallel sailing along the quay and common regular berthing approaches as plotted by Figure 11.

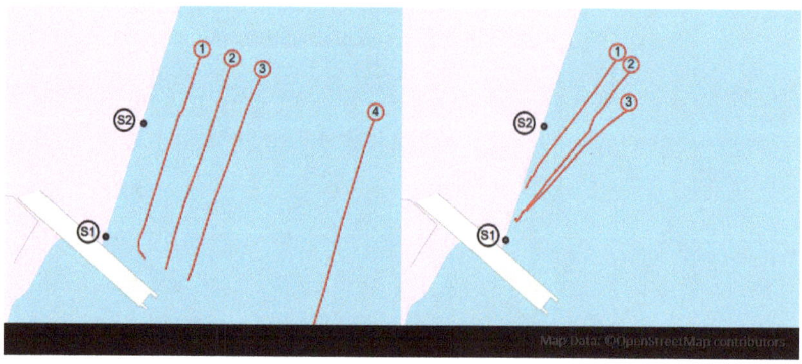

(a) (b)

Figure 11. Test campaign scenarios (**a**) parallel sailing (scenarios 1.1–1.4), (**b**) berthing maneuvers (scenarios 2.1–2.3).

Because vessels over 200 m approach with berthing angles < 0.5 degrees but the ship available for the test campaign was much smaller, we decided to sail parallel tracks in four different distances to the quay: 20 m, 40 m, 60 m, and 150 m. We did not consider closer distances as the detection range of the LD-LRS 3611 sensors started above 5 m and chose 150 m as the maximum distance based on weighting the requirements of the pilots that stated 120 m as the berthing initiation range with the technical characteristics of the LiDAR Sensors (we expected the best detection results until this distance even for bad weather situations). All these tracks had a similar length of around 125 m, were sailed with nearly 3 kn, and the track times were each between 92 and 98 s. For the 20 m parallel track, the vessel slowed down and initiated a turn towards the end of the track to avoid a collision with the adjacent quay wall. To have comparable parallel tracks, the 20 m track was therefore cut, starting from the point that the berthing angle was >5 degrees.

Moreover, three different berthing maneuvers were performed (c.f. Figure 11b). For berthing, the ship started north of our berthing location heading towards the quay. The ship is approaching the location with a berthing angle of 15 degrees and an initial speed of four knots and continuously reducing it to one knot. A berthing angle of 15 degrees was the maximum of what was typically observed for smaller vessels without tug boat assistance [16], such as the Argus ship. The shipmaster's target was to berth between the two LiDAR sensors. This scenario was also varied with a start speed of six knots. Because our LiDAR sensors have a minimum range of 5 m, they were clipped up to this distance. The track lengths vary between 38 and 54 s due to the different starting positions and velocities.

The Argus ship master performed all maneuvers by navigating manually with the help of the onboard bridge systems (ECIDS, compass, and GPS). For the experiment, we installed an additional IMO approved (transmitting heading and satellite navigation device) DGPS sensor (JRC JLR-21) centrally on the ship, which enabled us to record the sailed maneuver tracks with more precision and keep it in sync with the LiDAR sensor measurements. In our configuration, the sensor provides a minimum accuracy of 6 m by specification, which is less accurate than the LiDAR sensor. But there were no buildings or other infrastructure on the quay wall that could affect the DGPS measurements. Good weather conditions were also present during the test campaign. Without clouds, rain, or other influencing factors for the DGPS signal, and therefore with very good satellite coverage, we assume the DGPS fixes to be much more accurate and precise than the corresponding minima stated in the sensor specifications.

We defined Reference Points along the quay wall starting from the south-east sensor (c.f. Figure 9) every 5 m to reflect the Argus ship size for a total of 120 m resulting with 24 Reference Points. An additional Reference Point was placed at meter mark 0 m to indicate where the ship needs to stop. We also aligned a vertical one to the southern quay to communicate forward speed and distance from there as well. The length of the vertical Reference Points was increased to 150 m to cover the 150 m maneuver track. The same applies to the width of our horizontal Reference Point.

The following Section 6.2 covers the results of the parallel sailing maneuvers, whereas Section 6.3 focuses on presenting the results of the berthing scenarios with higher berthing angles.

6.2. Parallel Sailing Results

For the parallel tracks a maximum variance by half the width of the ship can be expected. This is because when entering the area of a Reference Point, the bow of the ship is detected first, followed by a gradual inclusion of the outer hull of the ship. Therefore, for the ship used for this paper, the expected maximum variance is 2.4 m.

Figure 12 depicts the four parallel scenarios and for the three sensor combinations the 95% interval together with the expected maximum variance of 2.4 m for the measured distances for each Reference Point (x-axis, numbered vertical Reference Points with 5 m spacing).

It can be seen that almost all measurements are within the expected variance, and also that with increasing distance the measurements fluctuate less. This can be explained by the higher point density especially at close ranges and the resulting better resolution of the bow. At higher distances, effects such as the lower point density and widening of the beam for better detection of the outer hull have to be considered.

Considering a Reference Point spacing of 5 m and a parallel sailing ship of 16 m length, on average 3.2 adjacent Reference Points are expected to communicate simultaneously a parallel sailing ship in 20 m, 40 m, 60 m, and 150 m. Table 3 shows the mean and standard deviation divided according to the parallel runs. Here, the entry and exit times were removed so that the ship is completely inside the BSA.

For the first three scenarios with minor distances, the average is above the expected number of 3.2. In the 150 m scenario the ship sails outside the calculated BSA that requires at least three simultaneous measuring Reference Points. A significant reduction of the mean of measuring Reference Points for the 150 m scenario confirms the expectation. The standard deviation is around 0.5 for all scenarios. With increasing distance to the quay wall, the number of Reference Points that simultaneously detect a ship is reduced nearly linearly. Based on the calculations, we can also conclude that three reference points per ship length cannot be achieved in the 60 m scenario due to the standard deviation. In the initial sailing phase of the scenario, only the bow of the ship is detected. After the ship has passed the second sensor, the stern is also illuminated so that three Reference Points detect the ship. The illumination is therefore dependent on the relative positioning of the ship to the sensors. However, it is necessary to investigate why the average within the BSA is above the expected value of three. Figure 13 shows for each timestamp the number of Reference Points that simultaneously detect a ship for the 20 m scenario including the entry and exit times.

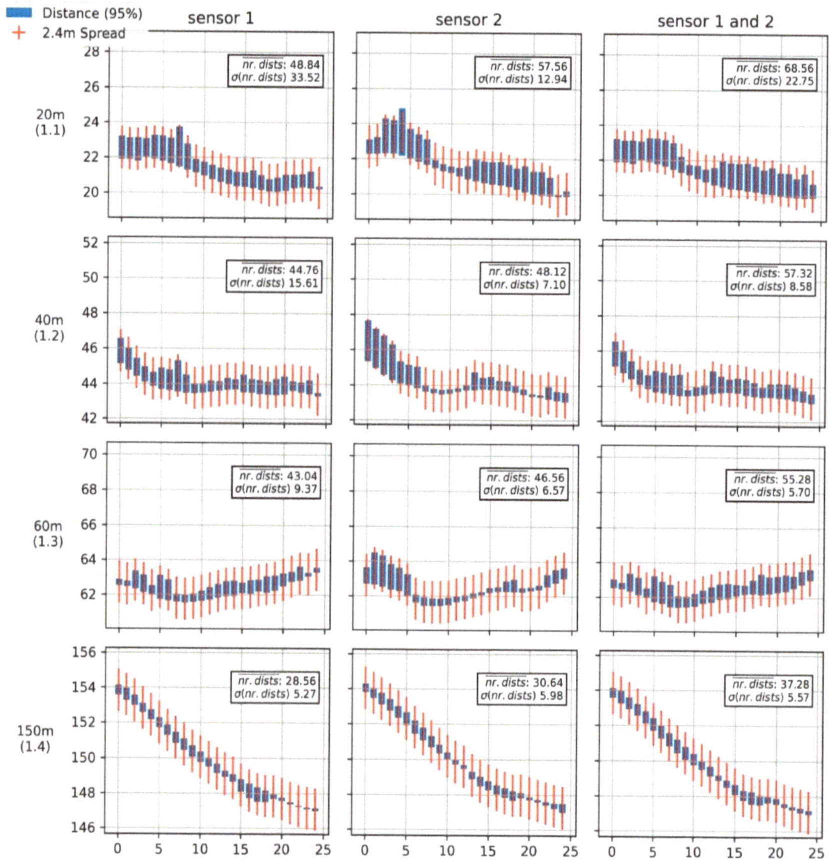

Figure 12. Measured distances against the maximum allowed spread for each sensor and scenario configuration.

Table 3. Number of Reference Points that detect a vessel simultaneously.

Scenario	Mean	STD
1.1 (20 m)	3.72	0.54
1.2 (40 m)	3.56	0.54
1.3 (60 m)	3.34	0.53
1.4 (150 m)	2.42	0.49

In the first few seconds the ship enters the BSA, the number increases. As soon as the ship has completely entered the BSA, at least three Reference Points are able to measure the ship (after 16 s). It is noticeable that in short time intervals the number of Reference Points which recognize a ship briefly sinks or rises (example: between 60 s and 80 s). In these moments the ship leaves one Reference Point and changes to the next one, so that the number changes. The fact that the number is above the expected value of three is therefore due to the fact that with a 16 m ship length divided by 5 m a surplus arises. If the ship is seen by three Reference Points, the bow and stern can be in the adjacent Reference Points so that it is seen by five.

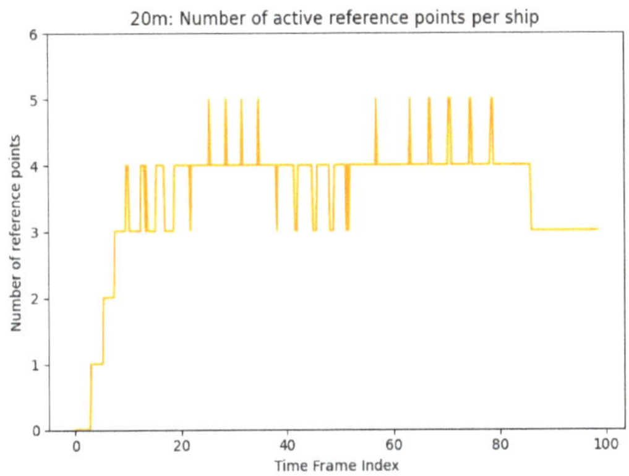

Figure 13. Number of active Reference Points during the 20 m scenario.

In addition to the number of active Reference Points, we also measured how often these were turned on and off. One would expect a ship to enter and exit each vertical Reference Point just once. A higher toggling number is an indicator that point outliers may occur, which would result in causing false alarms. We therefore measured how often a Reference Point is triggered, which is listed by Table 4.

Table 4. Number of activations per Reference Point.

Scenario	Mean	STD	Max
1.1 (20 m)	1.16	0.46	3
1.2 (40 m)	1.12	0.32	2
1.3 (60 m)	1.12	0.32	2
1.4 (150 m)	1.32	0.61	3

For none of the scenarios the expected value of one activation per scenario was reached. For the 20 m and the 150 m scenario a Reference Point was activated three times. In the 20 m scenario this was caused by point outliers caused by the relatively tiny hull size of the ship. Due to the fact that 2D LiDAR sensors were used, movements of the ship on the X or Z axis result in suddenly seeing other parts than the hull such as objects on the ship's deck. For the 150 m scenario, we realized that the point density of the ship is significantly lower than for the other scenarios. Due to the minimum number of points that must lie within a Reference Point in order to perform a distance measurement, measurements are discarded more often, and thus false alarms are produced more frequently. We therefore checked how often measurements were discarded. For this purpose, we collected the number of successful and unsuccessful measurements. Successful in this context means that there are at least five points per Reference Point measurement. Otherwise, measurements are not successful and were not considered. Figure 14 summarizes the number of successful and unsuccessful measurements for each scenario.

We recognized that with higher distance the point density decreases, and the number of unsuccessful measurements increases. Regarding the 150 m scenario, approximately 25% of our measurements were not considered because of the low point density. The still high amount of successful measurements indicates that the IMO positioning fix requirements of 1 Hz while berthing can be fulfilled with the 5 Hz sensor. To investigate this in more detail, we measured how many times per second a Reference Point takes a measurement. The factors that influence the calculation rate are the time synchronization of the LiDAR measurement and the failure rate due to too few points. Figure 15 shows the calculated average Hz number for each scenario.

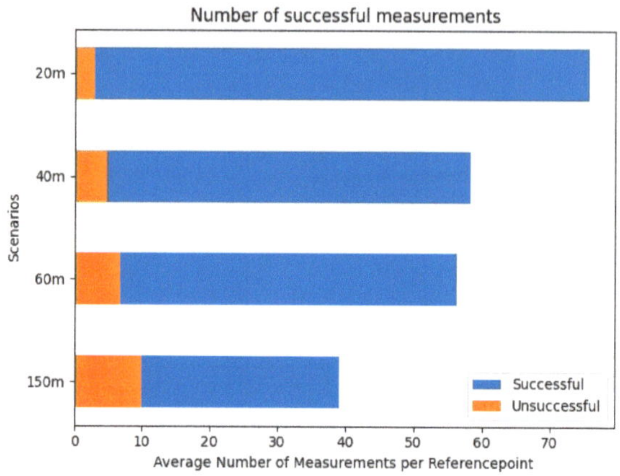

Figure 14. Number of successful and unsuccessful measurements of Reference Points. With a raising distance the number of successful measurements decreases.

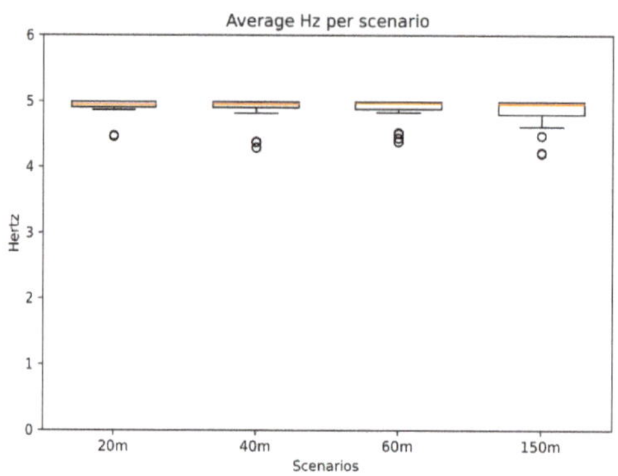

Figure 15. Average Hz of Reference Points per scenario.

Due to the refresh rate of the sensor, the upper limit is 5 Hz, and our results are very close to this value. However, it is noticeable that there are a few outliers. The lowest value can be observed for the 150 m scenario (4.2 Hz).

6.3. Berthing Maneuvers Results

Scenarios 2.1–2.3 define the other extreme of berthing maneuvers with an angle of 15 degrees (which is in fact a typical berthing angle for small ships such as the one we used for our experiments). We evaluated the distance to Reference Point, speed, forward speed, and distance to meter mark 0 m. For these metrics, we examined the deviations from the DGPS. Because in general we expected the accuracy and precision of the DGPS to be much lower when compared to LiDAR, we focused a comparison of the measurement deviations to verify the stability of the Reference Point measurements.

We will start with the comparison of the measured distances on vertical (distance to the quay wall) and horizontal (distance ahead) level. It is expected that the variance of the measurements is low and thus a constant deviation between both measurement methods is achieved. Because several Reference Points were defined along the quay wall, but a DGPS sensor only determines a single position and speed, the measurements of the Reference Points are combined. Therefore, the shortest distance of all Reference Points is used as a reference value to compare it with the DGPS. Based on the determined positions of the GPS sensor, we calculated the vertical distance to the quay wall and the horizontal distance to enable the comparison.

Figure 16 shows the vertical (a) and horizontal (b) deviations between the Reference Point measurements and the DGPS.

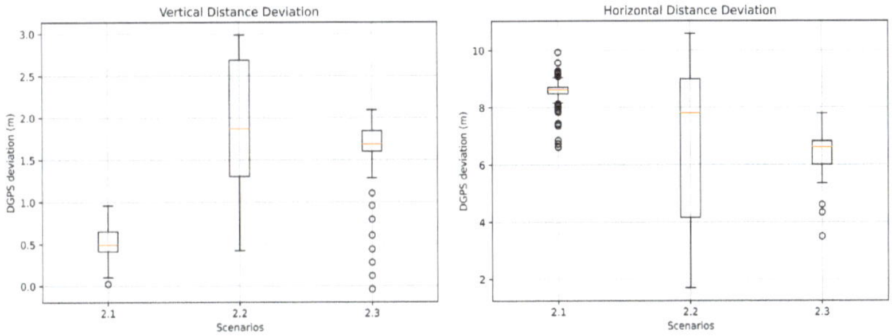

(a) (b)

Figure 16. Vertical ((**a**), left) and horizontal ((**b**), right) deviations between DGPS sensor and Reference Points.

In scenario 2.1, the variance of the measurements is small compared to other scenarios. However, with respect to the horizontal distance measurements, outliers are present. In the second scenario (2.2), the highest variance is present. Accordingly, the minimum and maximum values are also high. Due to the high range of values, there are also no outliers. In the third scenario, the variance is similar to that of scenario 2.1 (vertical and horizontal). However, outliers are also found here, especially below the minimum value. To first check for outliers in the first scenario, we first examine the horizontal distance measurements.

Figure 17 shows the horizontal distance measurements to the quay wall for scenario 2.1.

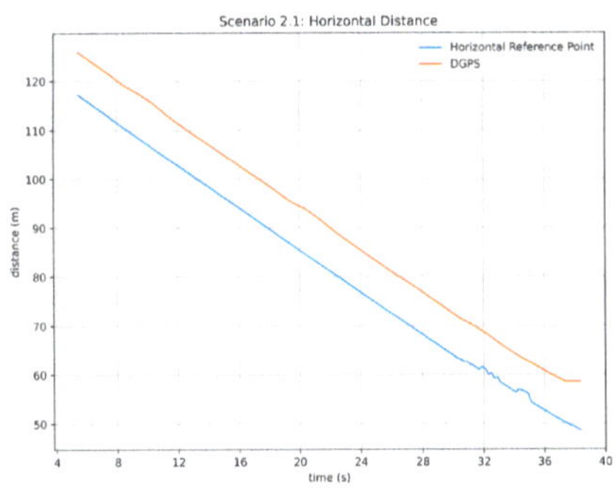

Figure 17. Scenario 2.1 horizontal distance to quay wall.

Up to second 32, the deviations between the measurement methods are relatively constant. From this point on, however, the horizontal Reference Point shows fluctuating distance measurements, which causes the outliers in Figure 16b. Regardless of the scenario considered, these anomalies are also found in other scenarios (cf. Figure 16 scenario 2.3), although we ascertained that the length of these phases is smaller.

The next step is to investigate the high variance in scenario 2.2. For this purpose, Figure 18 shows the horizontal distance to the quay wall (a) and the heading of the ship (b).

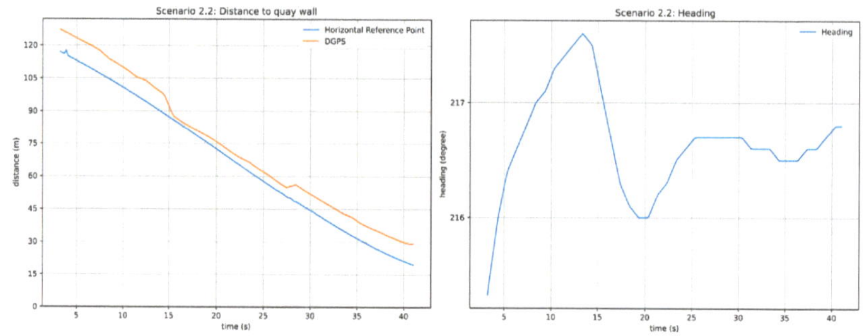

(**a**) (**b**)

Figure 18. Scenario 2.2 horizontal distance to quay wall (**a**) and ships heading (**b**).

It is particularly striking that the deviations between the two measurement methods decrease between 15 s and 27 s. This increases the variance of the measurements, which is reflected in the Figure 16b. However, the reason for this behavior is unknown. As can be seen from the right side of the figure, the ship did not change course. In the period under consideration, only a course change of <1° was made. We also checked the quality of the DGPS measurements. However, the number of satellites was constant at 8 and the Horizontal Dilution of precision (HDOP) was 1. Infrastructure at the mooring was not available, so the GPS measurements were not exposed to any influence. We therefore assume that this is an anomaly.

The next part is the comparison of measured speed between Reference Points and DGPS. For these we examine the average, standard deviation, mean error, and rooted mean squared error (RMSE) to be able to make a statement about the stability of the measurements. Due to deviations in the distance measurements (see Figure 16), we have removed all speed values greater than 10 m/s, as this leads to disproportionate deviations. However, they will be analyzed in more detail in the following. Table 5 shows the vertical velocity measurements (quay wall approach speed) of the respective scenarios.

Table 5. Comparison of vertical approach speed between Reference Points approach and DGPS measurements using berthing scenarios (Scenarios 2.1–2.3).

	Reference Points		**DGPS**		**Deviations**	
Scenario	**Mean**	**STD**	**Mean**	**STD**	**Mean Error**	**RMSE**
2.1	0.68 m/s	0.11 m/s	0.66 m/s	0.1 m/s	−0.01	0.11
2.2	0.89 m/s	0.19 m/s	0.83 m/s	0.19 m/s	−0.07	0.21
2.3	0.81 m/s	0.15 m/s	0.8 m/s	0.09 m/s	−0.01	0.14

The table shows that the measurements of the two methods are quite comparable. In all scenarios the absolute deviation is below 0.07 m/s (peak in Scenario 2.2). The highest deviation was recorded in scenario 2.2, in which the position anomalies on Figure 18 cause varying velocity anomalies to be measured. However, the standard deviation of the Reference Points is consistently higher than that of the DGPS, so that the measurements vary more strongly. Due to the higher measurement rate of the LiDAR sensors (5 Hz) compared to the DGPS (1 Hz), more velocity measurements were recorded, which led to higher deviations. Table 6 shows the measured forward velocities of the horizontal Reference Point and the DGPS.

Table 6. Comparison of forward speed between Reference Points approach and DGPS measurements using berthing scenarios (Scenarios 2.1–2.3).

	Reference Points		DGPS		Deviations	
Scenario	**Mean**	**STD**	**Mean**	**STD**	**Mean Error**	**RMSE**
2.1	2.09 m/s	0.98 m/s	2.08 m/s	0.25 m/s	−0.01	0.99
2.2	2.52 m/s	0.95 m/s	2.59 m/s	1.21 m/s	0.08	1.44
2.3	1.59 m/s	0.7 m/s	1.6 m/s	0.34 m/s	0.01	0.63

Again, the absolute deviation to the DGPS is relatively small and is below 0.08 (peak again in scenario 2.2). Again, however, the standard deviation of the Reference Points is higher in scenario 2.1 and 2.3 than for the DGPS. Especially in scenario 2.3, the standard deviation is two times higher.

The highest speed deviation between both systems was recorded for scenario 2.3 (peak deviation). We also observed a high standard deviation for the Reference Points in this scenario, compared to DGPS. Because the velocity is calculated based on the measured distance, we further investigated in the distance measurements of scenario 2.3. Figure 19 depicts the distance to meter mark 0 m and the resulting forward speed of the vessel.

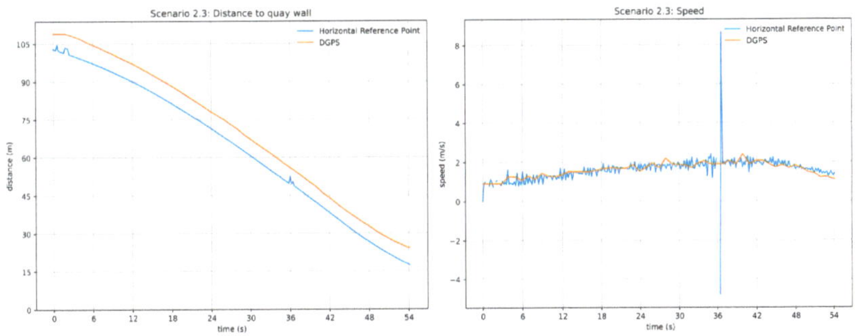

(**a**) (**b**)

Figure 19. Horizontal distance to meter mark 0 m (**a**) and forward speed for scenario 2.3 (**b**).

We observed that the deviation in distance between the DGPS sensor and the horizontal Reference Point is relatively constant between 5 and 8 m apart (cf. Figure 16b). However, between second 0 and 3 we measured jitter in the horizontal distance to meter mark 0 m. Because this generates incomparably high-speed measurements, they were removed. Velocity anomalies were also measured in the time range between 35 and 37. This results in speed deviations; thus, we calculated a maximum above ~8 m/s and a minimum below ~−4 m/s in forward speed. Due to the high update frequency of the LiDAR sensor (5 Hz) and distance change in a small time-window, the resulting forward speed is high. Compared to the DGPS, we can see that the reference points take varying velocity measurements, thus resulting in jitter. In comparison to the DGPS sensor, less frequent speed changes can be observed. High-speed deviations were also observed for the other scenarios. Therefore, to improve the stability of the measurements, a filter is required to smoothen the speed measurements for the nautical personnel.

7. Discussion

After considering the results of the previous section, it appears that the concept of Reference Points is promising. We applied the Reference Point method to four parallel and three berthing tracks with the port operation ship Argus and specifically focused on the corner cases (very small berthing and very high berthing angles).

Our setup is based on defining at least three Reference Points per ship length to ensure that a future system could also calculate berthing angles and rate of turn (ROT) based on the Reference Points. We showed that this can be ensured as long as the ship is within the specified sensor range. As soon as the ship is outside the BSA (for the 150 m scenario) and thus outside the specifications, only two reference points could simultaneously detect the ship. Running with 5 Hz, the IMO requirements that require 1 Hz fixations can be easily achieved (we reported at least 4.2 Hz with an average close to the maximum refresh rate of 5 Hz). Verification of the spread of measurements showed that almost all measurements at all reference points were within the range that we anticipated.

The comparison between LiDAR and GPS showed that deviations were within the expected range of half a ship width (with the GPS antenna assumed to be installed at the center of the ship). Nearly all of the deviations can be explained by a dependence on the heading, because the ship's hull is longer than it is wide and the LiDAR sensors detect the hull of the ship while the GPS measures the position at the ship-center. Thus, a change of the orientation has no effect on the GPS measurement, but directly one on the measured minimum distance via LiDAR.

Furthermore, when looking at the results, it is noticeable that many of the errors mentioned above occurred during fluctuations of the LiDAR measurements and thus individual measured values led to outliers. This can be explained to a large extent by the prototypical design of the system, where the LiDAR was positioned a little too high. It can be seen on the picture of the ship (Figure 7) that the shape of the hull in the middle of the side does not reach the same height as at the bow and stern of the ship. If the ship is now positioned unfavorably to the sensor, it is therefore possible that the hull was not ideally hit and thus the structure of the ship is briefly measured. Therefore, to improve detection for smaller ships, 3D LiDAR sensors must be applied. This would make it possible to reduce the outliers, because measurements are not only made on the horizontal plane. For bigger vessels (which are actually the targeted ones), these kind of LiDAR fluctuations would not be expected.

Regardless of the point measurement errors, it remains a problem to use the sensor values directly to determine the speed. Our velocity measurements have shown a high standard deviation and the reliability of these are therefore low. Due to the high frequency of the LiDAR sensors, small distance deviations provide high speeds. Therefore, these must be filtered before they are used. Assuming that the reason for the deviations is the frequency of the measurements, we decided to calculate them only once per second (1 Hz) to compensate for the fluctuations. Figure 20 plots the horizontal speed calculation for scenario 2.3.

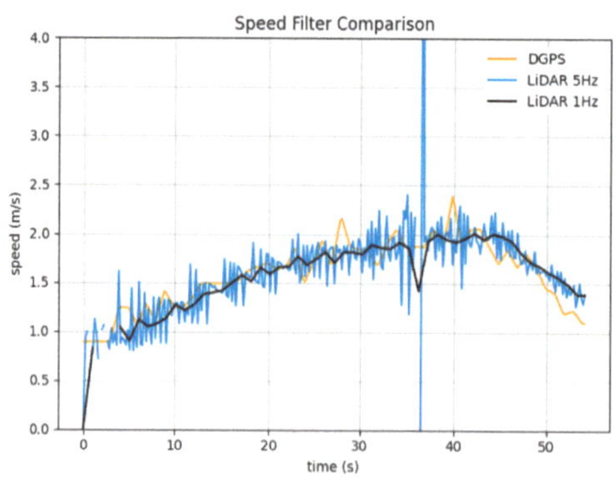

Figure 20. Comparison of speed filtering methods for scenario 2.3.

The DGPS measures velocity in 1 Hz intervals, with measurements based on LiDAR in 5 Hz (sensor specification) and in 1 Hz. The calculation in 5 Hz intervals shows high fluctuations compared to the DGPS. But if the velocity is calculated only once per second, the fluctuations are massively reduced. Only between second 34 and 37 are high deviations found in the LiDAR data, so that outliers are generated. However, these outliers are much lower than previous measurements. To further investigate the performance of these methods, Table 7 shows the mean, standard deviation, and RMSE compared to the DGPS of all methods.

Table 7. Comparison of speed filtering methods.

Method	Mean	STD	RMSE
DGPS	1.6	0.34	-
LiDAR 5 Hz	1.59	0.7	0.63
LiDAR 1 Hz	1.6	0.34	0.17

If the speed measurement is performed only once per second, the deviations are close to those of the DGPS and reduce the RMSE from 0.63 to 0.17. The precision and accuracy of the used DGPS sensor is not a sufficient ground truth for evaluating the LiDAR sensor to validate the accuracy of the velocity measurements. However, a reduction in fluctuations would lead to better stability to present more consistent measurements to nautical personnel. Nevertheless, because we do not have ground truth, it is also possible that more accurate velocity values were obtained from the LiDAR measurements. Future work should therefore select appropriate filtering techniques and collect ground truth measurements to validate the calculations.

Another point we noticed during the evaluation is that it is relatively difficult to evaluate LiDAR-based methods. Because these sensors can have high accuracy down to the millimeter range, only a few comparable sensors can be found. Therefore, better evaluation sensor technology must be used for ground truth measurements. It is also difficult to evaluate the robustness and reliability of the system. LiDAR-based techniques are dependent on various weather conditions, so long term testing is required. Alternatively, realistic ship and sensor simulations must be used to test all possible situations.

8. Conclusions

In this paper, we presented a ship-independent mooring assistance system based on LiDAR sensors. For this, we summarized interviews performed with pilots in the form of workshops and derived requirements for our proposed BAS. Then we presented the current state in shore-based assistant systems and checked if these can cover all of the pilots' requirements that we ascertained. We found that most systems make use of non-deterministic or black box algorithms, so that the functional safety is hard to ensure. We have therefore introduced the concept of a Reference Point. They can be arbitrarily placed on a mooring site e.g., to reflect typical landmarks that pilots typically used for orientation to during berthing. We have placed vertical Reference Points along the quay wall to measure the distance to the quay wall and the approach speed. Horizontal Reference Points are placed at the end of the berth to measure the approach distance and speed. To ensure functional safety, we defined a Berthing Support Area based on the port structure, sensor specifications, and pilotage requirements. The BSA was determined by a mathematical model and defines an area in which the support is provided by calculating the possible illumination of a target by LiDAR sensors.

We implemented the BSA within the SmartKai project and installed the system at a berth in Wilhelmshaven in Germany. For the evaluation of our BAS, we conducted a test campaign using this prototypical setup. We performed several scenarios (parallel sailing and berthing scenarios) to verify that our BAS conforms with the requirements derived based on pilot interviews. Our results show that we can fulfill most of our requirements. The BAS conforms to the IMO Resolution A.915 (22) with respect to the required update frequency of 4.2 Hz > 1 Hz. We were able to achieve a high measurement stability, in which only a few outliers could be found. Accuracy measurements require further investigations with more accurate sensors than the applied DGPS sensors. We observed that the biggest challenge for our system is the speed calculation. Because of the high update frequency of the LiDAR sensors, small distance deviations result in high-speed deviations. Thus, an approach is needed to filter the distance measurements beforehand to retrieve accurate speed calculations.

With the concept of Reference Points and the Berthing Support Area, we see additional use cases that can be covered. In the future, we want to support pilots not only during berthing maneuvers, but also during casting off. In situations involving tug assistance, such a system could also offer important data not only to the captain of the vessel but also to the skippers of the tugs, improving the coordination between vessel and tugboats. In our interviews, the pilots have reported that not only the distance to the quay wall, but also the rate of turn of a ship is an important information. Thus, using multiple Reference Points, the heading of a ship and the rate of turn could be determined. Warnings, for example, at high approach speeds or high rate of turns could also provide additional support. Therefore, captains and pilots should be warned about a possible danger to improve the safety of the berthing process. Further use cases include support during approaching lock entrances or bridge crossings. The BSA can additionally be transferred to other LiDAR-based systems.

Furthermore, it is also possible to integrate our system into Berthing Planning Systems [36]. Compared to other systems, Reference Points are only loosely coupled with each other. Therefore, our system can be dynamically split if needed (e.g., dividing one berth into several), so that this flexibility supports harbor operators in the dynamic environment of the port. This is especially useful for unintended berths, where berth locations are dynamically allocated [37]. This approach can also support the planning of berthing processes in the port. In [24] and [25], procedures were presented to solve the berth allocation and quay crane assignment problem. The authors partition the berth into segments and assign a slot to arriving vessels. This procedure can be supported by partitioning the reference points to ensure a safe and efficient berthing process.

Author Contributions: Conceptualization, J.M., H.W. and S.F.; data curation, J.M.; investigation, J.M.; methodology, J.M., H.W. and S.F.; project administration, S.F.; software, J.M. and H.W.; supervision, S.F.; validation, J.M., H.W. and S.F.; visualization, J.M., H.W. and S.F.; writing—original draft, J.M., H.W. and S.F.; writing—review and editing, J.M., H.W. and S.F. All authors have read and agreed to the published version of the manuscript.

Funding: This work has been funded by the German Federal Ministry of Transport and Digital Infrastructure (BMVI) within the funding guideline "Innovative Hafentechnologien" (IHATEC) under the project SmartKai with the funding code 19H19008E.

Conflicts of Interest: The authors declare no conflict of interest. The funders had no role in the design of the study; in the collection, analyses, or interpretation of data; in the writing of the manuscript; nor in the decision to publish the results.

References

1. Sánchez, R.J.; Perrotti, D.E.; Fort, A.G.P. Looking into the Future Ten Years Later: Big Full Containerships and Their Arrival to South American Ports. *J. Shipp. Trade* **2021**, *6*, 2. [CrossRef]
2. Forum, I.T. *The Impact of Mega-Ships*; International Transport Forum Policy Papers, No. 10; OECD Publishing: Paris, France, 2015. [CrossRef]
3. Sánchez, R.; Mouftier, L. *Reflections on the Future of Ports: From Current Strains to the Changes and Innovation of the Future*; CEPAL: Santiago, Chile, 2016.
4. Hsu, W.-K.K. Assessing the Safety Factors of Ship Berthing Operations. *J. Navig.* **2015**, *68*, 576–588. [CrossRef]
5. Bui, V.; Kawai, H.; Kim, Y.-B.; Lee, K. A Ship Berthing System Design with Four Tug Boats. *J. Mech. Sci. Technol.* **2011**, *25*, 1257–1264. [CrossRef]
6. Felski, A.; Naus, K.; Świerczyński, S.; Wąż, M.; Zwolan, P. Present Status And Tendencies In Docking Systems' Development. *Annu. Navig.* **2014**, *21*. [CrossRef]
7. Harati-Mokhtari, A.; Wall, A.; Brooks, P.; Wang, J. Automatic Identification System (AIS): Data Reliability and Human Error Implications. *J. Navig.* **2007**, *60*, 373–389. [CrossRef]
8. Bialer, O.; Jonas, A.; Tirer, T. Super Resolution Wide Aperture Automotive Radar. *IEEE Sens. J.* **2021**, *21*, 17846–17858. [CrossRef]

9. Athavale, R.; Ram, D.S.H.; Nair, B. Low Cost Solution for 3D Mapping of Environment Using 1D LIDAR for Autonomous Navigation. In *IOP Conference Series: Materials Science and Engineering*; IOP Publishing: Bristol, UK, 2019; Volume 561, p. 012104. [CrossRef]
10. Trelleborg Smart Dock Laser. Available online: https://www.trelleborg.com/en/marine-and-infrastructure/products-solutions-and-services/marine/docking-and-mooring/docking-aid-system/smart-dock-laser (accessed on 17 February 2021).
11. DockAssist®. *The Most Advanced Berthing Aid System in the World*; Metratek Telematics Ltd.: Nicosia, Cyprus. Available online: https://metratek.co.uk/dockassist (accessed on 10 February 2021).
12. Reiher, D.; Hahn, A. *Review on the Current State of Scenario-and Simulation-Based V&V in Application for Maritime Traffic Systems*; IEEE: Piscataway, NJ, USA, 2021.
13. Corso, A.; Moss, R.J.; Koren, M.; Lee, R.; Kochenderfer, M.J. A Survey of Algorithms for Black-Box Safety Validation. *J. Artif. Intell. Res.* **2021**. [CrossRef]
14. Czarnecki, K. *Operational Design Domain for Automated Driving Systems—Taxonomy of Basic Terms*; Waterloo Intelligent Systems Engineering (WISE) Lab.: Waterloo, ON, Canada, 2018.
15. Falk, M.; Saager, M.; Harre, M.-C.; Feuerstack, S. Augmented berthing support for maritime pilots using a shore-based sensor infrastructure. In *HCI International 2020—Late Breaking Posters*; Communications in Computer and Information, Science; Stephanidis, C., Antona, M., Ntoa, S., Eds.; Springer International Publishing: Cham, Switzerland, 2020; Volume 1294, pp. 553–559, ISBN 978-3-030-60702-9.
16. PIANC—International Navigation Association. Guidelines for the Design of Fender Systems—MarCom Report of WG 33. 2002. Available online: https://www.pianc.org/publications/marcom/guidelines-for-the-design-of-fender-systems (accessed on 1 February 2021).
17. Hein, C. Anlegegeschwindigkeiten von Großcontainerschiffen. In Proceedings of the PIANC Deutschland (Hg.): Deutsche Beiträge, 33. Internationaler Schifffahrtskongreß, San Francisco, CA, USA, 1–5 June 2014; PIANC Deutschland: Bonn, Germany, 2014; pp. 1–5.
18. Stadt Wilhelmshaven Besondere Hafenordnung Für Den Stadthafen Wilhelmshaven. Available online: https://www.wilhelmshaven.de/PDF/Stadtrecht/Sr32-09_Besondere_Hafenordnung_fuer_den_Stadthafen_WHV.pdf?m=1418987622& (accessed on 23 August 2021).
19. IMO Resolution, A. *915 (22) Revised Maritime Policy and Requirements for a Future Global Navigation Satellite System (GNSS)*; Adopted on 29 November 2001; International Maritime Organization: London, UK, 2001.
20. Perkovič, M.; Gucma, L.; Bilewski, M.; Muczynski, B.; Dimc, F.; Luin, B.; Vidmar, P.; Lorenčič, V.; Batista, M. Laser-Based Aid Systems for Berthing and Docking. *J. Mar. Sci. Eng.* **2020**, *8*, 346. [CrossRef]
21. Kim, H.; Kim, D.; Park, B.; Lee, S.-M. Artificial Intelligence Vision-Based Monitoring System for Ship Berthing. *IEEE Access* **2020**, *8*, 227014–227023. [CrossRef]
22. Chen, C.; Li, Y. Ship Berthing Information Extraction System Using Three-Dimensional Light Detection and Ranging Data. *J. Mar. Sci. Eng.* **2021**, *9*, 747. [CrossRef]
23. Kuzu, A.; Arslan, O. Analytic comparison of different mooring systems. In Proceedings of the Global Perspectives in MET: Towards Suistainable, Green and Integrated Maritime Transport, Varna, Bulgaria, 11–14 October 2017.
24. Iris, Ç.; Pacino, D.; Ropke, S.; Larsen, A. Integrated Berth Allocation and Quay Crane Assignment Problem: Set Partitioning Models and Computational Results. *Transp. Res. Part E Logist. Transp. Rev.* **2015**, *81*, 75–97. [CrossRef]
25. Iris, Ç.; Lam, J.S.L. Recoverable Robustness in Weekly Berth and Quay Crane Planning. *Transp. Res. Part B Methodol.* **2019**, *122*, 365–389. [CrossRef]
26. Iris, Ç.; Lam, J.S.L. A Review of Energy Efficiency in Ports: Operational Strategies, Technologies and Energy Management Systems. *Renew. Sustain. Energy Rev.* **2019**, *112*, 170–182. [CrossRef]
27. Automated Mooring. Available online: https://www.cavotec.com/en/your-applications/ports-maritime/automated-mooring (accessed on 21 January 2022).
28. AutoMoor—Automated Mooring System. Available online: http://www.trelleborg.com/en/marine-and-infrastructure/resources/videos/automoor-automated-mooring-system (accessed on 21 January 2022).
29. In Tallinn Old City Harbour Ships Are Now Served by Automated Mooring Equipment. *Port of Tallinn*. 2021. Available online: https://www.ts.ee/en/in-tallinn-old-city-harbour-ships-are-now-served-by-automated-mooring-equipment/ (accessed on 17 January 2022).
30. Fischer, Y. Wissensbasierte Probabilistische Modellierung für die Situationsanalyse am Beispiel der Maritimen Überwachung. Available online: https://publikationen.bibliothek.kit.edu/1000051065 (accessed on 5 February 2022).
31. Rossmann, J.; Schluse, M.; Bücken, A.; Krahwinkler, P. Using airborne laser-scanner-data in forestry management: A novel approach to single tree delineation. In Proceedings of the ISPRS Workshop on Laser Scanning, Espoo, Finland, 12–14 September 2007.
32. SmartKai—EMIR. Available online: https://www.emaritime.de/smartkai/ (accessed on 29 October 2021).
33. Goodin, C.; Carruth, D.; Doude, M.; Hudson, C. Predicting the Influence of Rain on LIDAR in ADAS. *Electronics* **2019**, *8*, 89. [CrossRef]
34. Montalban, K.; Reymann, C.; Atchuthan, D.; Dupouy, P.; Rivière, N.; Lacroix, S. A Quantitative Analysis of Point Clouds from Automotive Lidars Exposed to Artificial Rain and Fog. *Atmosphere* **2021**, *12*, 738. [CrossRef]

35. Stanley, M.; Laefer, D. Metrics for Aerial, Urban Lidar Point Clouds. *ISPRS J. Photogramm. Remote Sens.* **2021**, *175*, 268–281. [CrossRef]
36. Dai, J.; Lin, W.; Moorthy, R.; Teo, C.-P. Berth Allocation Planning Optimization in Container Terminals. In *Supply Chain Analysis*; Tang, C.S., Teo, C.-P., Wei, K.-K., Eds.; International Series in Operations Research & Mana; Springer: Boston, MA, USA, 2008; Volume 119, pp. 69–104, ISBN 978-0-387-75239-6.
37. Imai, A.; Nishimura, E.; Hattori, M.; Papadimitriou, S. Berth Allocation at Indented Berths for Mega-Containerships. *Eur. J. Oper. Res.* **2007**, *179*, 579–593. [CrossRef]

Journal of
Marine Science and Engineering

MDPI

Review

Ship Maneuvering in Shallow and Narrow Waters: Predictive Methods and Model Development Review

Mislav Maljković [1], Ivica Pavić [1], Toni Meštrović [1] and Marko Perkovič [2,]*

[1] Faculty of Maritime Studies, University of Split, 21000 Split, Croatia; mmaljkovi@pfst.hr (M.M.); ipavic71@pfst.hr (I.P.); tmestrovi@pfst.hr (T.M.)
[2] Faculty of Maritime Studies and Transport, University of Ljubljana, 6320 Portorož, Slovenia
* Correspondence: marko.perkovic@fpp.uni-lj.si

Abstract: The maneuverability of ships is influenced by several factors, including ship design, size, propulsion system, hull shape, and external conditions such as wind, waves, and currents. The size, shape, and arrangement of the hull, rudder, and propeller are decisive for maneuverability. Hydrodynamic forces such as bank effect and squat significantly impact the maneuverability of large ships in narrow channels. With the increasing trend of building ever-larger ships, the demand to evaluate the maneuvering performance of the ship at the design stage has become more critical than ever. Both experimental and computational methods are used to obtain accurate maneuvering characteristics of vessels. In this study, the methods for predicting ship maneuvering characteristics are analyzed using a systematic review based on the preferred reporting items for systematic reviews and meta-analyses (PRISMA). This article contributes to a deeper understanding of the hydrodynamic capabilities of ships and identifies possible future challenges in the field of ship hydrodynamics. The findings inform educators and the shipping industry about the importance of predicting the maneuvering performance of ships, with an emphasis on the education and training of seafarers needed to make timely decisions in critical situations.

Keywords: ship maneuverability; shallow water; hydrodynamic forces; bank effect; squat effect

Citation: Maljković, M.; Pavić, I.; Meštrović, T.; Perkovič, M. Ship Maneuvering in Shallow and Narrow Waters: Predictive Methods and Model Development Review. *J. Mar. Sci. Eng.* **2024**, *12*, 1450. https://doi.org/10.3390/jmse12081450

Academic Editor: Decheng Wan

Received: 27 June 2024
Revised: 9 August 2024
Accepted: 16 August 2024
Published: 21 August 2024

1. Introduction

Ship maneuvering has been the subject of extensive research for many years, particularly regarding the influence of hydrodynamic forces. There is extensive literature on this subject, which demonstrates the importance of understanding and analyzing the influence of these forces on ship maneuverability. The increasing size of ships, such as large tankers and large containers, has increased interest in researching the hydrodynamic performances of ships in confined waters. The increase in ship size is also accompanied by an increase in the operating speed of the ship, which is between 16 and 25 knots. While speed is not a priority for supertaskers with a capacity of 300,000 tons or more, which are no longer a rarity in shipping, it is crucial for container ships which have reached a size of 400 m and poses a challenge for Masters and pilots sailing in confined waters. Even before the Ever Given ran aground in the Suez Canal, some countries initiated studies on the passage of large container ships through their ports, as port pilots were increasingly concerned about the maneuvering difficulties of such ships in increased wind conditions. A serious incident in the Suez Canal, crucial to global trade between Asia and Europe, led to a six-day canal blockade and caused enormous costs to the global economy. The grounding of the Ever Given in the Suez Canal has brought focus to the safety and maneuverability of large vessels operating in narrow channels and shallow waters. This incident led to an in-depth study of the ability of large vessels to navigate in confined waters and renewed focus of shipbuilders and researchers on improving safety measures and maneuverability predictions [1–5].

1.1. Maneuvering Characteristic

Determining the maneuvering characteristics of ships remains a very important research topic in marine hydrodynamics. To gain a better understanding of ship maneuvers and the predictability of maneuvering characteristics, both experimental and computational tools must be used, with continuous improvement of existing models and development of new ones. Given the current trends and challenges, ship designers and builders are using advanced computer simulation and modeling tools to develop more accurate and reliable mathematical models to optimize ship design and performance [6]. Since the maneuverability of ships is considered very important for safety and efficiency, the International Maritime Organization (IMO) has set stringent standards. As a result, there is a growing demand for assessments of a ship's maneuverability early in the construction process [7]. These standards should be used to assess the maneuverability of ships, and those responsible for the design, construction, repair, and operation of ships should be aware of these standards [8]. Classification societies have established standards for testing and certifying a vessel's maneuverability, such as turning ability, course change and yaw checking ability, initial turning ability, stopping ability, straight line stability, and course keeping ability. All of these tests must be completed for the vessel to receive a seaworthiness certificate from the relevant classification society. Once the maneuverability tests of the ship's maneuverability have been completed, the test results must be entered into the wheelhouse poster, the maneuvering booklet, and the pilot card. It is important that the wheelhouse poster is displayed in a clearly visible place on the navigating bridge so that it is accessible to all navigating officers. The pilot card must be presented to the pilot on arrival on the navigating bridge and must contain a brief explanation of the maneuvering characteristics of the vessel. This is of utmost importance as all officers and pilots must be familiar with the maneuvering characteristics of the vessel [9].

Research into the maneuverability of ships has been ongoing for decades. In the mid-20th century, mathematical models were developed to better understand the resistance and propulsion of ships, considering both existing and new designs. The need to develop mathematical models arose not only from the need to better understand maneuvering, but also from the need to find appropriate mathematical tools to predict the maneuverability of ships. In the following years, as research in ship hydrodynamics progressed, mathematical models became increasingly important and contributed to improving the efficiency and safety of ships. With the construction of experimental tanks (towing tanks), new mathematical models were introduced to cover different types of ships. Sea tests in a towing tank focused on ships maneuvering in shallow and confined waters, and appropriate mathematical models were developed to predict ship performance under the influence of hydrodynamic forces. In parallel with the construction of the towing tank, the International Towing Tank Conference was established as a voluntary association of worldwide organizations whose goal is to predict a ship's hydrodynamic performance based on physical and mathematical results obtained using models of the towing tank. This is particularly important for larger ships, where the influence of hydrodynamic forces can be more significant when sailing in confined waters [10].

Sinkage and trim were not used in the mathematical models, as they were occasionally added for the calculation of the squat. Maneuvering a ship is a difficult task due to the influence of external factors such as wind, waves, and sea currents, and due to these factors the draft, trim, and heel of the ship change. For this reason, wind and waves are included in the maneuvering criteria. [11].

The factors that affect a ship's maneuverability can be characterized as a combination of external and internal factors. External factors depend on the area where the ship is sailing and cannot be controlled, while internal factors are related to the ship's design and can be adapted to the sailing conditions. External factors include the influence of shallow water and interaction with other vessels or the shoreline and the ship–ship and ship–shore interactions. Internal factors include speed, hull structure, propulsion, and the rudder. Adjusting these two factors can reduce the negative effects of external disturbances [12].

Based on experimental results on the quality of course keeping, a ship that remains stable in deep water can lose stability in a specific range of water depth when approaching shallow waters. Upon entering shallow waters, a ship tends to regain stability and can be more stable than in deep waters [13–15].

1.2. Methods Used to Predict Hydrodynamic Coefficient of a Ship

A hydrodynamic coefficient must be determined to predict the maneuvering motion of a ship. The methods used for this purpose are the captive model test, system identification techniques applied to the results of free-running model tests, and computational fluid dynamics (CFD). The methods are represented in Figure 1 [16–18].

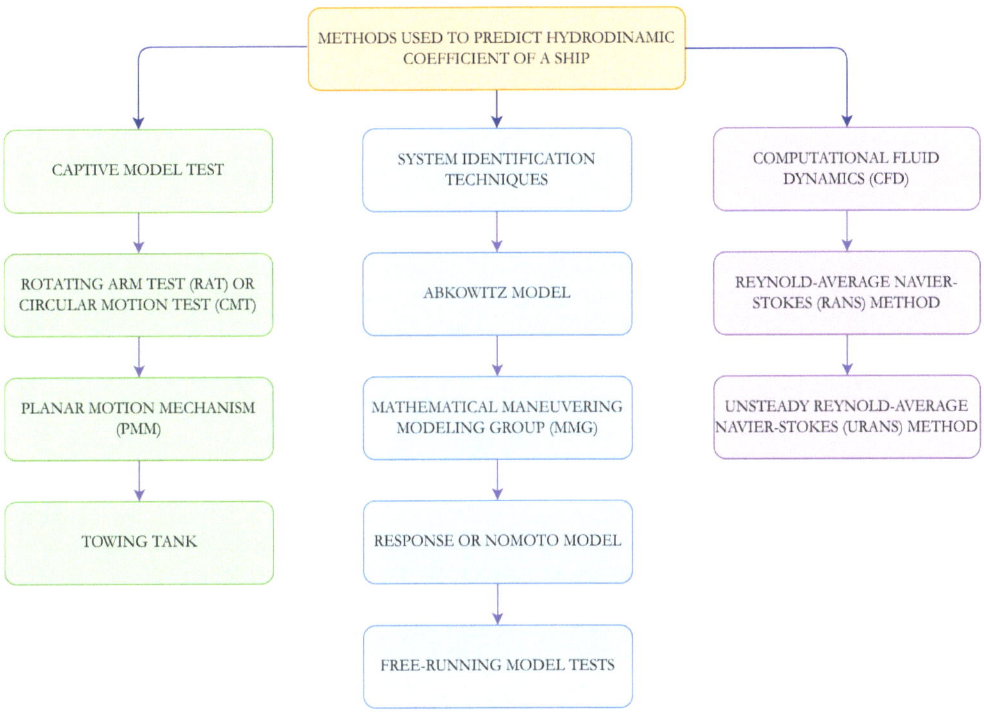

Figure 1. Methods used to predict the hydrodynamic coefficient of a ship.

The captive model test simulation is the most traditional and effective method for determining the hydrodynamic coefficient used in mathematical calculations to predict a ship's maneuverability. In this test, a model of a particular type of ship is placed in a closed tank called a towing tank, and certain parameters are measured, such as the change in rudder deflection and the speed of the ship. The most common measurements in this closed environment relate to the design of harbors and channels and to ship–shore and ship–ship interactions [5,19]. The tests used for these measurements are the Rotating Arm Test (RAT) or the Circular Motion Test (CMT), and the Planar Motion Mechanism (PMM) test [20]. In the RAT or CMT test, a major static captive model test, the models are tested in the towing tank using a rotating arm to obtain the hydrodynamic values of the yaw velocity function, which takes place in the horizontal plane. The test can also be performed in the vertical plane to obtain pitch velocity values [19,21].

The PMM test is a very common captive model test performed in the towing tank when the model has a drift angle with a straight flow. This steady-state test is called the Oblique Towing Test (OTT). In the dynamic test, the model is affected by forces of inertia.

The information obtained from these tests includes damping coefficients, acceleration, and velocity [19,20,22]. The main PMM tests are pure drift, sway, and yaw.

System identification techniques are tools developed in control engineering to create mathematical models of dynamic systems from measured data. The maneuver coefficient is often based on experimental data. System identification techniques can be applied to the results of free-running model tests with a ship model in the towing tank [6,23].

There are three basic types of mathematical models for predicting ship maneuvering: the Abkowitz model, the Mathematical Maneuvering Modeling Group (MMG) model, and the Response model [6,24].

The Abkowitz model is a hydrodynamic model that contains the equations for longitudinal, lateral, and yaw motion from which hydrodynamic factors for ship maneuvering are derived [24]. The MMG model proposed by the Japanese Maneuvering Modeling Group (JMMG) considers the characteristics of the hull, rudder, and propeller forces and their interactions [25].

A Response or Nomoto model is a simplified model that includes the input of simple parameters, the rudder angle, system output, and yaw rate, of a maneuvering vessel as a dynamic system [26,27].

CFD simulations became popular with the advancement of computer capabilities. They analyze ship behavior in shallow waters, including ship–shore and ship–ship interactions. The results obtained with the CFD method are very reliable and provide a detailed picture of surface elevations and velocity/pressure fields, leading to a better understanding of the hydrodynamic phenomena of a ship maneuvering in shallow waters. CFD calculations are an alternative to physical methods because they provide reliable data at a much lower cost [28,29].

Mathematical models for ship maneuvering consider ship motion in six degrees of freedom (DOF) known as surge, sway, heave, roll, pitch, and yaw. In the past, mathematical models with only three degrees of freedom (3 DOF) have been used to study maneuvering in shallow waters. However, to account for the effects of rolling motion, a fourth DOF was later added to these models.

With the introduction of IMO regulations on maneuvering criteria under the influence of wind and waves, a more comprehensive approach became necessary. As a result, mathematical models with all six DOF became mandatory for ship maneuvering [6,30–32]. By using all six DOF, these updated mathematical models can effectively estimate the longitudinal forces, pitching forces, and yawing moments acting on the ship during maneuvering [33].

1.3. Approach Channels and Waterways

Harbor pilots and ship Masters take sinking and trimming very seriously, especially when navigating large ships in channels and ports with limited depth [34]. The hydrodynamic forces acting on a ship in shallow waters differ in many ways from those in open water. In shallow waters, the ship becomes sluggish, causing reduced steerability. The consequences are a lower response to the rudder angle, a larger turning diameter, smaller drift angles, and a lower speed reduction when turning. The decisive pivot point for ship handling is the center of the forces acting on the ship. If the depth shifts from deep to shallow waters, this affects the pivot point by moving it backward, close to the ship's center of gravity. Adjusting the pivot point affects the ship's maneuvering characteristics. Given the reduced maneuverability in shallow waters, the vessel sails at a lower speed and reduced engine revolutions per minute (RPMs) [35–37]. Other disturbance factors that make navigating in shallow waters more difficult are the varying depths and sudden changes in the current direction followed by high and low tides. When transiting in shallow waters, the waves generated by the ship, including the effects of waterbed friction, must be considered, as the ship's wave resistance is present in shallow waters. The seabed also affects maneuverability, as the shallow water effect on large vessels is much stronger when the bottom consists of a muddy layer. All these factors must be considered before entering an area with shallow water. Continuous monitoring of the specified parameters is essential to be able to react in good time, so that adjustments can be made in the event of any deviations [38–40]. Squat depends not only on the speed of a ship but also on the speed of the sea current passing around it. Squats can also occur when the ship is alongside a current if a strong current is present. In addition to squat, an unstable turning moment can occur when passing close to the bank or in the fairway due to the formation of waves in shallow water, which makes it difficult for a ship to maintain a stable course; this becomes increasingly apparent as the ship's speed increases. Considering all these facts, shipping companies have introduced strict standard requirements for the minimum under-keel clearance and speed requirements in confined waters, which must be strictly adhered to by navigators to ensure safe navigation and protection of the environment.

In addition to the mathematical models used to calculate the ship's maneuverability, special attention is given to port infrastructure and the approach channels to accommodate large ships (PIANC). PIANC is a global organization working on this issue, providing guidance and technical advice on sustainable water transport infrastructure for ports, marinas, and waterways. The Maritime Navigation Commission, as the working group responsible for design guidelines for port access channels, is producing a PIANC report to review, update, and expand design recommendations for vertical and horizontal channels. As ship owners demand the testing of the maneuverability of vessels during the design phase, port authorities are also under pressure to provide deep enough approach channels for large vessels.

There are also canals and artificial waterways built for the passage of ships, where the strongest influence of geometric limits on sinkage and trim can be observed. At extremely shallow water levels, even the slightest change in the effective distance between a ship and the bottom of the fairway can have a significant impact on the behavior of the ship in the channel [41–44]. PIANC has also implemented an approach for selecting a group of seagoing vessels from the "IHS Sea-Web" database. The report contains the dimensions and characteristics of the vessels, categorized by size and type, with valuable data important for the planning and design of the port infrastructure so that it can accommodate the selected vessels [45].

This review article aims to provide a detailed look at the methods that have proven to be the best and most reliable in predicting ship maneuvering and to contributing to a better understanding of the problems associated with maneuvering ships in shallow waters and narrow channels. The results obtained with the methodology used in this article can be used in further research on mathematical models and other techniques for predicting the hydrodynamic coefficient when maneuvering a ship in confined waters. The information collected can be used to improve and enhance existing models to develop more efficient and accurate maneuvering prediction models. The guidelines of the Preferred Reporting Items for Systematic Reviews and Meta-Analyzes (PRISMA 2020) were followed throughout the review process to ensure the required standard of the entire article review process [46].

2. Materials and Methods

In writing this article and reviewing the literature on predicting the maneuvering behavior of large ships in shallow waters and narrow channels, a comprehensive examination of the existing literature revealed various methods and studies in this area of research. To ensure that we covered all the methods used for our research, it was necessary to find systematic and comprehensive research approaches in the extensive literature on the hydrodynamic capability of a ship when navigating through shallow waters. To ensure high-quality standards in the review process, we have adopted the PRISMA guidelines for this article. Figure 2 shows the information flow according to the PRISMA methodology, using four research steps which are explained in more detail in the following sections.

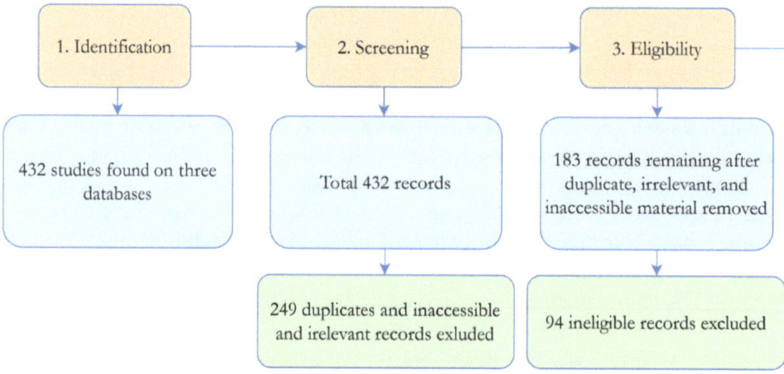

Figure 2. Literature retrieval process.

Step 1. Identification of research studies. The analysis and data collection process began with a literature search of the articles for the desired study. Search engines such as Scopus, WOS, and Sci Direct were used as they cover a sufficiently wide range of scientific articles. The research strategy included Scopus and WOS as large, multidisciplinary databases of the peer-reviewed literature: scientific journals, books, and conference proceedings. Sci Direct, on the other hand, as a third search engine, was included as we found a relatively large number of scientific papers, books, and conference papers related specifically to this field research. As we found many articles in Sci Direct, we had to narrow down the search to focus on the most relevant articles to our research. When entering a specific term into the search engines, it was found that the same terms were in the results, so it was difficult to determine which search engine would dominate. The search strategy included the use of general terms such as:

- ship maneuvering;
- restricted waters;
- shallow water effect;
- squat;
- bank effect;
- ship hydrodynamics.

When using general keywords in combination with these terms, such as "ship maneuvering in shallow waters", the accuracy of the search improved significantly. It helped identify many articles needed for the study. A total of 432 studies from all three databases were found (Figure 2). In Scopus, 201 (47%) studies were found; in WOS, 142 (33%); and in Sci Direct, 89 (21%).

Step 2. Screening research studies. Screening the articles aimed to determine the number of articles relevant to our study, including a thorough search of the articles, and excluding articles inconsistent with the study, with retention rate 249/432 (58%). Screening included reading the title of the article, the abstract, and the conclusion and skimming the article's content. The decision regarding reading the entire article depended on whether the article examined the maneuvering of large ships in narrow channels and shallow waters while also examining mathematical methods relevant to our study. In the screening phase, duplicate articles were identified and removed, and many of the relevant studies that were not accessible had to be excluded from the research.

Step 3. Eligibility assessment of research studies. In the process of acceptance of research studies, the articles were thoroughly analyzed, and the most important ones were selected for further processing. This was conducted according to certain criteria: the source of the article (whether it was published in a credible journal), the importance of the content of the article (were there significant efforts to find innovative solutions), and the validity of the research method (whether a sensible and rational method was used and whether the conclusions were rational). Publications that do not deal with ship maneuvering and are irrelevant to maritime education were excluded from the analysis. Research studies we included in the analysis were a ratio of 89/183 (49%). This step was important to ensure the broadest possible perspective in the research analysis and that sufficient material to conduct the bibliometric analysis was included.

Step 4. Research studies analysis. The final step included assessing the selected studies and analyzing them in detail. Only these studies were used to obtain specific answers in the analysis process and combined within the methodology best suited to overview the maneuvering characteristics of the vessel when navigating in narrow channels and shallow waters.

Figure 3, created with the WOS viewer, illustrates the interweaving of the keyword "ship maneuvers in shallow waters" with the 30 most frequently used keywords in the Scopus research database. At the center of this network is "shallow water", which is closely linked to "water depth" and "maneuverability", underlining their importance in the field. The visualization shows clusters around numerical studies, model testing, and the impact of physical geographies represented by terms such as "shore effect" and "interaction".

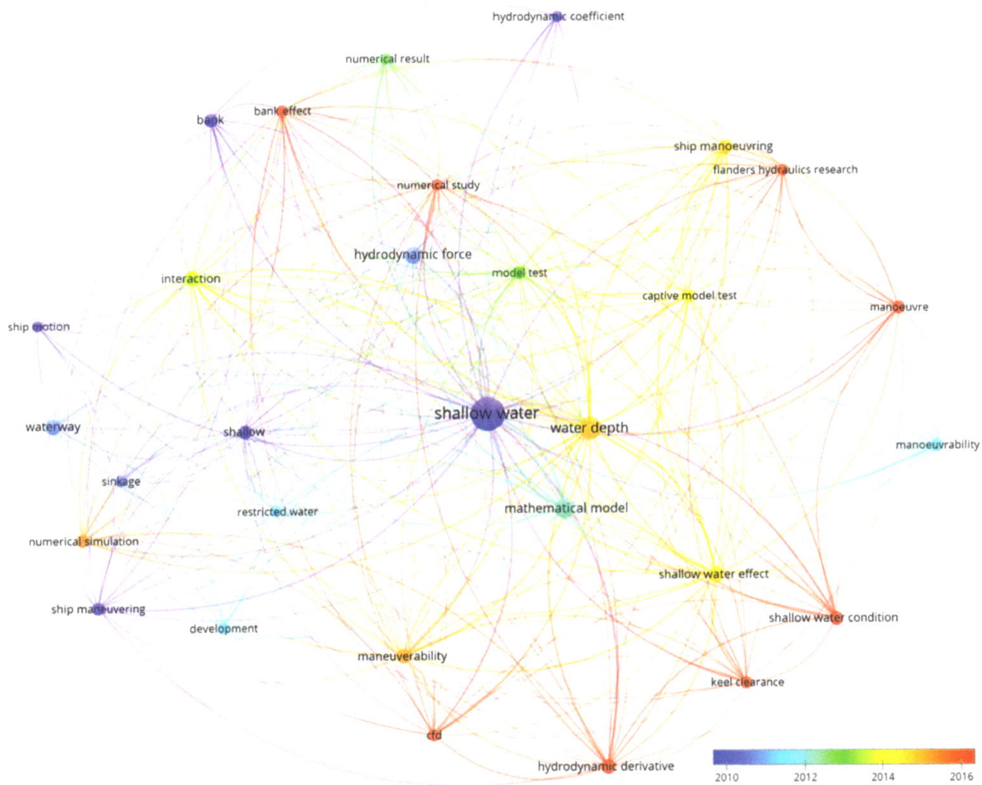

Figure 3. The linking of the keyword "ship maneuvering in shallow waters" with the 30 most used keywords in the research (Scopus).

The research covers various aspects including hydrodynamic forces, confined water environments, and waterway management. The color gradient shows the evolution of research trends from 2010 to 2016, reflecting the interdisciplinary and dynamic nature of studies to improve the safety and efficiency of vessel operations in shallow waters.

3. Results and Discussion

Bibliometric analysis considers studies by year of publication, leading countries, and leading authors. As seen in Figure 4, the publication of articles on ship maneuverability has increased significantly over the last ten years, particularly from 2009 onwards where a sudden increase can be observed. This trend is closely related to the greater demand for the construction of large ships due to transportation costs. From 2009 to the present, a series of maritime accidents have occurred that we believe have indirectly encouraged scientists to increase the number of research articles investigating the hydrodynamic capabilities of ships. The stranding of the Costa Concordia in 2012 encouraged experts to pay more attention to the safety of ships during coastal navigation, which is reflected in the increasing number of articles in the following years. Looking at the maritime accidents of the last ten years, it can be seen that the grounding of container ships in confined waters have characterized this period. The grounding of the Maersk Shams in the Suez Canal in 2016, followed by the grounding of the Vasco de Gama in the Thorn Canal in 2016, pointed to the increasing problems of large and ultra-large container ships navigating in shallow waters. The grounding of the Ever Given in the Suez Canal in 2021, which attracted public attention due to the week-long closure of the canal, was a sign to shipbuilders and scientists that they need to seriously consider the hydrodynamic capabilities of such vessels. Just one year later, in 2022, the Ever Forward ran aground in Chesapeake Bay. The latest example is the serious accident involving the container ship Dali on 26 March 2024, when the ship lost maneuverability due to a sudden power loss and collided with a bridge. Similarly, the incident involving the container ship MSC Michigan VII in Charleston on 5 June 2024, underscores the critical importance of understanding a ship's maneuverability to make timely decisions in emergencies. Due to a malfunction in the propulsion control systems, the ship sailed under a bridge at an uncontrolled speed. Based on a recent Baltimore bridge case, the crew could not reduce speed and hesitated to shut down the main engine because they were uncertain how the ship would respond without propulsion. They were particularly concerned about navigating the ship through the river's sharp turns, where a significant amount of critical infrastructure along the shore includes marinas, moored tankers, military vessels, and the bridge itself. Fortunately, there were no consequences in this specific scenario. In Figure 4, we can see the increase in the number of articles related to the hydrodynamic predictions of ships. A common link in these cases is that they involve large or ultra-large container ships that have run aground in narrow channels. The latest serious accident involving the container ship Dali in 2024, which destroyed the Baltimore bridge, will certainly lead to some changes in the maritime industry regarding the navigation of large ships in confined waters.

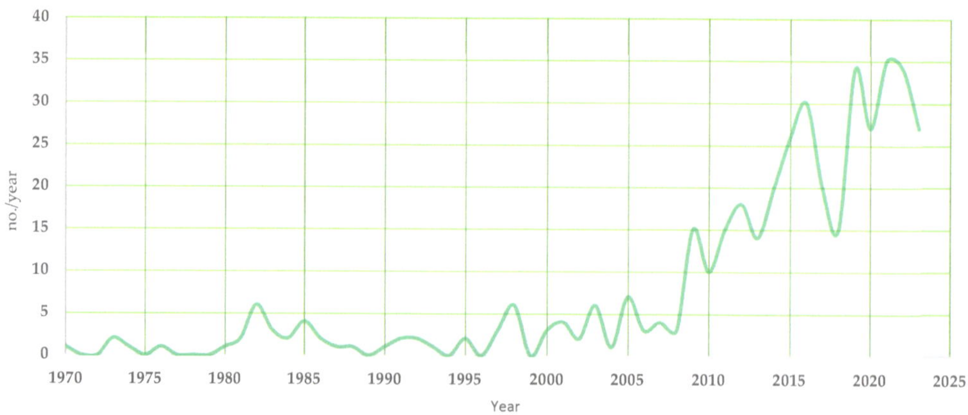

Figure 4. Annual publication count.

Container ships have experienced a real boom in increasing capacity and size. With ever-larger ships on the market, the question of the safety of navigating such ships, especially in shallow waters and narrow channels, has arisen. From the articles reviewed, it can be concluded that many articles focus on identifying the most reliable methods for predicting the maneuvering characteristics of ships before and during their construction.

Figure 5 lists the first ten countries (China, Belgium, the United States, the UK, Portugal, Germany, Japan, South Korea, The Netherlands, and India) that have contributed the greatest research into the maneuverability of large ships in shallow waters. China, which has a considerable number of scientists, has contributed the most scientific articles to this field of research. Slightly fewer, but still a large proportion of the contributions to this research come from Belgium, which we can attribute to the collaboration between Flanders Hydraulics Research and Ghent College of Applied Sciences. The Flanders institute installed a towing tank in 1992–1993, leading to numerous maneuvering tests being conducted, which resulted in new mathematical models for predicting ships' maneuvering characteristics and led to numerous scientific articles. The Flanders Institute has also contributed to many scientific articles from multiple European countries (the United Kingdom, Portugal, Germany, the Netherlands, Poland, Italy, and Norway), as many are based on maneuvering tests conducted in their towing tank. Japan has contributed a considerable number of articles through the efforts of the JMMG, whose work is based on this topic. Other countries with a significant number of scientific articles are the United States, South Korea, India, Malaysia, Brazil, and Australia. Based on the available research data, the study of the hydrodynamic properties of ships in confined waters is an interesting topic in most countries and continents.

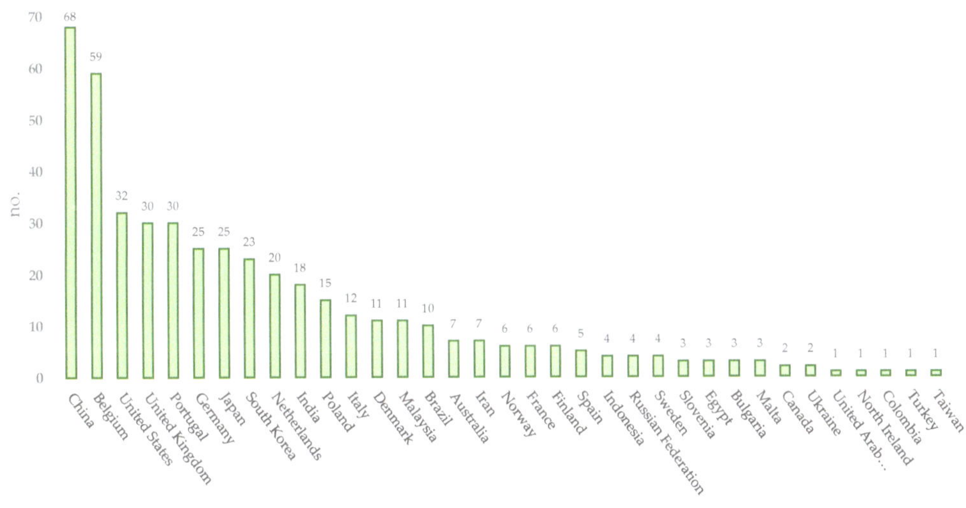

Figure 5. Number of publications per country.

Figure 6 was created using WOS Viewer. The figure displays the authors cited for their contribution to scientific articles on ship maneuvering in shallow waters with the prediction of hydrodynamic forces. The coloring of the grid represents different research groups, each focusing on specific aspects of ship hydrodynamics and characterized by their most important contributions to the field. This visualization highlights the diversity of research topics within the field and underscores the collaborative efforts that are driving progress in ship hydrodynamics. The figure shows that a certain number of researchers are responsible for many scientific articles that can be used for future research on this topic. Dele-Fortrie's research, for example, focuses on the navigation of large ships in shallow water and the forces acting on them.

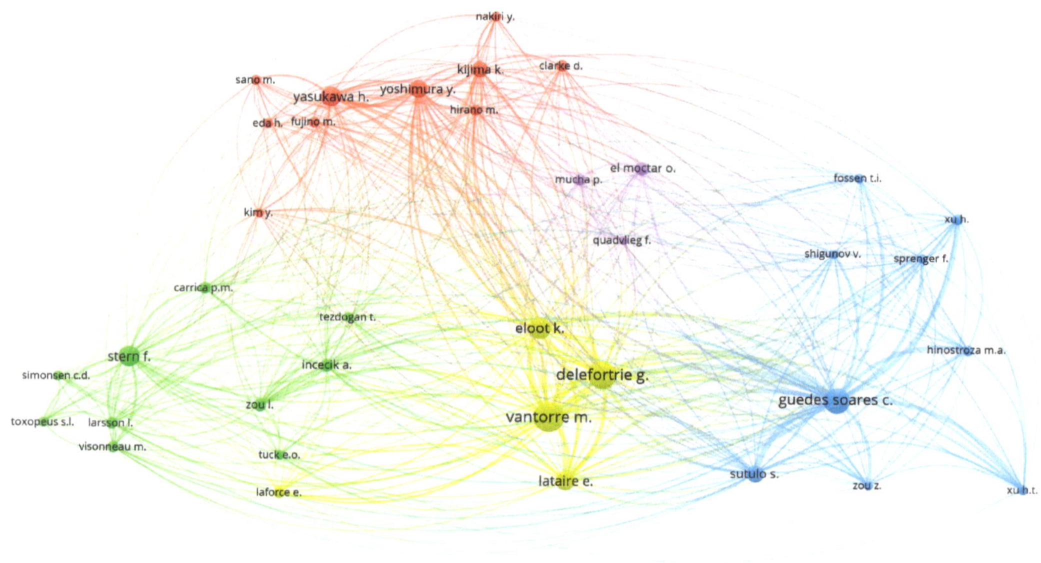

Figure 6. Authors with the most articles related to ship hydrodynamics.

Together with Lataire and Chen, they are investigating trajectory tracking controls, course keeping controls, and the validation of automatic steering algorithms for large ships navigating in shallow waters or narrow fairways [11]. In addition to the two other authors, he is also working with Vantorre to study the maneuvering behavior of Ultra-Large Container Vessels (ULCVs) in shallow waters.

In their articles, they investigate the effect of the heeling of the ship, the effect of water depth in the transition from medium to very shallow water, the wave forces when maneuvering the ship with effects on the propeller and rudder performance, and the turning ability. For most of their research, they use data obtained during free-running tests in the towing tank in Belgium [47]. Together with Eloot, they are engaged in research into the interactions between the banks, the forces that occur during lightering, the so-called ship-to-ship operation, and the effects of a muddy seabed on the maneuverability of ships [48]. Guedes Soares, in collaboration with Sutulo, is investigating the phenomenon of squats in coastal waters during ship–ship interactions. They are also investigating the effects of the bottom geometry in narrow channels on the maneuvering and seakeeping characteristics of ships. Their research aims to demonstrate the influence of different bottom types such as flat horizontal bottom, multi-level bottom, inclined bottom, or dredged channel, on the maneuverability of ships in narrow waters [49]. Guedes Soares, in collaboration with Haitong Xu, is investigating the maneuverability of large container ships in shallow waters and the necessary ship propulsion for ships of this size [50]. The results of towing tank tests are also used for certain measurements. Zou focuses his research on predicting maneuvers of large ships, especially container ships, using CFD methods and unsteady RANS equations, as well as predicting the hydrodynamic forces acting on a ship as it approaches a lock in shallow waters [22]. Yasukawa focuses on researching the maneuverability of ships with a single propeller shaft and two propellers/twin rudders ships when maneuvering in shallow waters and during berthing and unberthing operations using free model tests as part of the MMG. Yasukawa is investigating the maneuverability of ships under the influence of external forces such as wind and waves as well as the course stability of a ship near the bank or a channel wall in shallow water [51]. In their research papers, most authors investigate the influence of hydrodynamic forces acting on a ship when navigating in confined waters, including the squat and bank effects, using currently available methods.

In many research articles, a towing tank is used, which forms the basis for mathematical calculations. The towing tank models often refer to large container ships, which indicates a growing problem with ever-larger container ships in shallow waters, which are very susceptible to the influence of hydrodynamic forces, especially wind.

During our research, we identified many journals containing articles on ships' hydrodynamic properties. The most represented scientific journals on this topic are shown in Figure 7. The greatest number of authors published their work in the journal *Ocean Engineering*, followed by *Applied Ocean Research*. From the number of articles published, it can be concluded that these two are the leading journals for studying ship maneuvers and hydrodynamic properties. The other journals dealing with the hydrodynamic properties of a ship, such as the *Journal of Marine Science and Engineering*, and *TransNav* (the *International Journal of Marine Navigation and Safety of Sea Transportation*), have made a significant contribution to this research. The remaining number of relevant articles for this research are also listed in Figure 7.

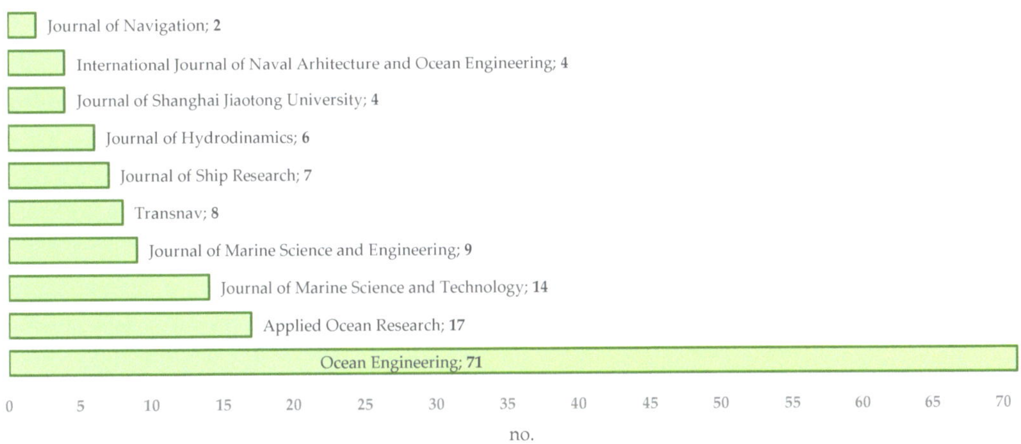

Figure 7. Distribution of articles related to ship hydrodynamics across journals.

Through a systematic review of the literature, original articles closely related to the study of maneuvering characteristics of ships in narrow channels and shallow waters, including the methods used to predict these characteristics, were selected for further analysis. Based on the analysis of the selected articles, we identified and categorized methods used for predicting the maneuverability of ships and classified them into three groups. These groups are the captive model test, computational fluid dynamics and system identification techniques. These three methods used in shipping industry are crucial for predicting a ship's hydrodynamic coefficient.

3.1. The Captive Model Test

The captive model test, which takes place in a controlled area such as a towing tank, where the ship model is towed at specific angles, measuring the sway force and yaw moment, is a good method for predicting the hydrodynamic forces acting on the ship. In this type of test, two mechanisms are used to conducted experiments: the rotating arm mechanism (in which the ship model is attached to a mechanical arm that rotates at a specific rate and speed), and the PMM (in which the ship model is attached to two columns, with one forward and the other aft, and the columns move the ship forward at a specific rate and create a variable yaw) [52].

The PMM test is a widely used method for the experimental investigation of the maneuvering characteristics of various ships, including catamarans, container ships, and large tankers. The PMM test includes maneuvering in shallow waters and narrow channels and the influence of the bank effect during maneuvering. Large tankers such as Very Large Crude Carriers (VLCCs) and Ultra-Large Crude Carriers (ULCCs) are not allowed to enter many ports due to their size and draught, so most operations take place further a sea from the port. These operations are known as ship-to-ship (STS) and lightering operations. In these operations, two (usually) large ships are positioned alongside each other at sea, with certain hydrodynamic forces occurring between the ships. The PMM test is used to understand better the hydrodynamic forces acting on both ships [53–57].

One of the most knowledgeable institutes for navigable ship model research is Flanders Hydraulic Research (FHR), operated by the Maritime Department of Ghent College. Their towing tank, built in 1992–1993, consists of a planar motion carriage, a wave generator, and ship-to-ship interaction equipment. The equipment was designed for captive model testing, and the experimental results were used to develop mathematical models for maneuvering simulations. Towing tank tests in confined waters are considered the most effective approach to understanding a vessel's hydrodynamic capabilities [58]. Since 2009, free-running model tests have been used to directly predict the maneuvering performance of full-scale ships, with the possibility of developing computer simulation models for further research [59]. Most captive model tests in shallow waters are conducted at FHR and based on a 6 DOF maneuvering model because of IMO-specific regulations for maneuvering in wind and waves [20].

The 6 DOF maneuvering model using FHR's towing tank can also be used to evaluate maneuvering tests on open water, in an attempt to find a mathematical model suitable for port maneuvering simulations [60]. Reynolds-averaged Navier-Stokes (RANS) simulations have proven to be a successful tool for determining hydrodynamic derivates for ships. RANS simulations simulate a series of model tests to predict ship maneuvering from two aspects: free surface flow and real rotating propeller [61,62]. With the trend of increasing ship tonnage, the demand for constructing twin-propeller double-rudder ships has increased significantly. The maneuvering characteristics of twin-propeller ships have a complicated area around the stern, so studying the interaction between the hull, propeller, and rudder is important. RANS simulations play a crucial role in simulating the hull, propeller, and rudder interaction in captive model tests for twin-screw ships [63]. Using a hexapod platform with the 6 DOF to perform captive model tests allows us to create independent or coordinated motions that are impossible with traditional methods [64]. The hexapod platform allows us to explore new types of tests, such as the vertical harmonic test, which are still in the research phase [65].

3.2. Computational Fluid Dynamics

Computational fluid dynamics (CFD) is one of the most reliable mathematical methods for accurately predicting and analyzing bank effects. To create a suitable mathematical model to study the bank effect, many intensive systems engineering model tests in the towing tank are required, which are quite expensive. Mathematical models, which are often used to predict the bank effect, have the limitation that they cannot provide detailed insight into the flow dynamics, which is necessary to understand the complex mechanisms of the bank effect. To solve this problem, researchers have turned to numerical techniques and CFD methods to study the phenomena of the bank effect. The most common method is the potential flow method, which allows detailed analysis of flow patterns, and most numerical research in this area relies on the theory of potential flow, which also accounts for free surface effects. However, this area of research does not consider important problematic factors in shallow water, such as breaking waves and viscous and turbulent effects. Therefore, the RANS method is used to predict the ship–bank hydrodynamic forces as a good alternative to obtain a more accurate fluid flow equation since it can include these important factors in its calculations. The CFD-based RANS method is a useful tool in marine hydrodynamics.

It is widely used, especially by ship designers, as they can test a ship's performance in the design phase and make the necessary corrections. This method can also be used for straight-ahead sailing and turning maneuvers tests when overloading of the propeller occurs. The CFD-based RANS can accurately predict flow details around the propeller in critical operating conditions. In addition, it is also used for shallow water problems, the squat effect, and ship–ship interactions [21,66–74]. In contrast to the static RANS methods, which are used for maneuvering problems in shallow waters, the CFD-based Unsteady Reynolds-Average Navier-Stokes (URANS) methods are used for predicting ship behavior in heavy weather, such as rolling and course-keeping ability in regular and irregular waves in low and high seas [75]. In this method, hybrid URANS solvers are often used to solve the time domain of hydrodynamic loads and motions when a ship performs dynamic maneuvers [76]. The CFD approach is also used for free-running simulations of maneuvers such as the zig-zag test, course keeping, and turning circle maneuvers, with highly satisfactory results. The results of these simulations include viscous and rotational effects that allow a clear and detailed analysis of the hydrodynamic interactions among hull, propeller, and rudder [21,77–81]. Trimarans, known for their high-performance capability with complex hull shapes and layouts, require hydrodynamic testing to be performed. CFD methods are very useful for the numerical simulation of the propulsion performance of a high-speed trimaran with waterjet propulsion [82]. With the advancement of computer technology, the CFD approach is increasingly prevalent in military shipbuilding and is applied to predict the hydrodynamic coefficient of submarines [83].

3.3. System Identification Techniques

The system identification method determines the hydrodynamic coefficient from a measured ship motion and the applied rudder angle. This method can be defined as a systematic approach to creating a mathematical model of an unknown system based on given input and output from measured data. Successful system identification depends on correctly selecting three key elements: a mathematical model representing the system, input–output data, and a parameter estimation scheme. It is a powerful technique that can correctly identify a vessel's movement, so it is often used to develop control and navigation systems [10,84].

System identification techniques with system-based (SB) and free-running CFD experiments are used to predict the maneuvering coefficient, as they are considered one of the most important simulation methods for predicting ship maneuverability. System-based simulations reduce the calculation time as they only solve the equations of motion within a given mathematical model. It requires only one minute per free-running test, whereas the CFD method requires several weeks or months, depending on the various factors of the ship's movement. Free-running model tests such as the zig-zag test, course keeping, and turning circle test are used to predict the maneuvering characteristics of a full-scale ship. The results of the free-running model tests are used to develop computer simulation models for further analysis [85,86]. The prediction of maneuvering trajectories of a newly designed ship without simulation, under the condition that the database of maneuvering parameters from many full-scale and free model tests is available, is the empirical "no simulation method" [84]. The simulation method is required for maneuvering parameters that are not included in the database. Figure 8 shows an overview of the different methods for predicting the maneuvering characteristics of ships.

Besides the system identification techniques, the Abkowitz and MMG models are the most recognized and widely used mathematical models for predicting ship maneuvers. The main difference between the Abkowitz and MMG models lies in determining the hydrodynamic forces acting on the maneuvering ship. The Abkowitz model examines all hydrodynamic forces simultaneously (longitudinal, surge, transverse sway, and yaw motion).

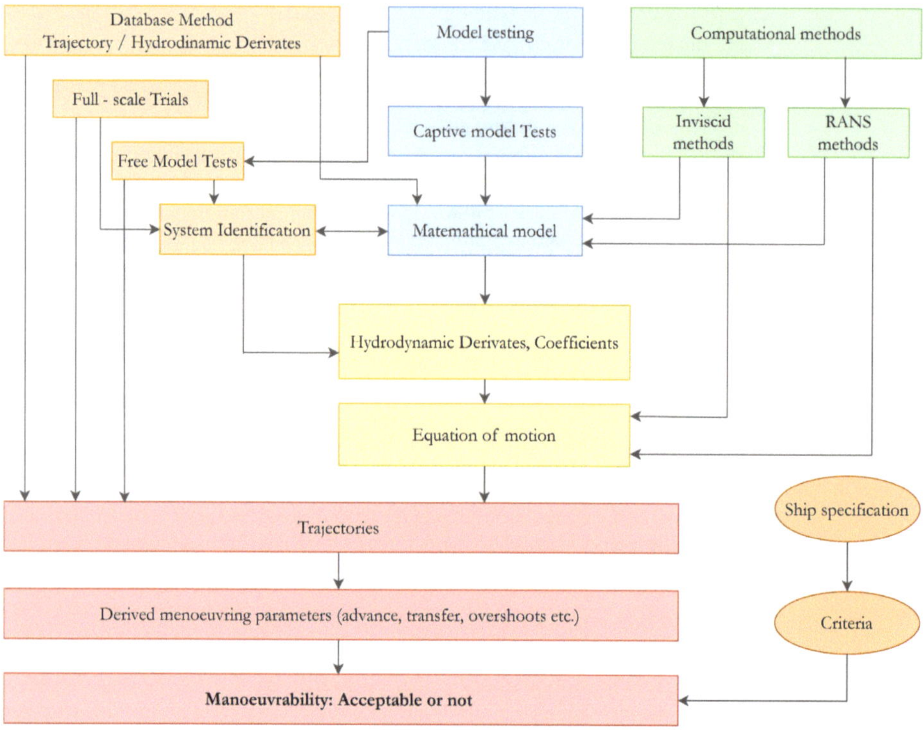

Figure 8. Overview of maneuvering prediction methods [85].

In contrast, the MMG model divides the hydrodynamic forces into categories such as hull, propeller, and rudder forces and their interactions [17,87–90]. The MMG model is also known as the "rudder to yaw response model" as it describes a ship's rate of turn response to the rudder actions. The MMG model can also be applied for shallow water maneuvering with reasonable accuracy in practical use, but with certain improvements to enhance its performance and by using the coefficient of hydrodynamic forces at every water depth [26,91]. Numerous simulation methods based on the MMG approach are in use. Still, there have been problems in fitting the hydrodynamic force data to the maneuver simulation because the different methods may not apply to each other. Therefore, it was concluded by the Japan Society of Naval Architects and Ocean Engineers that the basic parts of the method should be common. As a result, the MMG standard method was proposed, which consists of four elements: the maneuver simulation model, the procedure for the captive model test, the maneuver simulation analysis method, and the full-scale ship maneuver prediction method [92].

Semi-theoretical and semi-empirical methods are simplified mathematical models used for preliminary assessment of ship maneuvering characteristics using semi-empirical formulae derived from a database of conducted captive model tests. Empirical regressions are based upon the most prevalent ship hull forms like single-screw ships. Deviation from common hull forms, like in twin-screw ships, can exceed the parametric range of the experimental database and potentially lead to inaccurate predictions [93–96].

Artificial neural networks (ANNs), as an alternative approach, can be used to predict hydrodynamic ship parameters based on empirical information from experiments with a scale model and are a widely used tool to effectively predict the maneuverability of certain types of ships. In addition, ANNs are used to determine the optimal ship trajectory in narrow channels and shallow waters and for course alteration maneuvers. ANN applications have proven practical by using mathematical models to predict catamarans' and trimarans' pitching and heaving motions with unconventional underwater hulls. Research into the automatic mooring of ships is considered one of the most complex problems in ship control. The use of an ANN model has proven to be an extremely effective solution for the automatic docking of ships, as it can learn and imitate the actions of the human brain during docking maneuvers [96–99]. Neural network algorithms may have certain advantages since no structure of the mathematical model of the ship is required. However, at the same time, the lack of a physical structure can be a disadvantage because, without a physical basis, the ANN model cannot be extended, adapted, or modified [100,101]. ANNs are also used as a learning process in autonomous ship control. This can be described as a ship control system that simulates the learning process of an autonomous control unit that collects input signals and calculates the values of the necessary parameters for maneuvering the ship in confined waters [102]. With the development of Maritime Autonomous Surface Ships (MASS), predicting the maneuverability of these vessels has become an important issue, especially when navigating narrow channels and shallow waters. Autonomous navigation requires highly accurate maneuvering models to eliminate all possible uncertainties and inadequacies in the required actions. For this purpose, maneuver data from real ship maneuvers was systematically collected and analyzed. The comparison between real and simulated data showed the possibility of predicting the maneuvering characteristics of autonomous vessels using simulations in a controlled environment [103,104].

The performance of the ship's engine has always been an important factor in the construction of ships and their design. Achieving the desired speed with minimum fuel consumption makes a ship economical. Introducing the Energy Efficiency Design Index (EEDI) for new ships represents a significant step in setting energy-efficient regulations for different types of ships. However, fears have been raised that some ship designers will reduce engine power rather than develop new innovative propulsion systems to meet the new regulations. It is paramount to understand the vessel's maneuvering characteristics, especially in adverse weather conditions under the influence of waves, wind, and currents, as a reduction in engine power can lead to reduced maneuverability, compromising the vessel's safety in adverse weather. In addition, underpowering can affect the stopping distance, which may remain unchanged in unfavorable weather conditions. Given this new situation, the IMO adopted guidelines (Resolution MEPC) in 2011, which set out the minimum engine power requirements for different types of ships, also considering unfavorable weather conditions in the assessment. To meet the new challenges in the maritime industry, the EU has funded the SHOPERA (Energy Efficient Safe Ship Operation) project. The main objective of the project is to develop numerical methods and software tools and to carry out comprehensive studies on the propulsion and control systems of ships required for maneuvering in adverse weather conditions, including open sea, coastal waters, and confined waters [105–108].

During our research we have come to the conclusion that all three methods are used to predict the hydrodynamic forces of the ship and that the methods complement each other. The choice of method depends on the construction phase of the ship, the complexity of the ship design, and the construction budget. Often a combination of these methods is used to achieve the most reliable test results and reduce test costs, as certain methods are expensive and time-consuming or have certain limitations due to the influence of external factors such as wind and waves. Captive model tests provide very accurate and reliable data on hydrodynamic forces and moments. The results of these tests can be used to verify numerical models related to ship maneuvers, research projects (channel and port approaches), and the study of ship–shore and ship–ship interactions. The captive test is also used for rapid testing of ship designs to determine whether the vessel meets the maneuvering criteria set by the IMO. The measured forces from the research can be used as input data for ship simulators. Captive model tests are expensive and time-consuming and are therefore often used in conjunction with other methods such as CFD.

The advantage of CFD simulation, which has become popular with the development of computer capabilities and numerical techniques, is that it can provide detailed results of hydrodynamic forces in shallow waters. However, the CFD method requires a high level of computational expertise and the results need to be validated against empirical data. The combination of the CFD method with empirical formulas and model tests provides detailed test results.

The system identification technique is one of the most reliable techniques for improving mathematical models using collected data. It is a technique for creating mathematical models from measured data that can be applied to free-running model test results. However, the accuracy of the model depends on the quality and quantity of the measured data. Incomplete data can lead to inaccurate models. These models can also be integrated into ship simulators for the education and training of seafarers.

The CFD method is currently the standard method worldwide for solving hydrodynamic problems of ships, as it enables very accurate calculations of numerical simulations. For large ships, however, shipbuilders construct models for a towing tank after the CFD calculation, as these tests are considered to be the most reliable. The purpose of a towing tank test is to validate the numerical method used. Although towing tank tests are extremely expensive, this is not an issue for shipbuilders, as it is of great importance to obtain accurate hydrodynamic properties of the ship when it goes into series production.

4. Conclusions

This article aimed to identify and analyze the methods for predicting the hydrodynamic characteristics of large ships when passing through confined waters. Likewise, this article can guide future research toward the most effective and appropriate methods for determining the hydrodynamic coefficient to improve the safety of navigation at sea. In this article, we have identified the leading countries, the most frequent researchers, and the most used methods for predicting the hydrodynamic coefficient of large ships. After reviewing the most widely used methods in this field of research, we have also drawn attention to the leading countries, which account for a significant proportion of the number of scientific articles on this topic. The top ten countries—China, Belgium, the United States, the UK, Portugal, Germany, Japan, South Korea, The Netherlands, and India—are significant for the most research on this topic and for the results produced by the research articles. The tightening of IMO standards for assessing the maneuverability of ships has led to ship owners demanding maneuverability assessments at the initial stages of ship construction. This demand has encouraged scientists to develop more advanced and innovative methods to meet the new standards effectively. The significant increase in scientific articles over the last 15 years confirms the increased research efforts to meet the new requirements.

Given the considerable effort scientists have invested in developing innovative solutions for predicting maneuverability, it is essential not to ignore the human element, as the navigator makes the final decisions and steers the ship. Despite various methods for predicting the maneuverability of ships in confined waters, accidents such as groundings, collisions, and impacts still occur. A notable example is the serious accident involving the container ship Dali on 26 March 2024, when the ship lost maneuverability due to a sudden power loss and collided with a bridge. Similarly, the incident involving the container ship MSC Michigan VII in Charleston on 5 June 2024, underscores the critical importance of understanding a ship's maneuverability to make timely decisions in emergencies. Due to a malfunction in the propulsion control systems, the ship sailed under a bridge at an uncontrolled speed. Based on a recent Baltimore bridge case, the crew could not reduce speed and hesitated to shut down the main engine because they were uncertain how the ship would respond without propulsion. They were particularly concerned about navigating the ship through the river's sharp turns, where a significant amount of critical infrastructure along the shore includes marinas, moored tankers, military vessels, and the bridge itself. Fortunately, there were no consequences in this specific scenario. These cases highlight the need for effective training and education for seafarers to understand their ship's maneuverability and make timely decisions in critical situations. Training on bridge simulators is crucial in improving maneuvering skills and preparing seafarers for real-world scenarios. These simulators provide a realistic and controlled environment where mariners can practice handling various situations, including managing system malfunctions and navigating through challenging waterways.

As container ships continue to grow and reach lengths of up to 425 m, it is essential to thoroughly test the maneuverability of these vessels, especially when passing through narrow channels and shallow waters such as the Suez Canal. In addition to the influence of the hydrodynamic forces acting on the ship's size, the wind's influence must also be considered due to the large wind area of ships of this type and size. Studies have shown that the wind is the weakest link for large container ships. Especially in narrow channels with many bends, where the ship must turn frequently, the influence of the wind is most pronounced, and the danger of the influence of hydrodynamic forces is greatest. UKC decreases due to the ship's inclination caused by turning, which thus increases the effect of hydrodynamic forces, including the wind. Following this reason, many countries in the world have set criteria for wind restrictions in the port limits and canals where such ships can transit. Problems arise when testing container ship models in the towing tank because the tank has no curvatures. Towing tank tests provide the most accurate results on hydrodynamic forces but have the disadvantage that the tank is flat and has no curvatures, so the results on hydrodynamic forces cannot be determined when the ship tilts, and it is then that the ship is exposed to the strongest effects of hydrodynamic forces. The aim of future research would be to obtain accurate hydrodynamic data when the ship is sailing around bends in narrow channels, because this data would enable a complete analysis of hydrodynamic forces acting on a ship in confined waters. As the MASS industry is still in its infancy compared to human-crewed vessels, it is a concern when such vessels operate in confined waters. Future research should focus on the navigation of MASS vessels in shallow waters and narrow channels, as well as the docking and undocking, including a training program for human operators of such vessels.

Education and training on bridge simulators should become a mandatory seafarer training program to improve understanding of vessel performances. This training is equally important for crew-manned and MASS vessels. Finally, we must emphasize that this article has certain limitations. The lack of knowledge in this review article is due to the fact that we were not able to test the ship models with mathematical or other methods and draw concrete conclusions. Testing the maneuvering characteristics of the ship is a very sensitive area that is still under development and we were not able to determine which method would be most appropriate. Therefore, the information gathered in this review could serve as a guide and direction for future research on this topic, and we expect that it will support and encourage the improvement of existing methods and innovation in the development of new methods for predicting the hydrodynamic capabilities of ships.

Author Contributions: Conceptualization, M.M. and I.P.; methodology, M.M. and M.P; software, M.M.; validation, I.P., T.M. and M.P.; data curation, M.M.; writing—original draft preparation, M.M., I.P., T.M. and M.P; writing—review and editing, M.P.; visualization, M.M. and M.P; supervision, M.P.; project administration, M.P..; funding acquisition, I.P and M.P. All authors have read and agreed to the published version of the manuscript.

Funding: The publication of the paper is supported by the research group (P2-0394; Modelling and simulation in traffic and maritime engineering) at the Faculty of Maritime Studies and Transport, financed by the Slovenian National Research Agency.

Data Availability Statement: No new data were created or analyzed in this study. Data sharing is not applicable to this article.

Acknowledgments: In this section, you can acknowledge any support given which is not covered by the author contribution or funding sections. This may include administrative and technical support, or donations in kind (e.g., materials used for experiments).

Conflicts of Interest: The authors declare no conflict of interest.

Abbreviations

ANN	artificial neural networks
CFD	computational fluid dynamics
CMT	Circular Motion Test
DOF	degrees of freedom
EEDI	Energy Efficiency Design Index
FHR	Flanders Hydraulic Research
IMO	International Maritime Organization
JMMG	Japanese Maneuvering Modeling Group
MASS	Maritime Autonomous Surface Ships
MMG	Mathematical Maneuvering Modeling Group
OTT	Oblique Towing Test
PIANC	The World Association of Waterborne Transport Infrastructure
PMM	Planar Motion Mechanism
PRISMA	Preferred Reporting Items for Systematic Reviews and Meta-Analyzes
RANS	Reynolds-Average Navier-Stokes
RAT	Rotating Arm Test
RPMs	revolutions per minute
SB	system-based
SHOPERA	Energy Efficient Safe Ship Operation
STS	ship-to-ship
UKC	Under-Keel Clearance
ULCC	Ultra-Large Crude Carrier
ULCV	Ultra-Large Container Vessel
URANS	Unsteady Reynolds-Average Navier-Stokes
VLCC	Very Large Crude Carrier

References

1. Sjöberg, H.; Roos, H.; Edvall, A.; Hallbjörner, F. *Safe Handling of Ultra Large Container Ships in Strong Wind*; Gothenburg Pilot 2016-12-01; Final Report—Version 2.0; Swedish Maritime Administration: Norrköping, Sweden.
2. Fan, S.; Yang, Z.; Wang, J.; Marsland, J. Shipping accident analysis in restricted waters: Lesson from the Suez Canal blockage in 2021. *Ocean Eng.* **2022**, *266*, 113119.
3. Terziev, M.; Tezdogan, T.; Oguz, E.; Gourlay, T.; Demirel, Y.K.; Incecik, A. Numerical investigation of the behaviour and performance of ships advancing through restricted shallow waters. *J. Fluids Struct.* **2018**, *76*, 185–215.
4. Lee, S.; Hong, C. Study on the course stability of very large vessels in shallow water using CFD. *Ocean Eng.* **2017**, *145*, 395–405.
5. Gucma, L. *Risk Management in the Area of Bridges Situated on Waterways in the Aspect of Ship Collisions*, 1st ed.; Marine Traffic Engineering—MTE: Szczecin, Poland, 2024.
6. Özden, M.C.; Kurdoğlu, S.; Demir, E.; Sarıöz, K.; Gören, Ö. A compact motion controller-based planar motion mechanism for captive manoeuvring tests. *Ocean Eng.* **2021**, *220*, 108195.
7. Guo, H.; Zou, Z. System-based investigation on 4-DOF ship maneuvering with hydrodynamic derivatives determined by RANS simulation of captive model tests. *Appl. Ocean. Res.* **2017**, *68*, 11–25.
8. *MSC.137(76)*; I. Resolution MSC.137(76) (Adopted on 4 December 2002), Standards for Ship Manoeuvrability. International Maritime Organization: London, UK, 2002.
9. Atchison, J. *Guide for Vessel Maneuverability*; American Bureau of Shipping ABS: Houston, TX, USA, 2017.
10. Eloot, K.; Vantorre, M. NATO Unclassified + SWE Ship behaviour in Shallow and Confined Water: An Overview of Hydrodynamic Effects through EFD. *Assess. Stab. Control. Predict. Methods NATO Air Sea Veh.* **2011**, *20*.
11. Chen, C.; Verwilligen, J.; Mansuy, M.; Eloot, K.; Lataire, E.; Delefortrie, G. Tracking controller for ship manoeuvring in a shallow or confined fairway: Design, comparison and application. *Appl. Ocean. Res.* **2021**, *115*, 102823.
12. Liu, J.; Hekkenberg, R.; Rotteveel, E.; Hopman, H. Literature review on evaluation and prediction methods of inland vessel manoeuvrability. *Ocean Eng.* **2015**, *106*, 458–471.
13. Amin, O.M.; Hasegawa, K. Assessment of Ship Manoeuvrability in Shallow Waterways. In Proceedings of the MARTEC 2010, Osaka, Japan, 11 December 2010.
14. Osman, M.A.; Kazuhiko, H. Assessment of Ship Manoeuvrability in Shallow Waterways. In Proceedings of the International Conference on Marine Technology, Dhaka, Bangladesh, 11–12 December 2010.
15. Fujino, M. Studies on Manoeuvrability of Ships in Restricted Waters. *J. Soc. Nav. Arch. Jpn.* **1970**, *4*, 157–184.
16. Eloot, K.; Delefortrie, G.; Vantorre, M.; Quadvlieg, F. Validation of Ship Manoeuvring in Shallow Water Through Free-Running Tests. In Proceedings of the International Conference on Offshore Mechanics and Arctic Engineering, American Society of Mechanical Engineers, Hamburg, Germany, 31 May–5 June 2015.

17. Rev The I.T.T.C. Captive Model Test Procedure 7.5-02-06-02, Manual of the Recommended Procedures and Guidelines Maintenance, Manoeuvering Committee of the 29th ITTC, June 2021.
18. Xu, H.; Soares, C.G. Hydrodynamic coefficient estimation for ship manoeuvring in shallow water using an optimal truncated LS-SVM. *Ocean Eng.* **2019**, *191*, 106488.
19. Du, P.; Ouahsine, A.; Toan, K.T.; Sergent, P. Simulation of ship maneuvering in a confined waterway using a nonlinear model based on optimization techniques. *Ocean Eng.* **2017**, *142*, 194–203.
20. Testing and Extrapolation Methods Manoeuvrability Captive Model Test Procedures. In Proceedings of the ITTC International Towing Tank Conference ITTC, Edinburgh, Scotland, 1 September 2005; University of Newcastle upon Tyne: Newcastle upon Tyne, UK, 2005.
21. Hajivand, A.; Mousavizadegan, S.H. Virtual simulation of maneuvering captive tests for a surface vessel. *Int. J. Nav. Archit. Ocean Eng.* **2015**, *7*, 848–872.
22. Captive Model Test for Underwater Vehicles. In Proceedings of the ITTC International Towing Tank Conference ITTC, Virtual, 13–18 June 2021; Manoeuvering Committee: Wageningen, The Netherlands, 2021.
23. Liu, Y.; Zou, L.; Zou, Z.; Guoa, H. Predictions of ship maneuverability based on virtual captive model tests. *Eng. Appl. Comput. Fluid. Mech.* **2018**, *12*, 334–353.
24. Hajizadeh, S.; Seif, M.S.; Mehdigholi, H. Determination of ship maneuvering hydrodynamic coefficients using system identification technique based on free-running model test. *Sci. Iran.* **2016**, *23*, 2154–2165.
25. Zhang, X.G.; Zou, Z.J. Identification of Abkowitz model for ship manoeuvring motion using ε-support vector regression. *J. Hydrodyn.* **2011**, *23*, 353–360.
26. Ahmed, Y.A. Mathematical model of the manoeuvring motion of a ship. In *Advanced Structured Materials*; Springer: Berlin/Heidelberg, Germany, 2018; Volume 85, pp. 551–566.
27. Nomoto, K.; Taguchi, T.; Honda, K.; Hirano, S. *On The Steering Qualities of Ships*; Department of Naval Architecture; Osaka University: Osaka, Japan, 1957.
28. Luo, W.; Zou, Z. Identification of Response Models of Ship Maneuvering Motion Using Support Vector Machines. *J. Ship Mech.* **2007**, *11*, 6.
29. Kim, D.; Tezdogan, T.; Incecik, A. Hydrodynamic analysis of ship manoeuvrability in shallow water using high-fidelity URANS computations. *Appl. Ocean Res.* **2022**, *123*, 103176.
30. Mofidi, A.; Carrica, P.M. Simulations of zigzag maneuvers for a container ship with direct moving rudder and propeller. *Comput. Fluids* **2014**, *96*, 191–203.
31. Bertram, V. Ship Maneuvering. In *Practical Ship Hydrodynamics*; Elsevier: Amsterdam, The Netherlands, 2012; pp. 241–298.
32. Lataire, E. Captive model tests based 6 DOF shallow water manoeuvring model. In Proceedings of the 4th MASHCON, Karlsruhe, Germany, 23–25 May 2016.
33. Kazerooni, M.F.; Seif, M.S. Experimental evaluation of ship squat in shallow waters. *J. Braz. Soc. Mech. Sci. Eng.* **2014**, *36*, 559–569.
34. Taimuri, G.; Matusiak, J.; Mikkola, T.; Kujala, P.; Hirdaris, S. A 6-DoF maneuvering model for the rapid estimation of hydrodynamic actions in deep and shallow waters. *Ocean Eng.* **2020**, *218*, 108103.
35. Tuck, E.; Model Basin, D.T.; Washington, D.C. Shallow-water flows past slender bodies. *J. Fluid Mech.* **1966**, *26*, 81–95.
36. Hooft, J.P. Manoeuvring Large Ships in Shallow Water-I. *J. Navig.* **1973**, *26*, 189–201.
37. Carreño, J.E.; Mora, J.D.; Pérez, F.L. A Study of Shallow Water's Effect on a Ship's Pivot Point. *Ing. Investig.* **2012**, *32*, 27–31.
38. Xu, H.; Hinostroza, M.A.; Wang, Z.; Guedes Soares, C. Experimental investigation of shallow water effect on vessel steering model using system identification method. *Ocean Eng.* **2020**, *199*, 106940.
39. Hooft, J.P. Manoeuvring Large Ships in Shallow Water-II. *J. Navig.* **1973**, *26*, 311–319.
40. Zeng, Q.; Thill, C.; Hekkenberg, R. Shallow water effects on ship-generate d waves. In Proceedings of the 5th International Conference on Ship Manoeuvring in Shallow and Confined Water (MASHCON), Ostend, Belgium, 19–23 May 2019.
41. Sami, K.; Mohamed, A.; Emmanuel, L.; Hassan, S. Numerical Modeling of the Muddy Layer Effect on Ship Squat and Resistance. In Proceedings of the 5th International Conference on Ship Manoeuvring in Shallow and Confined Water (MASHCON), Ostend, Belgium, 19–23 May 2019.
42. The World Association for Waterborne Transport Infrastructure. *Pianc Report No 121-2014—Harbour Approach Channels Design Guidelines*; Maritime Navigation Commission: Basel, Switzerland, 2014.
43. *ROM 3.1-99*; Puertos del Estado (España) Recommendations for the Design of the Maritime Configuration of Ports, Approach Channels and Harbour Basins. Puertos del Estado: Madrid, Spain, 2007.
44. Bechthold, J.; Kastens, M. Robustness and quality of squat predictions in shallow water conditions based on rans-calculations. In Proceedings of the 5th International Conference on Ship Manoeuvring in Shallow and Confined Water (MASHCON), Ostend, Belgium, 19–23 May 2019; pp. 11–24.
45. Shevchuk, I.; Bottner, C.U.; Kornev, N. Numerical investigation of scale effects on squat in shallow water. In Proceedings of the 5th International Conference on Ship Manoeuvring in Shallow and Confined Water (MASHCON), Ostend, Belgium, 19–23 May 2019; pp. 410–422.
46. PIANC, The World Association for Waterborne Transport Infrastructure. *Ship Dimensions and Data for Design of Marine Infrastructure*; PIANC HQ Boulevard du Roi Albert II 20 B. 3, Ed.; No. 235-2022; General Secretariat of PIANC: Brussels, Belgium, 1990.

47. Liberati, A.; Altman, D.G.; Tetzlaff, J.; Mulrow, C.; Gøtzsche, P.C.; Ioannidis, J.P.A.; Clarke, M.; Devereaux, P.J.; Kleijnen, J.; Moher, D. The PRISMA statement for reporting systematic reviews and meta-analyses of studies that evaluate health care interventions: Explanation and elaboration. *J. Clin. Epidemiol.* **2009**, *62*, e1–e34. [PubMed]
48. Tello Ruiz, M.; Mansuy, M.; Delefortrie, G.; Vantorre, M. Modelling the manoeuvring behaviour of an ULCS in coastal waves. *Ocean Eng.* **2019**, *172*, 213–233.
49. Lataire, E.; Vantorre, M.; Vandenbroucke, J.; Eloot, K. Ship to Ship Interaction Forces During Lightering Operations. In Proceedings of the International Conference on Ship Manoeuvring in Shallow and Confined Water: Ship to Ship Interaction (Mashcon 2), Trodheim, Norway, 18–20 May 2011.
50. Sutulo, S.; Rodrigues, J.M.; Guedes Soares, C. Hydrodynamic characteristics of ship sections in shallow water with complex bottom geometry. *Ocean Eng.* **2010**, *37*, 947–958.
51. Xu, H.; Guedes Soares, C. Manoeuvring modelling of a containership in shallow water based on optimal truncated nonlinear kernel-based least square support vector machine and quantum-inspired evolutionary algorithm. *Ocean Eng.* **2020**, *195*, 106676.
52. Yasukawa, H.; Sakuno, R. Application of the MMG method for the prediction of steady sailing condition and course stability of a ship under external disturbances. *J. Mar. Sci. Technol.* **2020**, *25*, 196–220.
53. Michael, S.T.; Franz, S.H. *Maneuvering and Control of Surface and Underwater Vehicles*; Lecture Notes; OpenCourseWare: Cambridge, MA, USA, 2013.
54. Liu, H.; Ma, N.; Gu, X. Experimental Study on Ship-Bank Interaction of Very Large Crude Carrier in Shallow Water. *J. Shanghai Jiaotong Univ. Sci.* **2018**, *23*, 730–739.
55. Lataire, E.; Vantorre, M.; Delefortrie, G.; Candries, M. Mathematical modelling of forces acting on ships during lightering operations. *Ocean Eng.* **2012**, *55*, 101–115.
56. Milanov, E.; Zlatev, Z.; Chotukova, V.; Stern, F. Analysis of inherent course stability of a high-speed catamaran in deep and shallow water. *Int. Shipbuild. Progress* **2011**, *58*, 83–96.
57. Zhu, Z.; Kim, B.S.; Wang, S.; Kim, Y. Study on numerical PMM test and its application to KCS hull. *Appl. Ocean Res.* **2022**, *127*, 103327.
58. Hirano, M.; Nakamura', Y. *The Japan Society of Naval Architects and Ocean Engineers NII-Electronic Library Service an Experimental Study on Maneuvering Hydrodynamic Forces in Shallow Water*; The Japan Society of Naval Architects and Ocean Engineers: Osaka, Japan, 2010.
59. Delefortrie, G.; Geerts, S. The towing tank for manoeuvres in shallow water. In Proceedings of the 4th MASHCON, Karlsruhe, Germany, 23–25 May 2016.
60. The I.T.T.C. Free Running Model Tests Procedures 7.5-02-06-01, Manual of the Recommended Procedures and Guidelines Maintenance, Seakeeping Committee of the 29th ITTC, June 2021. 20 June.
61. Delefortrie, G.; Vantorre, M. 6DOF manoeuvring model of KCS with full roll coupling. *Ocean Eng.* **2021**, *235*, 109327.
62. Yao, J.; Liu, Z.; Song, X.; Su, Y. Ship manoeuvring prediction with hydrodynamic derivatives from RANS: Development and application. *Ocean Eng.* **2021**, *231*, 109036.
63. Lee, S.H.; Paik, K.J.; Hwang, H.S.; Eom, M.J.; Kim, S.H. A study on ship performance in waves using a RANS solver, part 1: Comparison of power prediction methods in regular waves. *Ocean Eng.* **2021**, *227*, 108900.
64. Guo, H.; Zou, Z.; Liu, Y.; Wang, F. Investigation on hull-propeller-rudder interaction by RANS simulation of captive model tests for a twin-screw ship. *Ocean Eng.* **2018**, *162*, 259–273.
65. Tu, H.; Song, L.; Xie, D.; Liu, Z.; Zhang, Z.; Sun, J. Performing captive model tests with a hexapod. *Ocean Eng.* **2019**, *171*, 49–58.
66. ITTC International Towing Tank Conference 2021 Rev 06 ITTC—Captive Model Test; 2021.
67. Ortolani, F.; Mauro, S.; Dubbioso, G. Investigation of the radial bearing force developed during actual ship operations. Part 1: Straight ahead sailing and turning maneuvers. *Ocean Eng.* **2015**, *94*, 67–87.
68. Zou, L.; Larsson, L. Computational fluid dynamics (CFD) prediction of bank effects including verification and validation. *J. Mar. Sci. Technol.* **2013**, *18*, 310–323.
69. Recommended Procedures and Guidelines Practical Guidelines for Ship CFD Applications. In Proceedings of the ITTC International Towing Tank Conference ITTC, Copenhagen, Denmark, 31 August–5 September 2014; Manoeuvering Committee: Wageningen, The Netherlands, 2014.
70. Practical Guidelines for RANS Calculation of Nominal Wakes. In Proceedings of the ITTC International Towing Tank Conference ITTC, Copenhagen, Denmark, 31 August–5 September 2014; Quality Systems Group: Copenhagen, Denmark, 2014.
71. Van Hoydonck, W.; Toxopeus, S.; Eloot, K.; Bhawsinka, K.; Queutey, P.; Visonneau, M. Bank effects for KVLCC2. *J. Mar. Sci. Technol.* **2019**, *24*, 174–199.
72. Hajivand, A.; Hossein Mousavizadegan, S. Virtual maneuvering test in CFD media in presence of free surface. *Int. J. Nav. Archit. Ocean Eng.* **2015**, *7*, 540–558.
73. Tezdogan, T.; Incecik, A.; Turan, O. Full-scale unsteady RANS simulations of vertical ship motions in shallow water. *Ocean Eng.* **2016**, *123*, 131–145.
74. DeMarco Muscat-Fenech, C.; Sant, T.; Zheku, V.V.; Villa, D.; Martelli, M. A Review of Ship-to-Ship Interactions in Calm Waters. *J. Mar. Sci. Eng.* **2022**, *10*, 1856. [CrossRef]
75. Kazerooni, M.F.; Seif, M.S. Experimental Study of Forces Exerted on Ships Due to the Vertical Walls of Navigation Channels. *Trans. Nav. Int. J. Mar. Navig. Saf. Sea Transp.* **2015**, *9*, 199–203.

76. Serani, A.; Diez, M.; van Walree, F.; Stern, F. URANS analysis of a free-running destroyer sailing in irregular stern-quartering waves at sea state 7. *Ocean Eng.* **2021**, *237*, 109600.
77. Jin, Y.; Duffy, J.; Chai, S.; Magee, A.R. DTMB 5415M dynamic manoeuvres with URANS computation using body-force and discretised propeller models. *Ocean Eng.* **2019**, *182*, 305–317.
78. Kim, D.; Song, S.; Tezdogan, T. Free running CFD simulations to investigate ship manoeuvrability in waves. *Ocean Eng.* **2021**, *236*, 109567.
79. Kim, D.; Tezdogan, T. CFD-based hydrodynamic analyses of ship course keeping control and turning performance in irregular waves. *Ocean Eng.* **2022**, *248*, 110808.
80. Broglia, R.; Dubbioso, G.; Durante, D.; Di Mascio, A. Turning ability analysis of a fully appended twin screw vessel by CFD. Part I: Single rudder configuration. *Ocean Eng.* **2015**, *105*, 275–286.
81. Dubbioso, G.; Durante, D.; Di Mascio, A.; Broglia, R. Turning ability analysis of a fully appended twin screw vessel by CFD. Part II: Single vs. twin rudder configuration. *Ocean Eng.* **2016**, *117*, 259–271.
82. Carrica, P.M.; Mofidi, A.; Eloot, K.; Delefortrie, G. Direct simulation and experimental study of zigzag maneuver of KCS in shallow water. *Ocean Eng.* **2016**, *112*, 117–133.
83. Jiang, F.; Li, Y.; Gong, J. Study on the manoeuvre characteristics of a trimaran under different layouts by water-jet self-propulsion model test. *Appl. Ocean Res.* **2021**, *108*, 102550.
84. Gao, T.; Wang, Y.; Pang, Y.; Chen, Q.; Tang, Y. A time-efficient CFD approach for hydrodynamic coefficient determination and model simplification of submarine. *Ocean Eng.* **2018**, *154*, 16–26.
85. Proposal, R.; Yoshimura, Y.; Pyo Rhee, K. *The Manoeuvring Committee Final Report and Recommendations to the 24th ITTC*; Manoeuvring Committee: Glasgow, UK, 2005.
86. Araki, M.; Sadat-Hosseini, H.; Sanada, Y.; Tanimoto, K.; Umeda, N.; Stern, F. Estimating maneuvering coefficients using system identification methods with experimental, system-based, and CFD free-running trial data. *Ocean Eng.* **2012**, *51*, 63–84.
87. Testing and Extrapolation Methods Manoeuvrability Free Running Model test. In Proceedings of the ITTC International Towing Tank Conference ITTC, Fukuoka, Japan, 14–20 September 2008; pp. 1–11.
88. Mohammadafzali, S. A Mathematical Model for the Maneuvering Simulation of a Propelled SPAR Vessel. Master's, Thesis, Memorial University of Newfoundland, Norris Point, NL, Canada, 2016.
89. A Abkowitz, B.M.T.U. *Hydro-Og Aerodynamisk Laboratorium Lectures on Ship Hydrodynamics-Steering and Manoeuvrability*; Faculty Mechanical, Maritime and Materials Engineering: Lyngby, Denmark, 1964.
90. Abkowitz, M.A. *Stability-and-Motion-Control-of-Ocean-Vehicles*; The MIT Press: Cambridge, MA, USA, 1969.
91. Vijay, A.; Somayajula, A. Identification of Hydrodynamic Coefficients using Support Vector Regression. In *Oceans Conference Record (IEEE)*; Institute of Electrical and Electronics Engineers Inc.: New York, NY, USA, 2022.
92. Yoshimura, Y. Mathematical model for the manoeuvring ship motion in shallow water. [Application of MMG mathematical model to shallow water]-41. *J. Kansai Soc. Nav. Archit. Jpn.* **1976**, *200*, 42–51.
93. Yasukawa, H.; Yoshimura, Y. Introduction of MMG standard method for ship maneuvering predictions. *J. Mar. Sci. Technol.* **2015**, *20*, 37–52.
94. Dr-Ing Zou Zaojian, B.; Jiao, S. *Lecture Notes on Ship Manoeuvring and Seakeeping*; School of Naval Architecture, Ocean and Civil Engineering: Shanghai, China, 2006.
95. Dubbioso, G.; Viviani, M. Aspects of twin screw ships semi-empirical maneuvering models. *Ocean Eng.* **2012**, *48*, 69–80.
96. Moreira, L.; Guedes Soares, C. Simulating Ship Manoeuvrability with Artificial Neural Networks Trained by a Short Noisy Data Set. *J. Mar. Sci. Eng.* **2023**, *11*, 15.
97. Moreira, L.; Guedes Soares, C. Dynamic model of manoeuvrability using recursive neural networks. *Ocean Eng.* **2003**, *30*, 1669–1697.
98. Im, N.; Hasegawa, K. A Study on Automatic Ship Berthing Using Parallel Neural Controller (2nd Report)—Motion Identification Considering Lateral Speed and Angular Velocity to cope with Disturbances—The Technology for Marine Traffic Hazard Control system using Ship navigation Big Data View project Ship Accident Analysis View project. *J. Kansai Soc. Nav. Archit. Jpn.* **2002**, *2002*, 237_127–237_132.
99. Im, N.K.; Nguyen, V.S. Artificial neural network controller for automatic ship berthing using head-up coordinate system. *Int. J. Nav. Archit. Ocean Eng.* **2018**, *10*, 235–249.
100. Sutulo, S.; Guedes Soares, C. An algorithm for offline identification of ship manoeuvring mathematical models from free-running tests. *Ocean Eng.* **2014**, *79*, 10–25.
101. Rajesh, G.; Bhattacharyya, S.K. System identification for nonlinear maneuvering of large tankers using artificial neural network. *Appl. Ocean Res.* **2008**, *30*, 256–263.
102. Lacki, M. Intelligent Prediction of Ship Maneuvering. *Trans. Nav. Int. J. Mar. Navig. Saf. Sea Transp.* **2017**, *10*, 511–516.
103. Gug, S.G.; Harshapriya, D.; Jeong, H.; Yun, J.H.; Kim, D.J.; Kim, Y.G. Analysis of manoeuvring characteristics through sea trials and simulations. In Proceedings of the IOP Conference Series: Materials Science and Engineering, Chennai, India, 16–17 September 2020; IOP Publishing Ltd.: Bristol, UK, 2020; Volume 929.
104. Xue, Y.; Liu, Y.; Xue, G.; Chen, G. Identification and prediction of ship maneuvering motion based on a gaussian process with uncertainty propagation. *J. Mar. Sci. Eng.* **2021**, *9*, 804. [CrossRef]
105. Sutulo, S.; Guedes Soares, C. Review on ship manoeuvrability criteria and standards. *J. Mar. Sci. Eng.* **2021**, *9*, 904. [CrossRef]

106. Papanikolaou, A.; Zaraphonitis, G.; Shigunov, V.; Papanikolaou, A.; Zaraphonitis, G.; Bitner-Gregersen, E.; Shigunov, V.; El Moctar, O.; Guedes Soares, C.; Reddy, D.N.; et al. Energy Efficient Safe Ship Operation (SHOPERA). In Proceedings of the SNAME Maritime Convention, Bellevue, WA, USA, 2–4 November 2016.
107. Van Zwijnsvoorde, T.; Tello Ruiz, M.; Delefortrie, G.; Lataire, E. Sailing in shallow water waves with the DTC container carrier: Open model test data for validation purposes. In *Knowledge Centre Manoeuvring in Shallow and Confined Water, Proceedings of the 5th International Conference on Ship Manoeuvring in Shallow and Confined Water (MASHCON), Ostend, Belgium, 19–23 May 2019*; Gent University: Gent, Belgium, 2019; pp. 411–422.
108. International Maritime Organization; Marine Environment Protection Committee (MEPC); Resolution MEPC.203(62) Amendments to the Annex of the Protocol of 1997 to Amend the International Convention for the Prevention of Pollution from Ships, 1973, as Modified by the Protocol of 1978 Relating thereto (Inclusion of Regulations on Energy Efficiency for Ships in MARPOL Annex VI); 2011.

Journal of
Marine Science and Engineering

MDPI

Article

Comprehensive Study on Optimizing Inland Waterway Vessel Routes Using AIS Data

Xiaoyu Yuan [1,2], Jiawei Wang [1], Guang Zhao [3,4] and Hongbo Wang [1,*]

1 State Key Laboratory on Integrated Optoelectronics, College of Electronic Science and Engineering, Jilin University, Changchun 130012, China; xyy22@mails.jlu.edu.cn (X.Y.); jwwang22@mails.jlu.edu.cn (J.W.)
2 Laboratory of Science and Technology on Marine Navigation and Control, China State Ship-Building Corporation, Tianjin 300131, China
3 Tianjin Navigation Instrument Research Institute, Tianjin 300131, China; zhaoguang2891@gmail.com
4 Tianjin Key Laboratory of Quantum Precision Measurement Technology, Tianjin 300051, China
* Correspondence: wang_hongbo@jlu.edu.cn; Tel.: +86-0431-8514-8242

Abstract: Inland waterway transport is an important mode of transportation for many countries and regions. Route planning optimization can reduce navigation time, avoid traffic congestion, and improve transportation efficiency. In actual operations, many vessels determine their navigation routes based on the experience of their shipowners. When the captain fails to obtain accurate information, experience-based routes may pose significant navigation risks and may not consider the overall economic efficiency. This study proposes a comprehensive method for optimizing inland waterway vessel routes using automatic identification system (AIS) data, considering the geographical characteristics of inland waterways and navigation constraints. First, AIS data from vessels in inland waters are collected, and the multi-objective Peak Douglas–Peucker (MPDP) algorithm is applied to compress the trajectory data. Compared to the traditional DP algorithm, the MPDP algorithm reduces the average compression rate by 5.27%, decreases length loss by 0.04%, optimizes Euclidean distance by 50.16%, and improves the mean deviations in heading and speed by 23.53% and 10.86%, respectively. Next, the Ordering Points to Identify the Clustering Structure (OPTICS) algorithm is used to perform cluster analysis on the compressed route points. Compared to the traditional DBSCAN algorithm, the OPTICS algorithm identifies more clusters that are both detailed and hierarchically structured, including some critical waypoints that DBSCAN may overlook. Based on the clustering results, the A* algorithm is used to determine the connectivity between clusters. Finally, the nondominated sorting genetic algorithm II is used to select suitable route points within the connected clusters, optimizing objectives, including path length and route congestion, to form an optimized complete route. Experiments using vessel data from the waters near Shuangshan Island indicate that, when compared to three classic original routes, the proposed method achieves path length optimizations of 4.28%, 1.67%, and 0.24%, respectively, and reduces congestion by 24.15%. These improvements significantly enhance the planning efficiency of inland waterway vessel routes. These findings provide a scientific basis and technical support for inland waterway transport.

Keywords: inland waterway transport; AIS data; trajectory compression; OPTICS clustering; NSGA-II; multi-objective optimization; route optimization

Citation: Yuan, X.; Wang, J.; Zhao, G.; Wang, H. Comprehensive Study on Optimizing Inland Waterway Vessel Routes Using AIS Data. *J. Mar. Sci. Eng.* **2024**, *12*, 1775. https://doi.org/10.3390/jmse12101775

Academic Editor: Marko Perkovic

Received: 11 August 2024
Revised: 26 September 2024
Accepted: 1 October 2024
Published: 6 October 2024

1. Introduction

In recent years, the computational science of ship routing has become a prominent topic in the field of maritime intelligence. The core research involves first inputting the starting point coordinates, ending point coordinates, and related parameters and then using efficient, intelligent optimization algorithms to explore and determine the optimal route. This process requires algorithms with excellent computational efficiency and accuracy while considering the safety, economic efficiency, and environmental sustainability of the route, thereby promoting the shipping industry toward more intelligent and refined development. Research on inland waterway route planning can be broadly divided into global and local route planning. Global route planning covers a larger area and ignores specific river information to construct a navigable inland network between ports. Using route planning algorithms, global route planning determines the ports and waterways to be traversed after inputting the start and end points. Conversely, local route planning covers a smaller area and typically relies on electronic nautical charts to consider non-navigable factors, such as obstacles. Using route planning algorithms, local route planning plots the route on a grid map. This study focuses on local route planning, addressing the route planning problems of cargo ships, such as container ships, within specific waterway sections.

Typically, the implementation of global route planning includes graph search algorithms, such as Dijkstra's algorithm [1], A* algorithm [2], and Rapidly exploring random tree (RRT) algorithm [3], and intelligent algorithms, such as ant colony optimization (ACO) [4] and genetic algorithm (GA) [5]. Local route planning involves dynamic planning algorithms, where the ship first senses the surrounding environment using its sensors and then plans a route necessary for safe navigation. This route planning is commonly applied in scenarios such as ship encounters and obstacle avoidance. The implementation of local route planning typically includes algorithms such as the dynamic window approach (DWA) [6] and artificial potential field (APF) [7]. In addition, some scholars have proposed intelligent algorithms, such as neural networks (NN) [8]. Various adaptive algorithms can be selected depending on different application scenarios and requirements. Some typical path-planning algorithms are listed in Table 1.

Table 1. Relevant path-planning algorithms.

Algorithm	Main Features	Advantages	Disadvantages
A* Algorithm	Efficient, shortest path finding	Quickly finds the shortest path	High memory usage
RRT	Good for complex spaces	Suitable for complex spaces	Generates suboptimal paths
ACO	Effective in dynamic environments	Highly effective in dynamic environments	Slow convergence
GA	Strong global search	Powerful global search capability	Risk of getting stuck in local optima
APF	Simple, easy implementation	Simple implementation	Local minima issues
Particle swarm optimization (PSO) [9]	Inspired by the social behavior of birds and fish	Fast convergence	Prone to local optima
NN	Excellent for complex patterns	Excellent at handling complex patterns	Long training time, data-intensive
Reinforcement learning (RL)	Learns from interactions with the environment	Adapts to dynamic environments	Long training time, resource-intensive

The A algorithm*, developed by Hart, P.E., enhances the Dijkstra algorithm by incorporating heuristic costs to expedite the search process. Singh et al. [10] introduced a constrained version of A* that reduces search time by defining a safety zone around a Maritime Autonomous Surface Ship (MASS), demonstrating its applicability in real-time local path planning with dynamic obstacles. Liu et al. [11] proposed the CTSPP method, combining Theta*-star and K-means algorithms, to generate safe and efficient inland vessel paths by dividing waterways into sub-segments and optimizing paths based on various navigational factors. Zhang et al. [12] proposed a global path-planning algorithm based on an improved A-star algorithm, utilizing bidirectional search and an enhanced evaluation function to reduce traversing nodes, with path smoothing to eliminate folding issues, improving the navigation efficiency of unmanned surface vehicles in environmental monitoring. The RRT algorithm is a sampling-based method where new nodes are created randomly until the goal is reached. Zaccone et al. [13] showcased RRT* in simulations for COLREG-compliant path planning, while Zhang et al. [14] developed a hybrid RRT solution for global path planning with dynamic adjustments for narrow areas. ACO algorithm draws inspiration from ant foraging behavior, using pheromones to guide optimal path selection. Lazarowska, A. et al. [15] applied ACO for dynamic path planning under COLREG compliance, and Wang, H. et al. [16] further evaluated its effectiveness in both static and dynamic maritime environments for collision avoidance. The GA is widely employed in optimization and path planning, either as a standalone method or integrated with other algorithms. While effective for complex problems, it requires careful parameter selection to ensure convergence. Kim, H. et al. [17] optimized path planning for minimal travel time using GA, and Xin, J. et al. [18] proposed GA variants with faster convergence and greater robustness. The APF algorithm relies on attractive and repulsive forces for path planning, with Lyu, H. et al. [19] proposing a COLREG-compliant APF for multiple ship encounters. Wang, Z. et al. [20] introduced an improved APF model based on the relative speed and acceleration between ships. The PSO algorithm is a global optimization method based on swarm intelligence, where particles collaborate to iteratively approach the global optimum. A hybrid approach combining ACO, PSO, and A* algorithms for multi-task path planning in AUVs is presented by Sui, F. et al. [21], achieving optimal and safe path planning. Xue, H. et al. [22] integrate the Sine Cosine Algorithm with PSO, proposing an SC-PSO variant that optimizes path planning by considering multiple constraints, including collision avoidance and path smoothness. Finally, machine learning methods, particularly RL, are gaining traction in maritime path planning. Zhang, X. et al. [23] proposed a two-layer Deep Reinforcement Learning (DRL) model for MASS, focusing on scene division and decision-making to avoid static and dynamic obstacles. Gao, M. et al. [24] developed a COLREG-compliant collision avoidance algorithm using AIS data, while Xu, Q.Y. et al. [25] utilized Convolutional Neural Networks (CNNs) to train systems for real-world scenario recognition, and Praczyk, T. et al. [26] explored evolutionary neural networks for developing autonomous systems in compliance with COLREG and collision avoidance.

Over the last decade, researchers have increasingly used AIS data in smart ocean construction. For example, in ship collision avoidance decision research [27–29], Shi et al. [30] identified collision risks using the ship domain and fed successful ship encounter data from AIS data into a double-gated recurrent unit neural network, generating collision avoidance decisions for unmanned ships through neural network learning. In ship trajectory prediction research [31–34], Volkova et al. [35] used AIS data to input ship positioning information into a neural network, solving the problem of weak satellite signal positioning caused by obstacles in inland waters. In ship emission research, Huang et al. [36] used AIS data extracted from ships to develop a quantitative model for estimating ship engine emissions under various operating conditions and applied spatiotemporal analysis to quantify these emissions. Moreover, AIS data are extensively used in areas such as ocean current detection, waterway traffic, environmental noise, marine fisheries, and marine trade. In Ref. [37], two indicators are defined to evaluate collision risk during navigation: the degree of velocity obstacle intrusion (DVOI) and the time of velocity obstacle intrusion (TVOI). These indicators measure collision risk based on two factors: speed and course. Ref. [38] develops an improved collision avoidance algorithm called Utility DWA (UDWA) for unmanned surface vessels (USVs), which incorporates maritime safety regulations (COLREGs) and accounts for wind and wave effects, enhancing navigational safety and efficiency in various conditions. Ref. [39] proposes the MDA-Traclus algorithm, an enhanced version of Density-Based Spatial Clustering of Applications with Noise (DBSCAN), designed to adaptively cluster ship trajectories with varying densities and directions, addressing the limitations of traditional DBSCAN in handling multi-density datasets and trajectory direction identification.

However, the AIS was initially designed to support communication for VTS, which poses several challenges to its use in scientific research. One of the most significant challenges is the storage and analysis of AIS data. The AIS transmits information every 2–12 s when a vessel is in motion and every 6 min in semistatic and static states.

In other words, the update rate of AIS data is occasionally significantly faster than the changes in the vessel speed, position, and heading during navigation. AIS data can be affected by various factors during transmission, potentially resulting in invalid or incorrect information. The main factors include equipment issues, such as malfunctions due to prolonged operation or lack of maintenance of signal transmission and reception devices; environmental factors, such as signal communication problems caused by obstructions, adverse weather, or waterway congestion; and data fusion problems, where data collected by different monitoring devices may encounter format mismatches or parsing errors during integration. Therefore, raw AIS data contain a significant amount of redundant information, which is not conducive to subsequent research and computation.

This study proposes an innovative and comprehensive method for optimizing inland waterway vessel routes using AIS data, addressing several challenges in existing technologies. The primary innovations and contributions of this study are as follows:

(1) AIS data preprocessing: The study introduces a robust AIS data preprocessing method that not only detects and corrects missing values, outliers, and inconsistencies but also establishes specific data validation criteria, such as MMSI length, geographical coordinates, and vessel dynamics (heading and speed). This preprocessing step is crucial for ensuring the accuracy and reliability of subsequent analyses.

Errors are classified as follows: (1) the MMSI length is not 9 digits or is otherwise invalid; (2) the longitude is less than 0° or greater than 180°, or the latitude is less than 0° or greater than 90°; (3) the vessel's heading or speed falls outside acceptable ranges (e.g., heading exceeding 360° or speed less than 0 kn). The corresponding data ranges are detailed in Table 2. Furthermore, incomplete trajectories with fewer than 20 points will be excluded from the analysis.

Table 2. Reasonable ranges for vessel trajectory data fields.

Field Name	Field Description	Data Range
LON	Longitude	[−180°, 180°]
LAT	Latitude	[−90°, 90°]
SOG	Speed over Ground	[0 kn, 102.2 kn]
COG	Course over Ground	[0°, 360°]
HDG	Heading	[0°, 360°]

(2) Data compression: A significant contribution of this research is the development of the MPDP algorithm, which addresses the limitations of traditional DP algorithms in multiturn route scenarios. By considering factors such as vessel speed, heading, and obstacle crossing, the MPDP algorithm significantly enhances the efficiency of trajectory data compression, reducing redundancy while maintaining the integrity of route information.

(3) Advanced clustering analysis using OPTICS: The study applies the OPTICS algorithm to the compressed data to improve the accuracy of route point extraction. Compared to the traditional DBSCAN algorithm, which struggles with parameter sensitivity and high noise levels in datasets, the OPTICS algorithm provides a more reliable clustering structure, thereby enhancing the effectiveness of subsequent path planning.

(4) Multi-Objective Path Optimization Based on NSGA-II: The research introduces a novel application of the NSGA-II algorithm, which simultaneously optimizes path length and congestion. This approach addresses the limitations of traditional algorithms that focus solely on path length, offering a more balanced and practical solution for inland waterway route planning.

By integrating these innovative methodologies, the study offers a comprehensive solution that improves the accuracy, efficiency, and practicality of route planning in complex waterway networks.

The proposed method was validated using real AIS data from waters near Shuangshan Island in the Yangtze River system. The experimental results show that the proposed method can significantly improve path planning efficiency, reduce path length, and alleviate congestion, providing important support for inland waterway transport. By systematically addressing data redundancy, path connectivity, and multi-objective optimization, this study provides a reliable framework for optimizing inland waterway vessel routes. Our findings provide a practical solution to enhance the efficiency and safety of inland navigation. The algorithm model is shown in Figure 1.

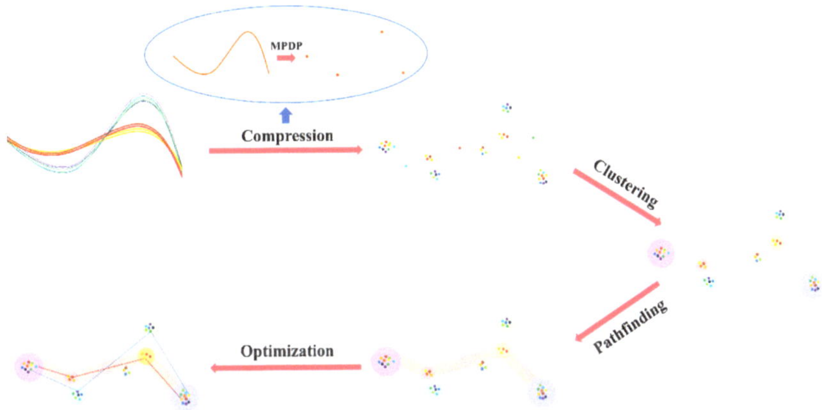

Figure 1. Schematic of algorithm model.

The remainder of this paper is organized as follows: Section 2 introduces the MPDP algorithm and describes comparative experiments with the conventional DP algorithm. Section 3 presents the OPTICS algorithm and performs comparison experiments with the DBSCAN algorithm in waters near Shuangshan Island in the Yangtze River system. In Section 4, the A* algorithm and NSGA-II algorithm are introduced for connectivity assessment and path optimization. Section 5 presents the path planning experiments, considering path length and traffic congestion as optimization objectives. Finally, Section 6 summarizes and concludes this study.

2. AIS Data Trajectory Compression

After acquiring the AIS data, the first step is to compress the trajectory for subsequent clustering. As shown in Figure 2, conventional algorithms typically use the DP algorithm [38] for trajectory compression. However, the approach of the DP algorithm for selecting the furthest distance as compression points misses some critical route points in multiturn waterways, making it unsuitable for winding inland waterways. This study employs a new MPDP algorithm for trajectory compression.

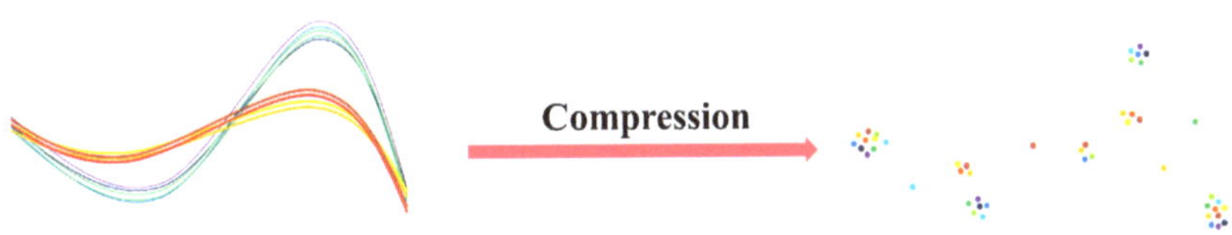

Figure 2. Trajectory compression.

Compressing AIS data is crucial for reducing storage and computational costs. Common compression methods include the Douglas–Peucker (DP) algorithm [40], Bellman algorithm [41], Spatio-Temporal Trajectory Clustering algorithm (STTrace) [42], sliding window algorithm [43], top-down time ratio algorithm, and opening window algorithm and its variants, such as the opening window time ratio and the opening window spatiotemporal algorithms. Data clustering algorithms are essential in big data mining, data analysis, artificial intelligence, and other fields. Clustering involves dividing a sample dataset into several subsets based on internal similarity characteristics, where the data within each subset have high similarity. Common clustering algorithms include the K-means algorithm [44], K-modes algorithm, hierarchical clustering algorithms, grid-based clustering algorithms, and density-based clustering algorithms. BIRCH [45] is a hierarchical clustering algorithm that builds a clustering feature tree by scanning data samples. This tree stores clustering features and clusters leaf nodes. STING [46] is a grid-based multiresolution clustering algorithm that divides spatial regions into hierarchical rectangular units. Higher-level units are subdivided into lower-level units, and a top-down query method with threshold filtering is employed. DBSCAN [47] is a density-based clustering algorithm that uses the search radius and the number of points within that radius as thresholds to assess the density relationship between samples and perform clustering.

2.1. Classic Douglas–Peucker Algorithm

The DP algorithm [38] is one of the most widely used compression algorithms. Its main task involves extracting a set of key trajectory points $\boldsymbol{KS} = \{K_1, \ldots K_j, \ldots K_M\}$ from the original trajectory point set $\boldsymbol{OS} = \{P_1, \ldots P_i, \ldots P_N\}$ that reflects the main shape characteristics of the original trajectory. The compression steps of the DP algorithm are as follows: First, set a compression threshold $\varepsilon(\varepsilon > 0)$ and add the start and end points of the original trajectory point set $\boldsymbol{OS}$ to the key trajectory points $\boldsymbol{KS}$. Then, divide $\boldsymbol{OS}$ into $M - 1$ subsets. For a subset $\boldsymbol{OS\_Sub_j}$ corresponding to two adjacent points K_j and K_{j+1}, use the line segment between K_j and K_{j+1} as the baseline and calculate the distance set $\boldsymbol{DS}$ from each point in $\boldsymbol{OS\_Sub_j}$ to the baseline. Finally, record the $Index$ corresponding to the maximum value $d_{\max}$ in $\boldsymbol{DS}$. If $d_{\max} \geq \varepsilon$, add $\boldsymbol{OS\_Sub_j}[Index]$ to the key trajectory point set $\boldsymbol{KS}$. The above process is repeated until the compression is complete. The compression process of the algorithm is shown in Figure 2, and the pseudocode is as follows (Algorithm 1):

Algorithm 1 Douglas-Peucker (DP) Algorithm

Input: Original trajectoiry points $OS = \{P_1, P_2, \ldots, P_N\}$
Output: Key trajectoty points $KS = \{K_1, K_2, \ldots, K_M\}$
Initialize $KS = \{P_1, P_N\}$
Set compression threshold ϵ
Divide OS into subsets between key points: $OS\_Sub_j = \{K_j, P_{j+1}, \ldots, K_{j+1}\}$
for each subset OS_Sub_j between K_j and K_{j+1} **do**
    Calculate distances $d(P_k, lineK_jK_{j+1})$ for all $P_k \in OS\_Sub_j$
    Find the maximum distance $d_{\max}$ and corresponding index k
    if $d_{\max} > \epsilon$ **then**
        Add P_k to KS
    end if
end for
Repeat the above steps until no more points are added to KS
return KS

The compression process of the DP algorithm is illustrated in Figure 3.

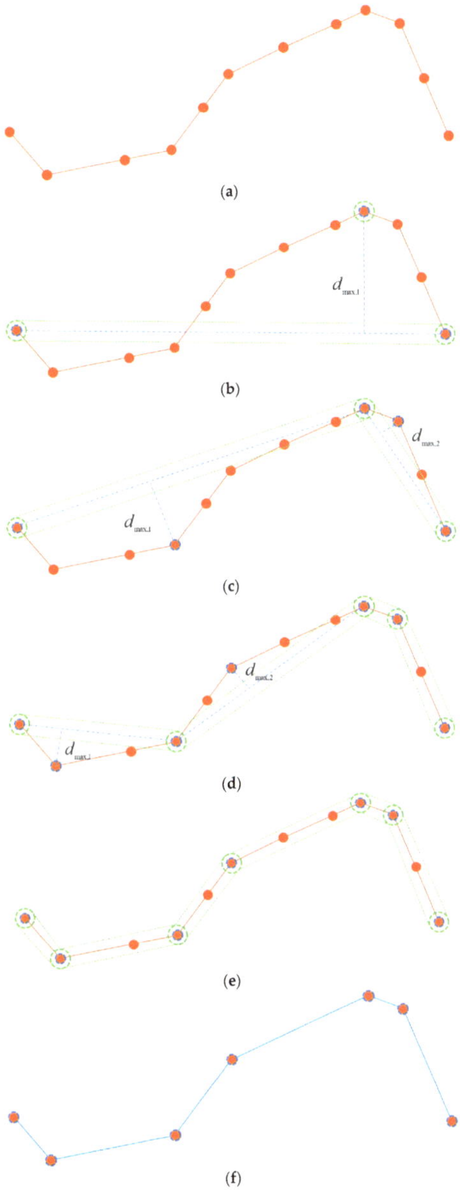

Figure 3. Schematic of the DP algorithm. (**a**) Original trajectory; (**b**) finding the farthest point; (**c**) evaluate and divide; (**d**) continue retaining points that do not meet the threshold; (**e**) all points meet the threshold; (**f**) compressed trajectory.

In Figure 3, the orange points represent the trajectory points, and the orange line represents the original trajectory. The green area represents the coverage area determined by the threshold ε. By continuously splitting and evaluating the relationship between $d_{\max}$ and ε, key route points are retained to obtain the final compressed trajectory. The conventional DP algorithm performs poorly in compressing multiturn trajectories of inland waterway vessels because it does not consider two important vessel indicators: heading and speed. Furthermore, it does not consider the accuracy of the compressed trajectory relative to the actual map. Therefore, this study adopts the MPDP algorithm to address these issues.

2.2. Multi-Objective Peak Douglas–Peucker Algorithm

The conventional DP algorithm [38] selects the farthest point as the split point at each compression level, decomposing the route chain and continuing the compression. However, the farthest point is not necessarily the point with the greatest degree of turning, and this is not ideal for winding inland waterways. Therefore, this study employs a multi-peak retention strategy that identifies local maximum points in the distance set ***DS***, generating a set of extreme points ***ES***. Each point in ***ES*** is compared with the threshold ε, and if it is greater than ε, it is stored in the set ***KS***. This method reduces the number of compression levels, thereby reducing the number of iterations.

To prevent the retention of too many trajectory points, a threshold Th_n for the number of peaks is set. When the number of extreme points in the set ***ES*** exceeds Th_n, the peaks in ***ES*** are re-evaluated, and the Th_n points with the most significant changes are retained. In the early stages of the compression process, many peaks appear in the distance set ***DS***. In the later stages, as the segments reduce, the local maximum value in ***DS*** equals the maximum value. Therefore, a threshold Th_l for the number of compression levels is set. When the number of compression levels is less than Th_l, multiple trajectory points are retained; otherwise, only a single trajectory point is retained.

Moreover, variations in speed and heading during navigation have a significant impact on the resulting trajectory. The MPDP algorithm employs a fitness function instead of the point-to-baseline distance used in the DP algorithm. This fitness function incorporates the distance from the point to the line, the rate of change in heading, and the rate of change in speed. Due to the substantial differences in units and dimensions among these three factors, they are not suitable for direct comparative analysis. Therefore, normalization is required to eliminate the effects of units and dimensions. The Min–Max method is applied to transform the original fitness function $f_{d,\eta,v}$, mapping the function value $F_{d,\eta,v}$ to the range $[0,1]$. The fitness function and the normalization method are defined as follows:

$$F_{d,\eta,v} = \frac{f_{d,\eta,v} - \min}{\max - \min}, \tag{1}$$

where $f_{d,\eta,v}$ represents the original fitness function, which includes the distance from the point to the line, the rate of change in heading, and the rate of change in speed; max denotes the maximum value of the data, and min denotes the minimum value of the data. The transformed fitness function is calculated according to the formulas shown in Equations (2) and (3).

$$fitness = \alpha \cdot d + \beta \cdot \eta + \gamma \cdot v, \tag{2}$$

$$\alpha + \beta + \gamma = 1, \tag{3}$$

where d represents the normalized distance from the point to the line, η represents the normalized rate of change in heading, and v represents the normalized rate of change in speed; α, β, and γ denote the corresponding coefficients.

In addition, the MPDP algorithm employs an obstacle-detection mechanism. When it detects that the line segment between two points $\boldsymbol{KS}(i)$ and $\boldsymbol{KS}(i+1)$ in the set ***KS*** crosses an obstacle, it re-compresses the original trajectory with $\boldsymbol{KS}(i)$ and $\boldsymbol{KS}(i+1)$ as endpoints, retaining key nodes until the restored trajectory successfully passes the obstacle.

Finally, when the vessel is engaged in back-and-forth movement, the trajectory points at different times may have overlapping coordinates. In such cases, the compression of the DP algorithm results in the baseline length approaching zero, which renders the point-to-baseline distance calculation ineffective. Therefore, a coordinate overlap detection mechanism is introduced: when two points have the same coordinates, the distance calculation changes from point-to-line to point-to-point. This ensures the correctness of the algorithm without changing its fundamental principles, which are expressed as follows:

$$d' = \sqrt{(x_{extreme} - x_{base})^2 + (y_{extreme} - y_{base})^2}, \tag{4}$$

where $x_{extreme}$ and $y_{extreme}$ denote the coordinates of the extreme point, and x_{base} and y_{base} denote the coordinates of the two overlapping baseline points. The pseudocode for the MPDP algorithm is as follows (Algorithms 2 and 3):

Algorithm 2 Trajectory Compression

```
Require: Original track points set OS, Threshold of the compression distance ϵ, Threshold of the
number of peak points Th_n, Threshold of compression layers Th_l
Ensure: Key track point set KS
L is the number of compression layers, and the initial value is 0
KS←MPDP(OS, L, c, Th_n, Th_l)
function MPDP(PS, Ztr, e, Th_n, Th_l)
    n is the size of point set PS
    repeat
      L←L + 1
      if PS[0] is same as PS[n− 1] then
        for i = 1 to n − 2 do
          Calculate the point-to-point distance d from PS[i] to PS[0] through Equation (3)
            DS[i] = d
        end for
      else
        for i = 1 to n − 2 do
          Calculate the point-to-line distance d from PS(i) to PS(0)PS(n − 1)
          DS[i] = d
        end for
      end if
      d is the set of distance set DS normalized
      η, v are the change rate of heading angle and speed of each track point after normalization
      FS = α ∗ d + β ∗ η + w ∗ v
      if L < Th_l then
        Index ← findpeaks(FS)
        m is the size of index set Index of points
        while m > Th_n do
          Index← lindpeaks(FS[Index])
          m ← size of index set Index of points
        end while
      else
        [fmax, Index] = max(FS)
      end if
      if Index = = [0, Index, n − 1] then
        k is the size of index set Index of points
        for i = 1 to k − 2 do
          if DS[Index[i]] >= ϵ then
            KS ← MPDP(PS[Index[i − 1] :Index[i] − 1], L, c, Th_n, Th_l − 1)
            KS ← MPDP(PS[Index[i]: Index[i + 1]], L, c, Th_n, Th_l)
          end if
        end for
      else
        Add PS[Index[i]] into KS
        E ←[E, DC[i]]
      end if
    until Index is full
    return KS
end function
return KS
```

Algorithm 3 Compressed Trajectory Obstacle Detection and Obstacle Avoidance

```
Require: Original track points set OS, Key track point set KS
Ensure: Key track point set KS, Recommended compression threshold ε_new
ε_new ← OBAvoid(OS, KS)
function OBAVOID(OS, KS)
    i = 1
    while i do
        if the obstacles between KS[i] and KS[i + 1] then
            PS is the subset of all trackpoints from KS[i] to KS[i +1] in OS
            n is the size of point set PS
            d_maz and Index are respectively the maximum distance between each point in PS and the line KS[i]KS[i + 1] and the index for obtaining the maximum distance
            if KS[i] is same as KS[i + 1] then
                for j = 0 to n − 1 do
                    Calculate the point-to-point distance d from PS[j] to KS[i] through Equation (3)
                    if d > d_max then
                        d_max = d
                        Index = i
                    end if
                end for
            else
                for j = 0 to n − 1 do
                    Calculate the point-to-line distance d from PS[j] to KS[i]KS[i + 1]
                    if d > d_max then
                        d_max = d
                        Index = i
                    end if
                end for
            end if
            Add PS[Index] into KS
            E = {E, d_max}
        else
            if KS[i + 1] is same as the last trajectory point in OS then
                break
            else
                i = i+ 1
            end if
        end if
    end while
    ε_new = min(E)
    return KS, ε_new
end function
```

3. OPTICS Algorithm

After extracting the key route points using the compression algorithm, clustering can be applied to group these key route points into clusters, thereby facilitating subsequent path optimization. This process is illustrated in Figure 4.

Figure 4. Route point clustering.

In inland waterway environments, navigation density and vessel distribution vary significantly across river sections, especially in narrow channels. The conventional DBSCAN clustering algorithm [45] clusters areas with similar densities, which can lead to inaccurate cluster identification and misjudgment. The selection of suitable parameters is challenging on datasets with varying densities.

In contrast, the OPTICS algorithm can discover nested clustering structures, making it suitable for complex clustering scenarios that may exist in inland waterway environments. For example, some areas may contain dense vessel clusters with smaller sub-clusters within them. The OPTICS algorithm generates a reachability plot that visually displays the density variation and clustering structure of the data points. This allows for a better understanding of the vessel distribution in inland environments and avoids boundary misjudgments that can occur with DBSCAN. Before elucidating the process of the OPTICS algorithm, it is imperative to define the following key concepts:

The core distance (*Core_Distance*) for a given point p with parameters ε_{OPTICS} and $MinPts$ is defined as the distance between the point p and its $MinPts$-th nearest neighbor within the ε_{OPTICS}-neighborhood. If the ε_{OPTICS}-neighborhood of point p contains fewer than $MinPts$ points, the core distance for point p is undefined.

Formally, the core distance can be expressed as

$$Core\_Distance(p) = \begin{cases} distance(p, p_{MinPts}), & if \left| N_{neigh}(p) \right| \geq MinPts \\ undefined, & if \left| N_{neigh}(p) \right| < MinPts \end{cases}, \tag{5}$$

where $N_{neigh}(p)$ denotes the set of points within the ε_{OPTICS}-neighborhood of point p, and p_{MinPts} is the $MinPts$-th nearest point within this neighborhood.

The reachability distance (***RD***) is the greater of the core distance of point p and the Euclidean distance from point p to point q, which can be calculated as follows:

$$Reachability\_Distance(q, p) = \max(Core\_Distance(p), distance(p, q)). \tag{6}$$

The neighborhood distance is defined as the Euclidean distance between point p and point q.

Formally, it is expressed as follows:

$$Neighborhood\_Distance = distance(p, q). \tag{7}$$

These definitions are critical for understanding the subsequent steps of the OPTICS algorithm. Below is a detailed explanation of the algorithmic steps:

Step 1: Initialize an empty ordered list to store processed data points and an empty priority queue to store candidate points and their reachability distances.

Step 2: Select an unprocessed point p from the dataset and calculate its core distance *Core_Distance* and reachability distance (***RD***).

Step 3: For the selected point p, calculate its core distance. The core distance is the smallest radius ε_{OPTICS} within which the neighborhood contains at least $MinPts$ points.

Step 4: Mark point p as processed and add it to the ordered list. Calculate ***RD*** values for all unprocessed points in the neighborhood of p and add these points, along with their ***RD*** values, to the priority queue.

Step 5: Select the point q with the smallest ***RD*** from the priority queue. If q has not been processed, calculate its core distance and ***RD***. Then, mark point q as processed and add it to the ordered list. Update the ***RD*** values for all unprocessed points in the neighborhood of q and add these points, along with their ***RD*** values, to the priority queue. Repeat the above steps until all points in the dataset have been visited.

The pseudocode is as follows (Algorithm 4):

Algorithm 4 OPTICS: Ordering Points To Identify the Clustering Structure

```
Require: Dataset D, Minimum points minPts, Neighborhood radius ϵ
Ensure: Ordered list of points with reachability distances
Initialize ordered list O as empty
Initialize priority queue Q as empty
for each unprocessed point p ∈ D do
    Mark p as processed
    N_p ← getNeighbors(p, ϵ)
    Compute core distance of p
    if core distance of p is defined then
        Update Q with N_p
        while Q is not empty do
            q ← extractMin(Q)
            if q is not processed then
                Mark q as processed
                N_q ← getNeighbors(q, ϵ)
                Compute core distance of q
                Update Q with N_q
            end if
        end while
    end if
end for
return ordered list O with reachability distances
function GETNEIGHBORS(p, ϵ)
    N ←{q ∈ D | distance(p, q) ≤ ϵ}
    return N
end function
function UPDATEQUEUE(Q, N_p)
    for each r ∈ N_p do
        Compute reachability distance of r
        Insert r into Q with priority as reachability distance
    end for
end function
function EXTRACTMIN(Q)
    q ← point in Q with smallest reachability distance
    Remove q from Q
    return q
end function
```

4. Connectivity Assessment and Path Optimization

After obtaining the clustering results, the A* algorithm is used to evaluate the connectivity between clusters, retaining the connected clusters. The NSGA-II [48] is then employed for path optimization. The procedure is illustrated in Figure 5.

Figure 5. Pathfinding and optimization.

In this study, the A* algorithm is utilized to find the shortest possible cluster path while ensuring connectivity. As a heuristic algorithm, A* efficiently identifies the shortest path between two points by employing a heuristic function. Due to its high search efficiency, we apply the A* algorithm to determine the shortest path between the cluster containing the starting point and the cluster containing the target point. Specifically, the Euclidean distance is used as the heuristic function to calculate the shortest distance from the starting cluster to the target cluster. This ensures that the A* algorithm effectively prioritizes adjacent clusters closest to the target during graph search, thereby enhancing the efficiency of the algorithm. The starting and target coordinates used by the A* algorithm are the central coordinates of the starting and target clusters, respectively.

The NSGA-II algorithm is highly suitable for solving multi-objective optimization problems. It effectively balances multiple conflicting objectives, such as path length and congestion, and provides a set of non-dominated solutions that form the Pareto front. In this study, the optimization objectives include path length and congestion, making NSGA-II an ideal choice as it can optimize these objectives simultaneously. Compared to other multi-objective optimization algorithms, NSGA-II efficiently handles large search spaces through fast, non-dominated sorting and elite preservation strategies. It employs an elite strategy to retain the best solutions and utilizes a rapid crowding distance calculation method to maintain solution diversity, ensuring a well-distributed set of solutions along the Pareto front. These features significantly improve the algorithm's convergence speed and global search capability.

The NSGA-II is used for path optimization with optimization objectives, such as the shortest distance and the lowest congestion. The fitness function is as follows:

$$f_1(path) = \sum_{i=1}^{N_{path}-1} \sqrt{(x_{i+1} - x_i)^2 + (y_{i+1} - y_i)^2}, \tag{8}$$

$$f_2(path) = \frac{1}{N_{path}} \sum_{i=1}^{N_{path}} C(t_i), \tag{9}$$

in Equation (8), N_{path} represents the number of points in the path and (x_i, y_i) denote the coordinates of the i-th point in the path. In Equation (9), t_i denotes the timestamp of the i-th point in the path and $C(t_i)$ denotes the number of points in the cluster at timestamp t_i.

The NSGA-II algorithm is configured with a maximum of 100 generations and a population size of 50. The function tolerance is set to 1×10^{-6}, meaning that the algorithm will terminate early if the change in the objective function is smaller than this value, controlling the precision of convergence. The crossover probability, which indicates the proportion of individuals involved in the crossover operation, is set to 0.8 in this study. The mutation method used is an adaptive feasible mutation, which adjusts the mutation probability adaptively based on the constraints of the problem and the diversity of the current solutions. Selection is performed using a binary tournament selection algorithm. The Pareto front fraction is set to 0.35, which determines the proportion of individuals in each generation that are retained on the Pareto front. A scatter crossover method is employed, where the values of decision variables for offspring are randomly selected from the parents. The migration ratio is set to 0.2, indicating that 20% of individuals will migrate to other sub-populations during each migration operation. The migration interval is set to 20 generations, meaning that every 20 generations, a portion of the population migrates to another sub-population to promote diversity. The parameter settings are shown in Table 3.

Table 3. NSGA-II parameter settings.

Parameters	Setting
Maximum Generations	100
Population Size	50
Function Tolerance	1×10^{-6}
Crossover Probability	0.9
Mutation Operation	adaptive mutation
Selection Operation	tournament selection
Pareto Front Fraction	0.35
Crossover Operation	scattered crossover

Table 3. *Cont.*

Parameters	Setting
Migration Fraction	0.2
Migration Interval	20

5. Experiments and Validation

To verify the effectiveness and feasibility of the proposed algorithm, this study uses a section of the Yangtze River near Shuangshan Island (120.32878 E–120.70191 E; 31.98000 N–32.09305 N) as the experimental background. Section 5.1 preprocesses the acquired AIS data. Section 5.2 compares the MPDP and DP algorithms. Section 5.3 compares the experimental results of the DBSCAN and OPTICS algorithms. Finally, Section 5.4 describes the use of the NSGA-II algorithm for path optimization to obtain the final recommended route. These algorithms are implemented and run on a computer with Windows 64-bit with 12 cores (Intel(R) Core (TM) i7-12700F@2.10 GHz) and 16 GB of RAM using MATLAB R2022a.

5.1. AIS Data Preprocessing

AIS data processing requires calculating the distance from a point to a line. However, the original AIS data contain vessel coordinate information based on a geographic coordinate system, which complicates the direct calculation of distances between two points on a sphere and the distance from a point to a line. To simplify the data calculations, this study converts the trajectory points $P(\lambda, \varphi)$ to Mercator projection coordinates $P'(x,y)$. The conversion process is as follows:

$$x_{Merca} = r_0 \cdot \lambda, \tag{10}$$

$$y_{Merca} = r_0 \cdot q_{iso}, \tag{11}$$

$$r_0 = \frac{l \cdot \cos(\varphi_0)}{\sqrt{1 - \left(e_{ell}{}^2 \cdot \sin^2(\varphi_0)\right)}}, \tag{12}$$

$$q_{iso} = \ln\left(\tan\left(\frac{\pi}{4} + \frac{\varphi}{2}\right) \cdot \left(\frac{1 - e_{ell} \cdot \sin\varphi}{1 + e_{ell} \cdot \sin\varphi}\right)^{\frac{e}{2}}\right), \tag{13}$$

where x_{Merca} and y_{Merca} represent the horizontal and vertical coordinates of the trajectory point, respectively; λ and φ represent the longitude and latitude of the trajectory point, respectively; r_0 represents the radius of the parallel circle at the standard latitude; q_{iso} represents the isometric latitude; l represents the semimajor axis of the Earth's ellipsoid; φ_0 represents the standard latitude in the Mercator projection; e_{ell} represents the first eccentricity of the Earth's ellipsoid.

In this study, the grid method is used for modeling. The map is divided into grids of the same size based on ground resolution, and the environmental characteristics within the grids are converted into binary information. The formula for calculating ground resolution is as follows:

$$Resolution = \frac{\left(\cos\left(\frac{\pi}{180}\varphi\right) \cdot 2\pi R_0\right)}{256 \cdot 2 \cdot Level}, \tag{14}$$

where R_0 denotes the equatorial radius, with $R_0 = 6378140$ (m), and *Level* denotes the zoom level. This study uses an electronic nautical chart with a zoom level of 13 and a resolution of 19.1093 (m). The electronic nautical chart and the binarized grid map are shown in Figure 6.

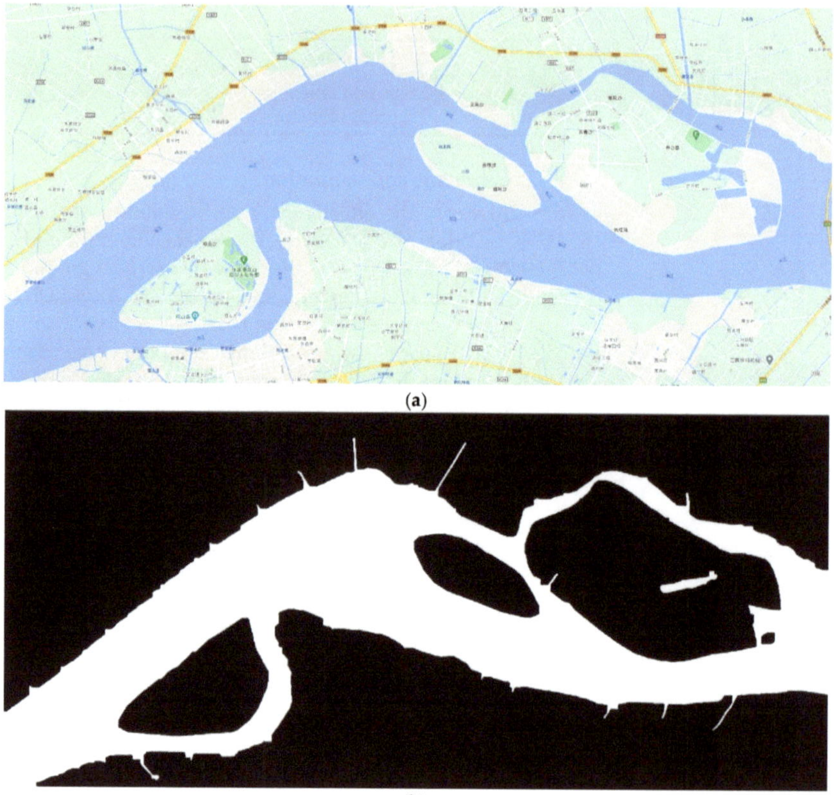

Figure 6. Inland waterway map generated using the grid method. (**a**) The electronic nautical chart; (**b**) the binarized grid map.

AIS data can be affected by various factors during transmission, potentially resulting in invalid or incorrect information. The main factors include equipment issues, such as malfunctions due to prolonged operation or lack of maintenance of signal transmission and reception devices, and environmental factors, such as obstructions, adverse weather, and waterway congestion, causing signal communication problems. Therefore, before proceeding to the next processing steps, it is necessary to clean unreasonable data based on the reasonable range of each field. The data preprocessing consists of the following steps:

(1) Data acquisition: The raw AIS data stream is acquired from AIS devices or base stations. These data are typically transmitted in the NMEA format and include information such as the vessel's position, speed, and heading.

(2) Data parsing: The cleaned data are parsed, and the NMEA format data are converted into structured data that can be analyzed and processed. This step typically involves splitting data into different fields and extracting key information such as the vessel's maritime mobile service identity (MMSI), position, and timestamp, as shown in Figure 7.

(3) Data filtering: Based on requirements, the parsed data are filtered to select specific periods, regions, or types of vessel data. This helps reduce the data volume and improves the efficiency of subsequent analyses.

(4) Data cleaning: The data were cleaned to remove any duplicate, invalid, or incorrect data. This includes detecting and correcting missing values, outliers, and inconsistent data.

MMSI	latitude	longitude	beam	course	draft	ship name	destination	heading	ship length	speed	sailing status
413810431	32.09419	120.5389	7	14	0	BAO GUANG JI 688		1.4	37	0.1	
413780458	32.09469	120.5391	7	260.4	0	HAI AN JI 00808			42	0.1	
413888877	32.09782	120.6937	6	0	11	JIANGSUTUO2380	0		29	0	
413793442	32.08897	120.5378	15	34.3	5.4	HUA YUAN 79	NANTONG		86	0	
413554250	32.07043	120.4736	16	249.5	6	MIN HANG 169	JIANG YIN		97	7.3	0
412767260	32.07613	120.4765	29	352.4	3.5	TONG BAO SHI JIE	JING JIANG		85	0	5
413826915	32.0769	120.4878	7	16.5	0.5	XUGANGTUO9006	B	1.6	27	2.1	0
413831205	32.07007	120.4857	19	95.6	4.2	HEXIE169	NANTONG	9.5	113	0.3	1
413832209	32.07228	120.4877	11	78.8	3.3	WAN TAI LUN 168	HUZHOU		65	2.4	
413288870	32.0752	120.4837	40	221	5	HAI FENG NENG YUAN	JING JIANG	3.7	149	0	5
413386570	32.07472	120.4823	12	109.1	3	GANGHUAGONG9	ZHOUSHANG		45	0	5
414110000	32.07604	120.4921	32	2.7	10.9	PING AN DA 60	JING JIANG	26.9	199	0	5
413832803	32.07092	120.4938	17		1.6	YU XIN HUO 17099	TAIZHOU		99	0	1
413797675	32.07673	120.4983	13	268.4	1.8	XINSHUN 66	JINGJIANG		79	0.1	5
413263780	32.07765	120.4999	12	24.2	3.8	HONGYUANTUO 1	JINGJIANG		38	0	5
413849285	32.07006	120.4912	16	91.3	4.3	JIANGRUN918	JINGJIANG		105	0	
413841988	32.07108	120.5036	16	354	4	HENGYANG19	CHONQIN		130	0	
413859506	32.07585	120.516	16	115	1.6	XINGLONGHESHENG	FENGJIE	0	99	0	
413843374	32.07922	120.5157	16	359.7	0	MENGYUN77	0	0	130	0	
413763886	32.06623	120.4729	15	0	2.5	JIN XIANG 18	TAIZHOU	0	98	0	0
413982287	32.06433	120.474	19	267.1	6	XINGUANGHUIHUANG	WUHU		111	0	
413802195	32.06969	120.4858	19	187.3	0	XIN YI SHUN HANG	A	18.7	118	0	
413829398	32.06849	120.4813	19	354.2	5.7	XINLONGCHANGSHENG	NANTONG	35.4	109	0	
413832048	32.06957	120.4861	16	276.5	4.3	HENGYANG16	CHONGQ		130	0	
413816927	32.06933	120.4859	16	325.1	2.2	YANGZIJIANGWULIU688	TAIZHOU		105	0.1	
413965823	32.0687	120.4836	17		4	XINGYUAN1666	JIANGYIN		99	0.1	0

Figure 7. Processed AIS data.

Errors are categorized as follows: (1) the MMSI length is not nine digits or is otherwise unreasonable; (2) the longitude is less than −180 or greater than 180, or the latitude is less than −90 or greater than 90; (3) the vessel heading or speed is outside reasonable ranges (e.g., heading greater than 360° or speed less than 0 knots). In addition, incomplete trajectories with fewer than 20 points are excluded. This study preprocesses 5 h of AIS data from the Shuangshan Island area in the Yangtze River system, containing 656 navigation trajectories with 145,359 trajectory points. The final processed AIS data are shown in Figure 8.

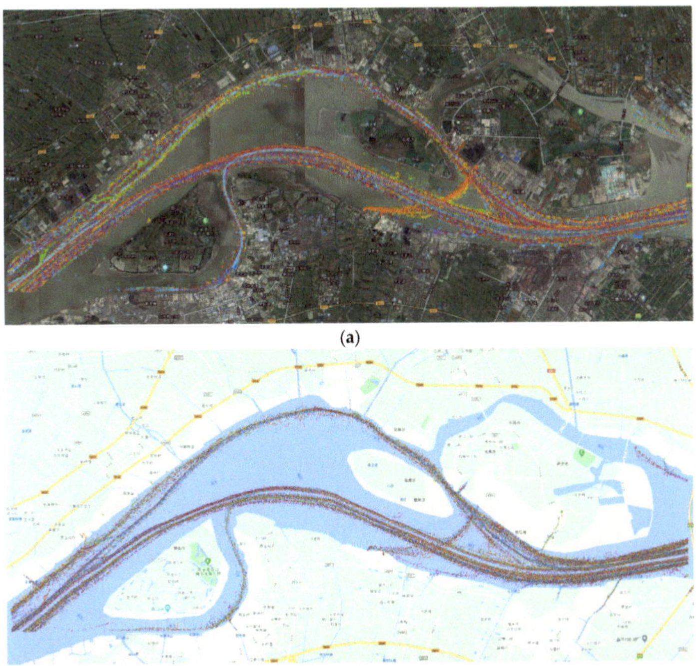

(**a**)

(**b**)

Figure 8. Preprocessed AIS trajectories. (**a**) Original navigation trajectories in the waters near Shuangshan Island in the Yangtze River system (satellite imagery); (**b**) original navigation trajectories in the waters near Shuangshan Island in the Yangtze River system (simplified map).

5.2. MPDP Trajectory Compression Algorithm Experiment

This study uses five performance metrics to measure the compression effect: compression rate R_c, length loss rate R_l, synchronized Euclidean distance D_{SE}, average speed deviation R_v, and average heading deviation R_θ. These metrics are as follows:

Compression rate: This is the ratio of the number of trajectory points discarded during compression to the number of original trajectory points and can be calculated as follows:

$$R_c = \frac{n_c}{N}, \tag{15}$$

where n_c represents the number of discarded trajectory points and N represents the total number of original trajectory points for each vessel.

Length loss rate R_l: This is the ratio of the difference between the original trajectory length and the reconstructed length after compression to the original trajectory length and can be calculated as follows:

$$R_l = \frac{\sum_{i=1}^{N-1} \overline{P_i P_{i+1}} - \sum_{j=1}^{M-1} \overline{K_j K_{j+1}}}{\sum_{i=1}^{N-1} \overline{P_i P_{i+1}}}, \tag{16}$$

where $\overline{P_i P_{i+1}}$ denotes the distance between two adjacent points in the original trajectory, M represents the number of retained trajectory points after compression for each vessel, and $\overline{K_j K_{j+1}}$ denotes the distance between two adjacent points in the compressed trajectory.

Synchronized Euclidean distance D_{SE}: Let $P_i = (x_i, y_i, t_i)$ be a point in the original trajectory; $K_j = (x_j, y_j, t_j)$ and $K_{j+1} = (x_{j+1}, y_{j+1}, t_{j+1})$ be the points in the compressed trajectory that are sequentially before and after $P_i (t_j \le t_i \le t_{j+1})$, respectively; and $P_i' = (x'_i, y'_i, t_i)$ be the reconstructed trajectory point, which can be expressed as follows:

$$\begin{cases} x'_i = x_j + \left(\frac{t_i - t_j}{t_{j+1} - t_j}\right)(x_{j+1} - x_j) \\ y'_i = y_j + \left(\frac{t_i - t_j}{t_{j+1} - t_j}\right)(y_{j+1} - y_j) \end{cases}. \tag{17}$$

Let the distance D_{SE}^i be the distance between the reconstructed point P_i' and the original point P_i. The synchronized Euclidean distance D_{SE} is the average distance between the original and reconstructed trajectories and is commonly used to reflect the compression effect. D_{SE} can be expressed as follows:

$$\begin{cases} D_{SE}^i = \sqrt{(x_i - x'_i)^2 + (y_i - y'_i)^2} \\ D_{SE} = \frac{1}{N}\sum_{i=1}^{N} D_{SE}^i \end{cases}. \tag{18}$$

Average speed deviation R_v: This metric reflects the average speed difference between the reconstructed trajectory points P_i' and the original trajectory points P_i. It can be expressed as follows:

$$\begin{cases} v'_i = v_j \\ R_v^i = |v'_i - v_i| \\ R_v = \frac{1}{N}\sum_{i=1}^{N} R_v^i \end{cases}, \tag{19}$$

where v_i denotes the speed of the vessel at the original trajectory point P_i; $v\prime_i$ denotes the speed of the vessel at the reconstructed trajectory point P_i'; v_j represents the speed of the vessel at key trajectory point K_j.

Average heading deviation R_θ: The heading restoration rate is used to reflect the average heading difference between the reconstructed point P_i' and the original point P_i, as shown in the following equation:

$$\begin{cases} \theta'_i = \arctan\left(\frac{y'_{j+1}-y'_j}{x'_{j+1}-x'_j}\right) \\ R^i_\theta = |\theta'_i - \theta_i| \\ R_\theta = \frac{1}{N}\sum_{i=1}^{N} R^i_\theta \end{cases}, \tag{20}$$

where θ_i denotes the heading angle of the vessel at the original trajectory point P_i; θ'_i denotes the heading angle of the vessel at the reconstructed trajectory point P'_i; $\left(x'_j, y'_j\right)$ and $\left(x'_{j+1}, y'_{j+1}\right)$ represent the coordinates of the vessel at the reconstructed trajectory points P'_i and P'_{i+1}, respectively.

To verify the effectiveness of the algorithm, a specific trajectory in the waters near Shuangshan Island in the Yangtze River system was selected for a compression comparison experiment between the MPDP and DP algorithms. The parameter settings and experimental results are shown in Table 4 and Figure 9, respectively.

Table 4. Algorithm parameter settings.

Algorithm	Compression Threshold ε (m)	Number of Peaks Threshold *Th_n*	Compression Level Threshold *Th_l*	α	β	γ
DP	42.40	-	-	-	-	-
MPDP	42.40	4	2	0.5	0.3	0.2

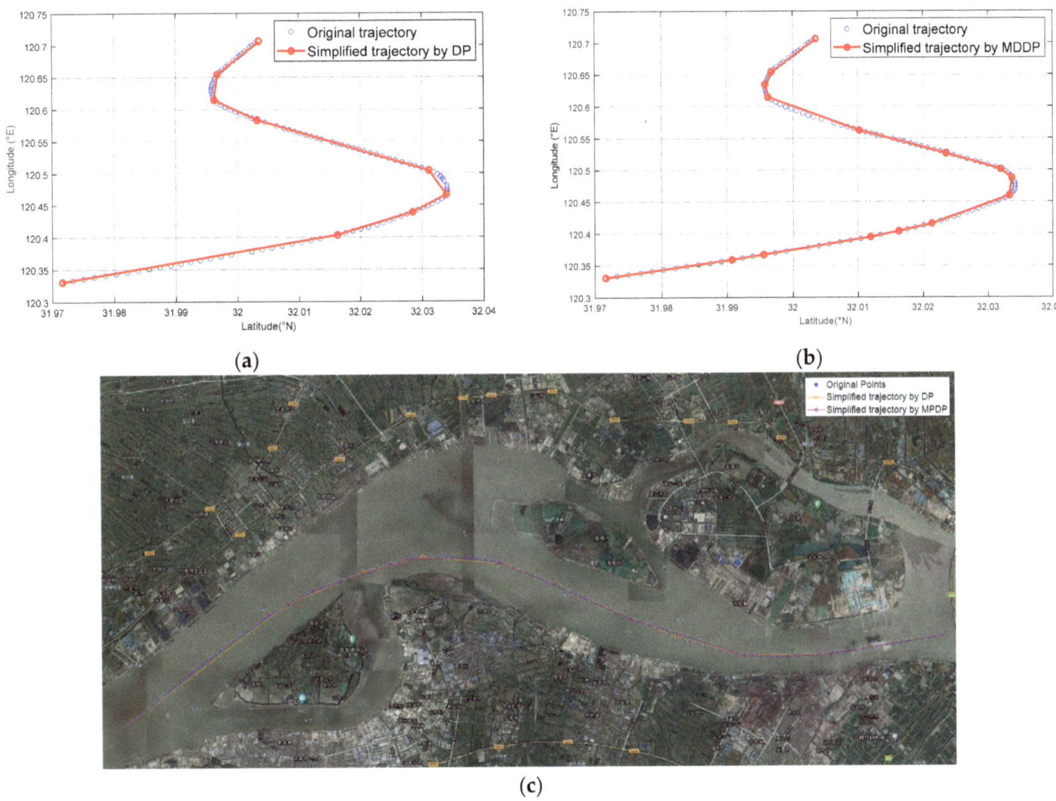

Figure 9. Trajectory compression results of DP and MPDP algorithms. (**a**) DP; (**b**) MPDP; (**c**) Comparison of trajectory compression between DP and MPDP algorithms.

The MPDP algorithm reduced the compression rate by 5.27% compared with the DP algorithm. As shown in Figure 10 and Table 5, the MPDP algorithm retained seven more trajectory points than the DP algorithm, reducing the length loss rate by 0.04%, representing an improvement of 26.67%. The synchronized Euclidean distance decreased by 82.6876 m, representing an improvement of 50.16%. The average heading deviation decreased by 0.0104, representing an improvement of 23.53%. The average speed deviation decreased by 0.0886 knots, representing an improvement of approximately 10.86%. As observed from Figure 11, the mean deviation in heading angle has decreased by 0.0104, reflecting a 23.46% improvement. Similarly, the mean deviation in speed has reduced by 0.0886 knots, corresponding to a 10.86% improvement.

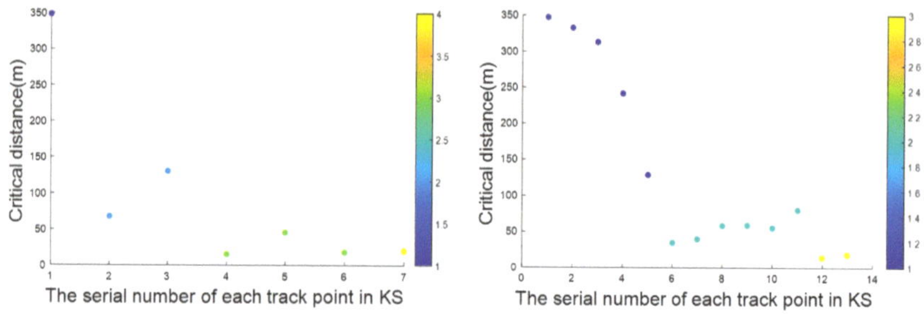

(a) (b)

Figure 10. Changes in critical distance. (**a**) DP; (**b**) MPDP.

Table 5. Trajectory compression results.

Algorithm	R_c	R_l	D_{SE} (m)	R_θ (rad)	R_v (kn)
DP	92.11%	0.15%	164.8631	0.0442	0.8158
MPDP	86.84%	0.11%	82.1755	0.0338	0.7272

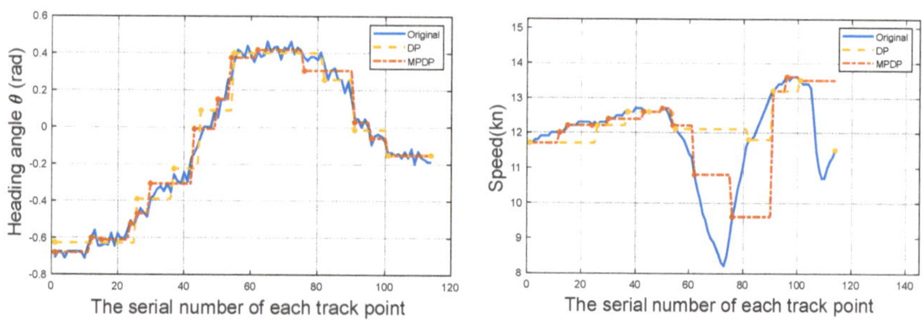

(a) (b)

Figure 11. Comparison of compressed heading angles and speeds. (**a**) Comparison of heading angles; (**b**) Comparison of speeds.

Based on the above experimental results, only minor differences in the compression rate and length loss rate are observed between the MPDP and DP algorithms. The retained trajectory points show a slight improvement in the average heading deviation R_θ. However, the MPDP algorithm significantly improves the synchronized Euclidean distance D_{SE} and average speed deviation R_v. In addition, the inclusion of an obstacle detection mechanism ensures thorough route connectivity.

Next, a compression experiment using the DP and MPDP algorithms was conducted on 5 h of AIS data from the waters near Shuangshan Island. The experimental results are shown in Figures 12 and 13.

Figure 12. Compression results of the DP algorithm.

Figure 13. Compression results of the MPDP algorithm.

Using the MPDP algorithm, the average compression rate R_c decreased by 2.11%; the average length loss rate R_l decreased by 0.14%; the average synchronized Euclidean distance D_{SE} decreased by 11.5922 m, an improvement of approximately 25.70%; the average heading deviation R_θ decreased by 0.0133 rad, an improvement of approximately 11.73%; the average speed deviation R_v decreased by 0.059 knots, an improvement of approximately 25.13%. The experimental results are presented in Table 6.

Table 6. Compression metrics for Shuangshan Island ship tracks.

Algorithm	R_c	R_l	D_{SE} (m)	R_θ (rad)	R_v
DP	96.08%	0.65%	145.8896	0.1271	0.1946
MPDP	94.28%	0.59%	125.4893	0.1103	0.1525

5.3. Trajectory Point Clustering Experiment Based on OPTICS Algorithm

After obtaining the key trajectory points, this section describes the use of the DBSCAN and OPTICS algorithms to cluster the waterway data. The DBSCAN algorithm generated 46 clusters, whereas the OPTICS algorithm generated 111 clusters. The reachability plot obtained by the OPTICS algorithm is shown in Figure 14.

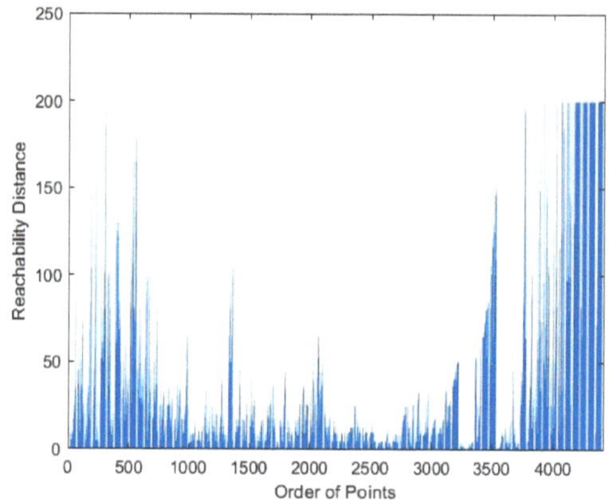

(**a**)

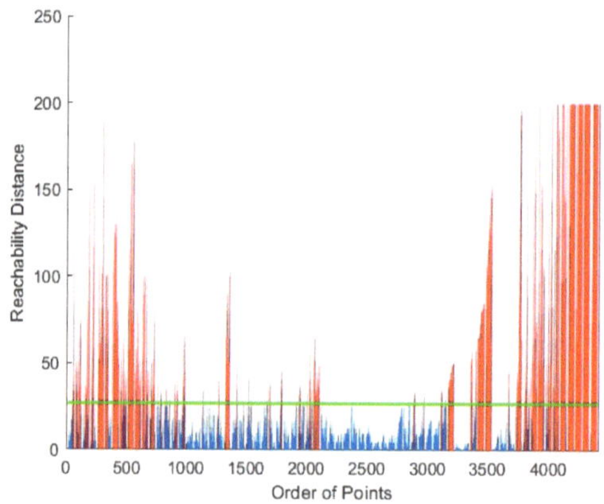

(**b**)

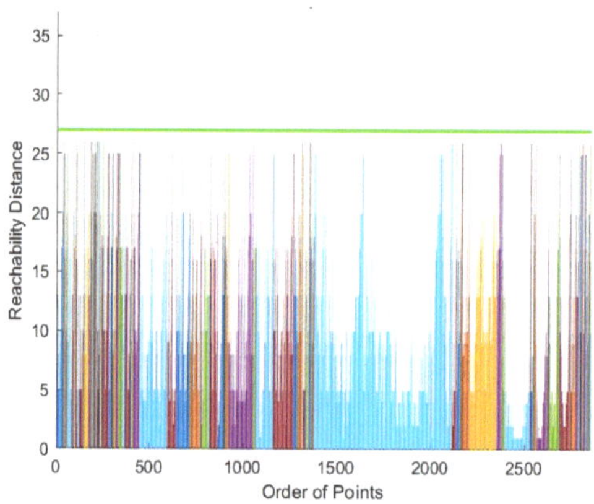

(**c**)

Figure 14. Reachability plot. (**a**) Distribution of reachability distance for each point; (**b**) clusters categorized with a threshold of 27, with red points indicating noise; (**c**) clustering results after classification, where different colors represent different clusters.

After obtaining the reachability plot, it becomes easy to understand the density and structure of the dataset. Low-value regions (valleys) in the reachability plot represent dense areas of data points; the larger the range of the low-value region on the x-axis, the larger the cluster and the more points it contains. High-value regions (peaks) indicate sparse areas of data points; the higher the peak, the sparser the area. In this study, the threshold is set to 27, resulting in the experimental results shown in Figure 14b, where the red points represent noise points with reachability distances greater than the threshold. The final classification results are shown in Figure 14c, where different colors represent different clusters. The final classification results of the AIS data for the waters near Shuangshan Island in the Yangtze River system are shown in Figures 15 and 16.

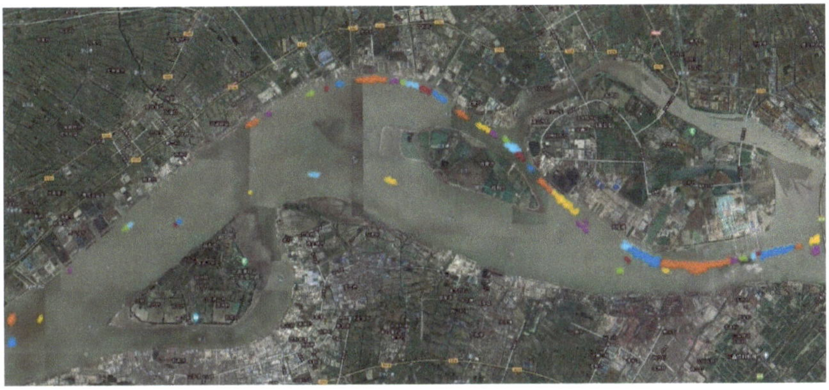

Figure 15. Clustering results of the DBSCAN algorithm in waters near Shuangshan Island.

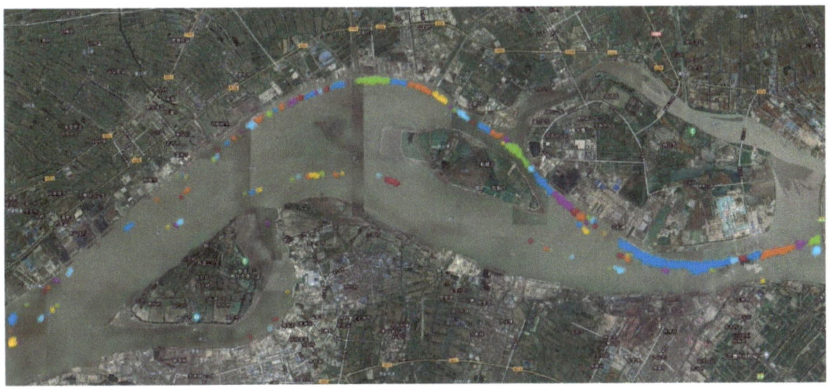

Figure 16. Clustering results of the OPTICS algorithm in waters near Shuangshan Island.

As shown in Figures 15 and 16, the OPTICS algorithm produces more and finer clusters than the DBSCAN algorithm. These clusters include key waypoint clusters, such as the tributary in the lower left corner, which the DBSCAN algorithm misses. Such information loss can hinder subsequent connectivity assessment and path optimization.

5.4. Route Planning Experiment Based on A* Algorithm and NSGA-II

This study uses the shortest navigation distance and minimum congestion as optimization objectives and employs the A* algorithm and NSGA-II for route searching and optimization, respectively, among clustered waypoints obtained using the OPTICS algorithm. First, the A* algorithm is used to determine the connectivity between each cluster of waypoints. The NSGA-II is then used to select the optimal route points within the connected clusters. The experimental results are shown in Figure 17.

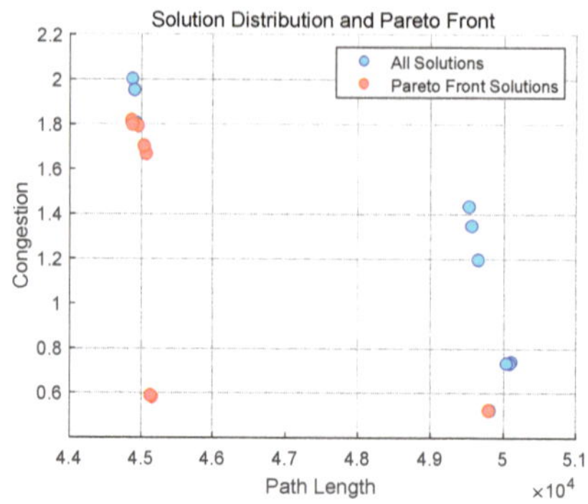

Figure 17. Pareto front distribution of optimized routes.

Figure 18a shows the experimental results obtained after the final iteration of the NSGA-II. Figure 18b shows the routes corresponding to the Pareto front solutions. To verify the feasibility of the proposed algorithm, three typical trajectories were selected from historical AIS data for the Yangtze River system. The actual route information was input (Table 7) and compared with the optimized routes. The experimental results are shown in Figures 19 and 20.

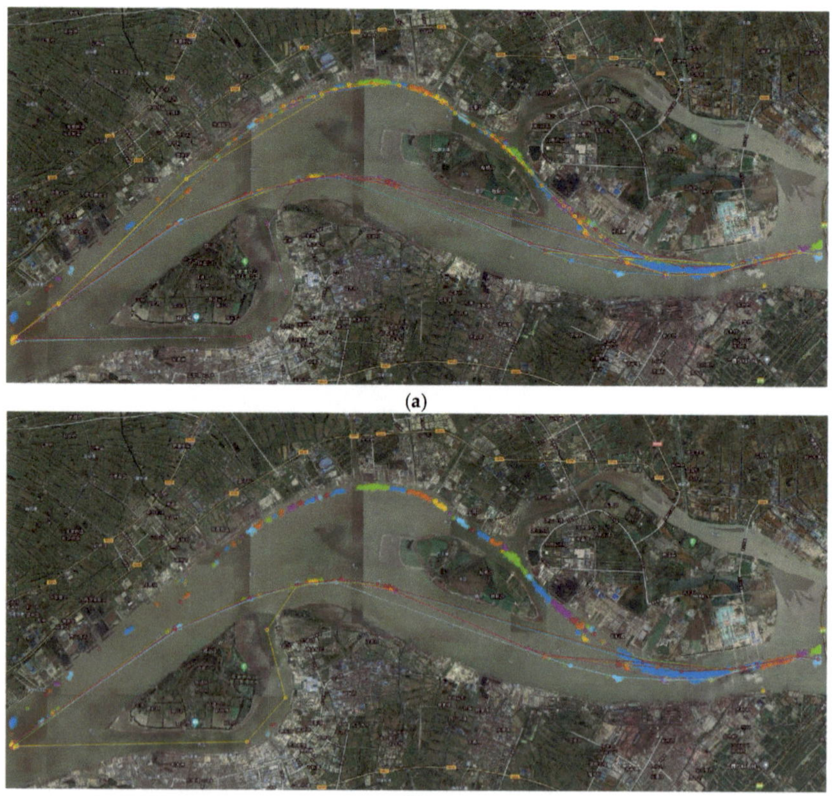

(a)

(b)

Figure 18. Optimized routes and Pareto front routes obtained using the NSGA-II. (**a**) NSGA-II-optimized routes; (**b**) routes along the Pareto front.

Table 7. Original data for route planning experiment.

ID Number	Starting Point Longitude	Starting Point Latitude	Ending Point Longitude	Ending Point Latitude
Case 1	120.7099	32.0102	120.3204	31.9718
Case 2	120.7085	32.0066	120.3233	31.9710
Case 3	120.7098	32.0068	120.3210	31.9707

Figure 19. Original AIS trajectory map (three typical trajectories).

Figure 20. Optimized routes (three typical trajectories).

The results of the route planning experiment are summarized in Table 8. As shown in Figures 17 and 18, the Pareto front solutions clearly avoid routes passing through the northern channel, as these routes are not only longer but also exhibit higher congestion levels.

Table 8. Results of route planning experiment.

Optimization Objective	Path Length (km)		Congestion	
ID Number	Before Optimization	After Optimization	Before Optimization	After Optimization
Case 1	51.776	49.559	1.559	1.346
Case 2	45.812	45.049	0.887	0.586
Case 3	49.926	49.808	0.788	0.521

From Figures 19 and 20 and Table 8, it is evident that compared to the original trajectories, the optimized routes successfully eliminated many unnecessary turning points, resulting in a total route length reduction of 3.098 km, with an improvement rate of 2.10%. Simultaneously, the congestion level decreased by 0.781, reflecting a 24.15% improvement in congestion. These results demonstrate that the optimized routes are significantly superior to the original trajectories in terms of reducing total route length and mitigating congestion.

Through experiments, the research results demonstrate that the proposed method achieves significant optimizations in both path length and congestion compared to classical original routes, validating the effectiveness and feasibility of this approach in practical applications. This provides a scientific basis and technical support for inland waterway transportation, effectively enhancing route planning efficiency, reducing traffic congestion, and ultimately improving transportation efficiency.

6. Summary

Inland waterway transport is a crucial mode of transportation for many countries and regions. Route planning optimization can reduce navigation time, avoid traffic congestion, and improve transportation efficiency. This study proposes an inland vessel route optimization method based on AIS data, combining the MPDP compression algorithm, OPTICS clustering algorithm, A* algorithm, and NSGA-II multi-objective optimization algorithm. Experiments conducted in waters near Shuangshan Island in the Yangtze River system demonstrated that the proposed method significantly improves path planning efficiency, reducing both path length and route congestion. Specifically, the MPDP algorithm effectively compressed AIS trajectory data, optimizing the length loss rate by 26.67%, the synchronized Euclidean distance by 50.16%, the average heading deviation by 23.53%, and the average speed deviation by 10.86%, thereby improving data processing efficiency. The OPTICS algorithm accurately identified clusters of route points in environments with uneven densities, including key waypoint clusters missed by the DBSCAN algorithm, which is crucial for subsequent path optimization. The A* algorithm ensured cluster connectivity, and the NSGA-II provided the optimal route under multi-objective optimization. These results indicate that the proposed method is highly feasible and effective in practical applications, providing a scientific basis and technical support for inland waterway route planning. The experimental results indicate that, compared to the original routes, the path lengths of the three typical trajectories were optimized by 4.28%, 1.67%, and 0.24%, respectively, leading to an overall path length optimization rate of 2.10%. Moreover, congestion was reduced by 24.15%. These data demonstrate that the proposed method is highly feasible and effective in practical applications, providing a scientific basis and technical support for inland waterway route planning.

However, this study has certain limitations. Firstly, the sample size and geographic scope are limited, as the experiments were conducted in terrains with fewer than five different turning angles. In reality, inland waterways may present more varied turns and angles. Future research will focus on validating the algorithm across different inland waterway terrains to enhance its generalizability and applicability. Secondly, this study only considers the impact of path length and congestion on route planning. In actual scenarios, varying weather conditions and water flows (e.g., during flood seasons) can significantly affect the feasibility of the planned routes. This study did not account for these factors, which represents another area for improvement in future research.

Author Contributions: Conceptualization, X.Y. and H.W.; methodology, X.Y.; software, X.Y.; validation, X.Y., H.W., J.W. and G.Z.; formal analysis, X.Y., G.Z. and H.W; investigation, X.Y., G.Z. and H.W.; resources, X.Y. and H.W; data curation, J.W.; writing—original draft preparation, X.Y.; writing—review and editing, X.Y., J.W. and H.W.; visualization, X.Y. and G.Z.; supervision, H.W.; project administration, H.W.; funding acquisition, H.W. All authors have read and agreed to the published version of the manuscript.

Funding: This study was funded by the Laboratory of Science and Technology on Marine Navigation and Control, China State Shipbuilding Corporation, grant number 2023010302.

Institutional Review Board Statement: Not applicable.

Informed Consent Statement: Not applicable.

Data Availability Statement: The original contributions presented in the study are included in the article, further inquiries can be directed to the corresponding author.

Acknowledgments: The authors are grateful to the anonymous reviewers for their valuable comments and suggestions that helped improve the quality of this manuscript.

Conflicts of Interest: Author Xiaoyu Yuan was employed by the company Laboratory of Science and Technology on Marine Navigation and Control, China State Ship-Building Corporation. The remaining authors declare that the research was conducted in the absence of any commercial or financial relationships that could be construed as a potential conflict of interest.

Abbreviations

ACO	Ant Colony Optimization
AIS	Automatic Identification System
APF	Artificial Potential Field
CNN	Convolutional Neural Networks
DBSCAN	Density-Based Spatial Clustering of Applications with Noise
DP	Douglas–Peucker
DRL	Deep Reinforcement Learning
DVOI	Degree of Velocity Obstacle Intrusion
DWA	Dynamic Window Approach
GA	Genetic Algorithm
MASS	Maritime Autonomous Surface Ship
MPDP	Multi-objective Peak Douglas–Peucker
PSO	Particle Swarm Optimization
RL	Reinforcement Learning
RRT	Rapidly exploring Random Tree
STTrace	Spatio-Temporal Trajectory Clustering
TVOI	Time of Velocity Obstacle Intrusion
UDWA	Utility DWA
USV	Unmanned Surface Vessels
VTS	Vessel Traffic Services

References

1. Dijkstra, E.W. A note on two problems in connexion with graphs. *Numer. Math.* **1959**, *1*, 269–271. [CrossRef]
2. Hart, P.E.; Nilsson, N.J.; Raphael, B. A Formal Basis for the Heuristic Determination of Minimum Cost Paths. *IEEE Trans. Syst. Sci. Cybern.* **1968**, *4*, 100–107. [CrossRef]
3. LaValle, S. Rapidly-exploring random trees: A new tool for path planning. *Res. Rep. 9811* **1998**.
4. Dorigo, M.; Maniezzo, V.; Colorni, A. Ant system: Optimization by a colony of cooperating agents. *IEEE Trans. Syst. Man Cybern. Part B (Cybern.)* **1996**, *26*, 29–41. [CrossRef]
5. Holland, J.H. *Adaptation in Natural and Artificial Systems: An Introductory Analysis with Applications to Biology, Control, and Artificial Intelligence*; MIT Press: Cambridge, MA, USA, 1992.
6. Fox, D.; Burgard, W.; Thrun, S. The dynamic window approach to collision avoidance. *IEEE Robot. Autom. Mag.* **1997**, *4*, 23–33. [CrossRef]
7. Khatib, O. Real-time obstacle avoidance for manipulators and mobile robots. *Int. J. Robot. Res.* **1986**, *5*, 90–98. [CrossRef]
8. McCulloch, W.S.; Pitts, W. A logical calculus of the ideas immanent in nervous activity. *Bull. Math. Biophys.* **1943**, *5*, 115–133. [CrossRef]
9. Kennedy, J.; Eberhart, R. Particle swarm optimization. In Proceedings of the ICNN'95-International Conference on Neural Networks, Perth, WA, Australia, 27 November–1 December 1995; pp. 1942–1948.
10. Singh, Y.; Sharma, S.; Sutton, R.; Hatton, D.; Khan, A. A constrained A* approach towards optimal path planning for an unmanned surface vehicle in a maritime environment containing dynamic obstacles and ocean currents. *Ocean Eng.* **2018**, *169*, 187–201. [CrossRef]

11. Liu, C.G.; Zhang, K.; He, Z.B.; Lai, L.H.; Chu, X.M. Clustering Theta* based segmented path planning method for vessels in inland. *Ocean Eng.* **2024**, *309*, 118249. [CrossRef]
12. Zhang, H.X.; Tao, Y.D.; Zhu, W.L. Global Path Planning of Unmanned Surface Vehicle Based on Improved A-Star Algorithm. *Sensors* **2023**, *23*, 6647. [CrossRef]
13. Zaccone, R.; Martelli, M.; Figari, M. A COLREG-Compliant Ship Collision Avoidance Algorithm. In Proceedings of the 2019 18th European Control Conference (ECC), Naples, Italy, 25–28 June 2019; pp. 2530–2535. [CrossRef]
14. Zhang, Z.; Wu, D.; Gu, J.; Li, F. A Path-Planning Strategy for Unmanned Surface Vehicles Based on an Adaptive Hybrid Dynamic Stepsize and Target Attractive Force-RRT Algorithm. *J. Mar. Sci. Eng.* **2019**, *7*, 132. [CrossRef]
15. Lazarowska, A. Ship's Trajectory Planning for Collision Avoidance at Sea Based on Ant Colony Optimisation. *J. Navig.* **2015**, *68*, 291–307. [CrossRef]
16. Wang, H.; Guo, F.; Yao, H.; He, S.; Xu, X. Collision Avoidance Planning Method of USV Based on Improved Ant Colony Optimization Algorithm. *IEEE Access* **2019**, *7*, 52964–52975. [CrossRef]
17. Kim, H.; Kim, S.-H.; Jeon, M.; Kim, J.; Song, S.; Paik, K.-J. A study on path optimization method of an unmanned surface vehicle under environmental loads using genetic algorithm. *Ocean Eng.* **2017**, *142*, 616–624. [CrossRef]
18. Xin, J.; Zhong, J.; Yang, F.; Cui, Y.; Sheng, J. An Improved Genetic Algorithm for Path-Planning of Unmanned Surface Vehicle. *Sensors* **2019**, *19*, 2640. [CrossRef]
19. Lyu, H.; Yin, Y. COLREGS-Constrained Real-time Path Planning for Autonomous Ships Using Modified Artificial Potential Fields. *J. Navig.* **2019**, *72*, 588–608. [CrossRef]
20. Wang, Z.; Li, G.; Ren, J. Dynamic path planning for unmanned surface vehicle in complex offshore areas based on hybrid algorithm. *Comput. Commun.* **2021**, *166*, 49–56. [CrossRef]
21. Sui, F.; Tang, X.; Dong, Z.; Gan, X.; Luo, P.; Sun, J. ACO plus PSO plus A*: A bi-layer hybrid algorithm for multi-task path planning of an AUV. *Comput. Ind. Eng.* **2023**, *175*, 108905. [CrossRef]
22. Xue, H. A quasi-reflection based SC-PSO for ship path planning with grounding avoidance. *Ocean Eng.* **2022**, *247*, 110772. [CrossRef]
23. Zhang, X.; Wang, C.; Liu, Y.; Chen, X. Decision-Making for the Autonomous Navigation of Maritime Autonomous Surface Ships Based on Scene Division and Deep Reinforcement Learning. *Sensors* **2019**, *19*, 4055. [CrossRef]
24. Gao, M.; Shi, G.-Y. Ship-Collision Avoidance Decision-Making Learning of Unmanned Surface Vehicles with Automatic Identification System Data Based on Encoder-Decoder Automatic-Response Neural Networks. *J. Mar. Sci. Eng.* **2020**, *8*, 754. [CrossRef]
25. Xu, Q.Y.; Zhang, C.J.; Zhang, L. Deep Convolutional Neural Network Based Unmanned Surface Vehicle Maneuvering. In Proceedings of the 2017 Chinese Automation Congress (CAC), Jinan, China, 20–22 October 2017; pp. 878–881.
26. Praczyk, T. Neural anti-collision system for Autonomous Surface Vehicle. *Neurocomputing* **2015**, *149*, 559–572. [CrossRef]
27. Zhang, W.; Goerlandt, F.; Montewka, J.; Kujala, P. A method for detecting possible near miss ship collisions from AIS data. *Ocean Eng.* **2015**, *107*, 60–69. [CrossRef]
28. Rong, H.; Teixeira, A.P.; Soares, C.G. Ship collision avoidance behaviour recognition and analysis based on AIS data. *Ocean Eng.* **2022**, *245*, 110479. [CrossRef]
29. Horteborn, A.; Ringsberg, J.W. A method for risk analysis of ship collisions with stationary infrastructure using AIS data and a ship manoeuvring simulator. *Ocean Eng.* **2021**, *235*, 109396. [CrossRef]
30. Shi, J.H.; Liu, Z.J. Deep Learning in Unmanned Surface Vehicles Collision-Avoidance Pattern Based on AIS Big Data with Double GRU-RNN. *J. Mar. Sci. Eng.* **2020**, *8*, 682. [CrossRef]
31. Pallotta, G.; Vespe, M.; Bryan, K. Vessel Pattern Knowledge Discovery from AIS Data: A Framework for Anomaly Detection and Route Prediction. *Entropy* **2013**, *15*, 2218–2245. [CrossRef]
32. Liu, T.; Ma, J. Ship Navigation Behavior Prediction Based on AIS Data. *IEEE Access* **2022**, *10*, 47997–48008. [CrossRef]
33. Bao, K.; Bi, J.; Gao, M.; Sun, Y.; Zhang, X.; Zhang, W. An Improved Ship Trajectory Prediction Based on AIS Data Using MHA-BiGRU. *J. Mar. Sci. Eng.* **2022**, *10*, 804. [CrossRef]
34. Xiao, Y.; Li, X.; Yao, W.; Chen, J.; Hu, Y. Bidirectional Data-Driven Trajectory Prediction for Intelligent Maritime Traffic. *IEEE Trans. Intell. Transp. Syst.* **2023**, *24*, 1773–1785. [CrossRef]
35. Volkova, T.A.; Balykina, Y.E.; Bespalov, A. Predicting Ship Trajectory Based on Neural Networks Using AIS Data. *J. Mar. Sci. Eng.* **2021**, *9*, 254. [CrossRef]
36. Huang, L.; Wen, Y.; Geng, X.; Zhou, C.; Xiao, C.; Zhang, F. Estimation and spatio-temporal analysis of ship exhaust emission in a port area. *Ocean Eng.* **2017**, *140*, 401–411. [CrossRef]
37. Wang, Y.; Zhang, Y.; Zhao, H.C.; Wang, H.B. Assessment Method Based on AIS Data Combining the Velocity Obstacle Method and Pareto Selection for the Collision Risk of Inland Ships. *J. Mar. Sci. Eng.* **2022**, *10*, 1723. [CrossRef]
38. Yuan, X.Y.; Tong, C.C.; He, G.X.; Wang, H.B. Unmanned Vessel Collision Avoidance Algorithm by Dynamic Window Approach Based on COLREGs Considering the Effects of the Wind and Wave. *J. Mar. Sci. Eng.* **2023**, *11*, 23. [CrossRef]
39. Zhang, Y.J.; Yuan, X.Y.; Li, M.; Zhao, G.; Wang, H.B. Multi-Density Adaptive Trajectory Clustering Algorithm for Ships Based on AIS Data. *IEEE Access* **2023**, *11*, 108198–108210. [CrossRef]
40. Douglas, D.H.; Peucker, T.K. Algorithms for the reduction of the number of points required to represent a digitized line or its caricature. *Cartogr. Int. J. Geogr. Inf. Geovis.* **1973**, *10*, 112–122. [CrossRef]

41. Bellman, R. On the approximation of curves by line segments using dynamic programming. *Commun. ACM* **1961**, *4*, 284. [CrossRef]
42. Potamias, M.; Patroumpas, K.; Sellis, T. Sampling trajectory streams with spatiotemporal criteria. In Proceedings of the 18th International Conference on Scientific and Statistical Database Management (SSDBM'06), Vienna, Austria, 3–5 July 2006; pp. 275–284.
43. Keogh, E.; Chu, S.; Hart, D.; Pazzani, M. An online algorithm for segmenting time series. In Proceedings of the 2001 IEEE International Conference on Data Mining, San Jose, CA, USA, 29 November–2 December 2001; pp. 289–296.
44. MacQueen, J. Some methods for classification and analysis of multivariate observations. In Proceedings of the Fifth Berkeley Symposium on Mathematical Statistics and Probability; University of California Press: Berkeley, CA, USA, 1967; pp. 281–297.
45. Zhang, T.; Ramakrishnan, R.; Livny, M. BIRCH: An efficient data clustering method for very large databases. *ACM Sigmod Rec.* **1996**, *25*, 103–114. [CrossRef]
46. Wang, W.; Yang, J.; Muntz, R. STING+: An approach to active spatial data mining. In Proceedings of the 15th International Conference on Data Engineering (Cat. No. 99CB36337), Sydney, NSW, Australia, 23–26 March 1999; pp. 116–125.
47. Marin, M.; Moonen, L.; van Deursen, A. An integrated crosscutting concern migration strategy and its application to JHotDraw. In Proceedings of the Seventh IEEE International Working Conference on Source Code Analysis and Manipulation (SCAM 2007), Paris, France, 30 September–1 October 2007; pp. 101–110.
48. Deb, K.; Pratap, A.; Agarwal, S.; Meyarivan, T. A fast and elitist multiobjective genetic algorithm: NSGA-II. *IEEE Trans. Evol. Comput.* **2002**, *6*, 182–197. [CrossRef]

MDPI AG
Grosspeteranlage 5
4052 Basel
Switzerland
Tel.: +41 61 683 77 34

Journal of Marine Science and Engineering Editorial Office
E-mail: jmse@mdpi.com
www.mdpi.com/journal/jmse

www.ingramcontent.com/pod-product-compliance
Lightning Source LLC
Chambersburg PA
CBHW041422210926
51850CB00013B/3330